DISCARD

Sampling in Education
and the
Social Sciences

Sampling in Education and the Social Sciences

RICHARD M. JAEGER

University of North Carolina at Greensboro

Longman
New York and London

Sampling in Education and the Social Sciences

Longman Inc., 1560 Broadway, New York, N.Y. 10036
Associated companies, branches, and representatives
throughout the world.

Developmental Editor: Nicole Benevento
Editorial and Design Supervisor: James Fields
Production/Manufacturing: Ferne Y. Kawahara
Composition: Bi-Comp, Inc.
Printing and Binding: Maple-Vail Book Manufacturing Group

Library of Congress Cataloging in Publication Data

Jaeger, Richard M.
    Sampling in education and the social sciences.
    Bibliography: p.
    Includes index.
    1. Sampling (Statistics)   2. Social sciences—
Statistical methods.   3. Educational statistics.
I. Title.
HA31.2.J33   1984   519.5′2   83-19945
ISBN 0-582-28440-6

Manufactured in the United States of America
Printing: 9 8 7 6 5 4 3 2 1   Year: 92 91 90 89 88 87 86 85 84

To LJC for teaching me how,
and to JMJ for making it possible

Contents

CHAPTER 8
IMPLICATIONS FOR PRACTICE: SUMMING UP 226

APPENDIX A
THE USE OF RANDOM NUMBER TABLES 237

APPENDIX B
THE STRASAMP-I COMPUTER PROGRAM 241

APPENDIX C
THE SYSAMP-I COMPUTER PROGRAM 253

APPENDIX D-1
THE CLUSAMP-I COMPUTER PROGRAM 285

List of Tables

Preface

A cursory review of programs for recent meetings of the American Educational Research Association or the American Sociological Association would convince even the most skeptical observer that educational researchers and social scientists are concerning themselves with an ever-broadening array of topics and problems. Although social scientists have long built their craft around the collection and analysis of data collected in situ, the primary methodological orientations of educational researchers have been derived from experimental psychology. Despite the dominance of field research in the social sciences, policy-related experiments, such as educational voucher plans and negative income tax investigations, are becoming commonplace. Current empirical studies in education delve into such problem areas as educational finance, the politics of education, the sociology of education, the economics of education, and educational policy formulation, to name but a few popular areas of inquiry. Each of these areas requires the collection of data under extralaboratory conditions, often through large-scale field studies of the sort that are traditional in sociological research.

The move to nonexperimental, empirical research in education and the persistence of this tradition in the social sciences create the need for training in procedures for efficient collection of data under field conditions. The theory of finite-population sampling is an important part of such training. Through the use of efficient sampling and estimation procedures, educational researchers and social scientists can realize significant economies and, more important, conduct studies that would not otherwise be feasible.

This book is designed to provide an introduction to the theory and practice of sampling from finite populations for students of educational and social science research; the topics treated and the examples used have been selected with these students and research workers in mind. No doubt, those in allied fields, such as economics, business administration, and historiography, will also find it useful and understandable.

The book is suitable for a one-semester or one-quarter course in sampling methods in a graduate applied research program or in a similar program for advanced undergraduates with strong methodological backgrounds. It might also be used as a supplementary text in courses that require knowledge of sampling procedures, such as demography, field research, or educational research methods. Although some mathematical and statistical background is assumed, the book is not at all demanding in these areas. To understand the topics discussed, the student should have a level of facility in basic algebra typically gained through a two-semester high school sequence. In addition, familiarity with the basic concepts of statistical inference, at the level of an introductory applied statistics course, is presumed. However, the student whose statistical skills have diminished through lack of use will find several elementary topics—for example, the meaning of confidence intervals—reviewed in the book.

Seventeen different sampling and estimation procedures are discussed. Even so, the book does not provide exhaustive coverage of finite-population sampling methods. We have been purposefully selective in choosing those procedures most likely to be of use to educational researchers and investigators in social science fields. Each procedure is introduced by defining its properties and discussing its advantages and disadvantages. Following a review of relevant estimation formulas, the procedure is applied to an important set of applied problems. The illustrated applications are of two kinds. First, the same set of data is used to illustrate the application of various sampling and estimation procedures to problems involving the academic achievement of elementary school students. Second, a variety of data sets are used to illustrate other applications of sampling to problems in the social sciences and education. By using a common data base in one set of examples, we have accomplished several objectives. We have provided a concrete example of the power of sampling methods in handling an educational research problem of near-universal concern. In addition, the reader has the opportunity to compare directly the results that might be expected from the entire array of sampling and estimation procedures considered in this book. These examples incorporate completely worked problems, including problem conception, choice of procedure, and complete numerical solutions. Given these examples, the reader can readily verify his or her understanding of the ways in which sampling and estimation methods are used in practice.

Various aspects of the theory and practice of sampling from finite populations are discussed in eight chapters and five appendixes. Chapter 1 can be viewed as an orientation to the use of survey research methods in education and the social sciences. Before discussing the details of any sampling method, it is useful to have a clear picture of the elements of survey research, since it is in the survey research context that sampling methods are most often applied. This is the role of Chapter 1. Chapter 2 presents definitions and illustrations of some fundamental concepts of sampling theory and introduces the basic technical vocabulary of the field. Chapters 3 through 7 are parallel in form. Each presents procedures, estimators, and applications of one or more methods of sampling and estimation, including several completely worked problems. The methods presented are simple random sampling, stratified sampling, systematic sampling, and single-stage cluster sampling. Many variations of the latter procedures are included. Chapter 8 contains a summary of the numerical results for each sampling and estimation method, when these methods are applied to the same achievement test data, and discusses the implications of these results. The value of this summary chapter goes beyond the specific numerical results presented. It serves as an illustration of the way in which data on alternative sampling procedures can be compared and provides an example of rational selection among competing sampling techniques.

Many sampling procedures require long hours of laborious computations. In fact, the complexity of required computations may seriously hamper the use of otherwise efficient sampling and estimation techniques. To overcome this problem, a series of novel computer programs has been developed for this book. The programs permit the reader to apply sampling and estimation procedures with a minimum of hand calculation. The programs are flexible and efficient, and should facilitate the use of sampling procedures in practice. Each program is completely documented and illustrated in an appendix. Sample input and output data are provided with each description. An additional appendix contains instructions on the use of random number tables in drawing samples by hand.

This book grew out of work begun during my doctoral studies at Stanford University in the late 1960s. I was assisted immeasurably in that work by Lee J. Cronbach, Ingram Olkin, and Robert Heath who have, through the years, provided the kind of stimulation, challenge, and encouragement that defines outstanding education. In this endeavor, as in so many professional pursuits, they have contributed greatly. Earlier versions of this book have been used as a text in survey research methods courses at the University of South Florida and the University of North Carolina. Dozens of students contributed to the revision of the text by pointing out opaque language, apparently inconsistent results, and the minute problems that always plague early drafts. I am grateful to every one of them. Finally, it is a pleasure to acknowledge the constant encouragement and assistance of my wife, Judy, without whose help this book would surely have been abandoned long since.

Richard M. Jaeger
Greensboro, North Carolina

1

Social Survey Research as a Context for Sampling

Introduction

Although this is a book on sampling methods, we thought it unwise to proceed directly to the details of sampling without a brief diversion to set the stage. Sampling procedures do not take place in a vacuum. Rather, they are an important component of survey research in education and the social sciences (the areas of application that are central to this book) and in other disciplines as well. So we will begin with an introduction to survey research methods that will describe the role of sampling in its larger research context. Of necessity, our discussion will be incomplete. Many volumes have been devoted to survey research methods, and we will recommend several for those who take the wise course of learning more about survey methodology than can be gleaned from this introductory chapter.

The objectives of survey research are to describe and to explain selected characteristics of some definable group of persons, institutions, or objects. These objectives are realized through the collection and analysis of carefully structured data from a relatively small group. The small group is intended to represent a larger group that constitutes the object of research.

On the surface, conducting a survey might appear to be a straightforward task. Just write a number of questions, assemble them to form a questionnaire or interview schedule, and ask some people what you want to know. What could be simpler? Doing surveys badly *is* easy. Examples abound in policy studies, educational research, and social inquiry. Unfortunately, collecting survey data that are trustworthy, appropriate to one's research purposes, and capable of supporting the conclusions typically advanced in educational and social research is both complicated and difficult. Notice that we did not say "impossible." When they are planned and conducted with thoughtfulness and diligence, surveys can be informative, reliable, and valid. In this chapter, we will give you some hints on making yours come out that way.

1

The balance of this chapter consists of three sections. In the next section, we list the major steps in planning and conducting a successful survey and illustrate the role of sampling in those steps. We next examine the objectives and assumptions of survey research more generally, as a prelude to a discussion of threats to the validity of survey results. We offer some practical advice on dealing with those threats. Finally, we provide a brief guide to further reading on survey research methods and some examples of their successful application in education and the social sciences.

Steps in Planning and Conducting a Survey

PROBLEM DEFINITION

One of the great temptations in survey research is to begin with the construction of a questionnaire. It is then easy to think of things that would be "nice to know" and add questions accordingly. Suppose, for example, that a researcher was investigating the perceived factors that lead students to terminate their formal education prior to graduation from high school (the school dropout problem). A questionnaire on this topic would surely include items on the age and grade at which the respondent quit school and basic demographic information such as the respondent's sex and ethnic origin. Beyond these questions, the field is wide open. One might seek data on the education of the respondent's siblings (if any) and parents, whether the siblings finished high school, whether the respondent's closest friends finished high school, the respondent's self-perceived academic ability, the respondent's report of school performance just prior to quitting school, the respondent's opinions about the value of formal schooling, the respondent's perceptions of peer and family pressures to leave school, the respondent's feelings about relationships with teachers and school officials, and so on. The range of possible issues related to dropping out of school is seemingly endless, and a questionnaire that dealt with all the issues would be endless as well. How does one draw the line? How can a survey researcher ensure that the truly important questions are asked and that needless questions are avoided? The answer lies in explicit problem definition.

The first step in designing a successful survey is carefully and completely specifying the research questions of interest. In our own work, we have found it useful to do this in a hierarchical fashion, beginning with the most general questions and moving in stages to the most specific. For example, we might structure several research questions on the school dropout problem as follows:

I. What social pressures are associated with dropping out of school?
   A. What pressures from peer groups are associated with dropping out of school?
      1. Are there higher dropout rates among students whose closest friends have recently dropped out of school?
      2. Are there higher dropout rates among students whose closest friends are perceived to devalue education?
   B. What family pressures are associated with dropping out of school?
      1. Are there higher dropout rates among students whose families are perceived to devalue education?

    2. Are there higher dropout rates among students whose parents did not complete a high school education?

    3. Are there higher dropout rates among students who feel that their parents wanted them to quit school?

II. What perceived economic pressures are associated with dropping out of school?

    A. Are there higher dropout rates among students who perceive family economic pressures?

        1. Are there higher dropout rates among students who feel they must work to support their families?

        2. Are there higher dropout rates among students who feel they must work to support themselves?

    B. Are there higher dropout rates among students who feel that school offers them no economic incentives?

        1. Are there higher dropout rates among students who feel that school will not equip them to get a better job?

        2. Are there higher dropout rates among students who feel that job markets will be worse the longer they stay in school?

Clearly, these two research questions do not exhaust the potential factors associated with a student's decision to quit school prior to high school graduation. They are offered as an illustration of the way major areas of research interest can be identified and elaborated. With hierarchically organized research questions, it is far easier to include issues that are of critical interest and exclude those that are not. For example, if the researcher had no interest in the perceived economic pressures associated with a decision to drop out of school, none of the more specific questions listed under Question II, our second major question, would have to be addressed. In turn, an entire set of questionnaire items on family economic status, students' beliefs about the economic value of education, students' beliefs about the current and future job market, etc., would be unnecessary.

The process of specifying the research questions that are of interest in a survey is often iterative. A preliminary list of major research questions is identified, and that list is discussed and refined several times. Previous research studies, discussions with experts in the field, and consultations with the ultimate users of survey results (if they can be identified) are often used to refine a set of research questions. As more specific research questions are proposed, the process of iteration and revision is repeated until the study is defined and bounded on the basis of required information and perceived feasibility. When other elements of the survey have been designed and preliminary data-collection activities have been evaluated, it may be necessary to further limit the focus of the survey by eliminating or recasting some research questions.

IDENTIFICATION OF THE TARGET POPULATION

In any empirical research study, there is some set of individuals, institutions, or other entities that constitutes the object of investigation. This object is called the *target population*. In survey research, perhaps more than with any other research methodology, the target population must be carefully and explicitly defined.

You will find in later chapters of this book that all the sampling methods presented share several critical requirements. One of these is that the population being sampled must be unequivocally defined. There must be no uncertainty as to whether a particular person, object, or institution is or is not a part of the population.

Defining a population might appear to be a trivially simple task. An example will illustrate the converse. Suppose you were interested in doing research on juvenile delinquency in a small city and that you wanted to sample the city's population of youthful offenders. What would your target population be? You would have to define the boundaries of your population very carefully. For example, what is a juvenile? Someone no more than 18 years old on his or her most recent birthday? Someone who has not yet had his or her eighteenth birthday? Someone who was no more than 16 years old on his or her most recent birthday? What is an offender? A person convicted of a crime by a court of law? The subject of a complaint to the city police department? What categories of crime should be included? Only crimes against persons? Property crimes? What about traffic violations? Assuming good answers to all these questions, what constitutes the city's youthful offender population? Youth convicted of committing a crime while within the boundaries of the city? Youth who reside in the city but have been convicted of committing a crime anywhere? This list of questions is extensive but certainly not exhaustive. The important point is that every combination of answers to these questions defines a different target population. And the results of a survey would differ, perhaps markedly, depending on how the target population was defined.

Often, there is no single "right answer" to the definition of a target population. The definition must depend on the purposes of the research and the interests of the ultimate users. In all cases, the definition will have to be discussed, modified, and, perhaps, negotiated before a satisfactory conclusion is reached. A crucial test of a satisfactory definition is whether the researcher can determine, unequivocally, whether any given person, object, or institution is, or is not, a member of the target population.

LITERATURE REVIEW

Once the purposes of a survey and the target population have been clearly defined, the next logical step is to determine whether one can build on the work of others in designing a survey. A review of literature on the general topic of interest can be extremely helpful in refining detailed research questions, suggesting additional research questions and procedures for securing specific information, and suggesting methods for analysis of data that might facilitate a comparison of current findings with those of previous researchers.

Sources of appropriate literature vary, depending on the topic of a survey. However, professional books and journals and computerized search services such as DIALOG and, in education, ERIC, are almost always appropriate and useful.

It is important to realize that surveys only rarely plow entirely new ground. If the purpose of a survey is what Cronbach and Suppes (1969) have termed "conclusion-oriented research," it is essential that the survey be solidly grounded in a body of relevant theory. Such grounding is possible only through a careful review of related disciplinary literature. Even if the purpose of the survey is "decision-oriented research," that is, research intended to serve a specific policy objective, consulting appropriate literature can prevent needless reinvention of previously used techniques and has the potential of facilitating comparison of results from the current survey with those of the past. Such comparisons are *strictly* valid only if the questions used in a previous survey are repeated, in exactly the same form, in a current survey. A simple change in the form of a question can evoke a dramatic change in the range of answers that result, rendering comparisons uninterpretable.

SELECTION OF SURVEY METHOD

Three basic methods of collecting data are available to the survey researcher: face-to-face interviews, telephone interviews, and mail surveys. Each has advantages and disadvantages, and selection of a method for a specific research study is a critical decision. Several books on survey research, including Moser and Kalton (1972), provide an extensive review of the characteristics of these three types of survey. We will consider only a few major points.

Face-to-face interviews have a multiplicity of advantages. They permit an interviewer to modify the sequence of questions, depending on the respondent's earlier responses. They ensure that a respondent only hears one question at a time, thus guaranteeing that later questions will not influence responses to earlier ones. This can be an important advantage in surveys designed to elicit opinions and judgments on subjects that range from the very general to the specific. Face-to-face interviews permit the unequivocal identification of the respondent (if a mail survey is sent to a business or residence, there is usually no way to tell who completed the survey questionnaire). Face-to-face interviews typically result in far higher rates of response than do mail surveys, since a respondent is less likely to refuse to cooperate with an interviewer on his or her doorstep than to toss a mailed survey questionnaire into a wastebasket. However, the response rates of mail surveys can often be increased markedly through careful planning and the use of appropriate follow-up procedures.

On the debit side, face-to-face interview surveys are often very costly compared to mail surveys—in many cases so costly as to be infeasible. Face-to-face interview surveys often require large expenditures for trained personnel, supervisory staff, and transportation. These costs are nearly avoided in mail surveys. Another problem with face-to-face interviews is that some respondents are more willing to provide confidential or embarrassing information through an anonymous questionnaire than to an on-site interviewer.

Telephone interview surveys enjoy many of the advantages of face-to-face interview surveys while avoiding many of their costs. Telephone costs are almost invariably lower than transportation costs. A telephone interview can be just as flexible as a face-to-face interview. However, it takes special skill to convince a respondent to provide information over the phone. It is almost as easy to say "no thanks" and hang up as to throw a mailed questionnaire in a trash basket. In general, rates of cooperation are lower in telephone surveys than in face-to-face interview surveys. Another advantage a face-to-face interview survey offers that is missed in a telephone survey is the opportunity for the interviewer to observe the respondent and his or her environment. Body language can be an important source of information in detecting a respondent's confusion, reluctance to respond, or veracity. In addition, through direct observation, a face-to-face interviewer can report on the type and quality of a respondent's living quarters and neighborhood, all of which can be important in social surveys.

The most obvious advantage of mail surveys is their relative economy. Because of budget limitations or the limited availability of trained or trainable interviewers, a mail survey may be the only feasible choice. In these cases, it is essential to evaluate the likelihood of obtaining accurate information through a mailed survey questionnaire. Remember that an interviewer can explain the purposes of a survey, the type of information requested, and the definitions of specialized terms. The mail survey questionnaire must stand on its own, regardless of the complexity of the desired information or the literacy and education level of the respondent. When the type of information sought is readily available fact or simple opinion, a mail survey should be totally appropriate. When complex judgments are required or the facts sought are difficult to obtain or recall, those who use mail surveys run the risk of collecting worthless information. Simplicity and clarity are essential in a mail survey questionnaire. When the subject matter of the survey makes these qualities impossible to achieve, an alternative form of data collection is indicated.

SECURE OR CONSTRUCT A SAMPLING FRAME

In order to select a sample of persons, objects, or institutions, one must have a list from which to sample. Such a list is called a *sampling frame*. In some studies, sampling frames are readily available. For example, a survey of major city governments in the state of California would make use of a sampling frame that could be obtained with little difficulty from any of a number of state agencies. A survey that required sampling of public schools in the state of North Carolina that enrolled fourth-graders could also make use of a readily obtainable sampling frame. A list of such schools is maintained on computer-readable magnetic tape by the state's Department of Public Instruction. In contrast, a survey that required a sampling frame of all fourth-grade students in a state would be very difficult, since it is unlikely that such lists exist. A sampling frame would have to be constructed, at great expense and, perhaps, with great difficulty if nonpublic as well as public school fourth-graders were to be included.

An available or constructable sampling frame provides an operational definition of the population to which survey results may be generalized. For example, even though the target population of interest might be all schools in New York state that enroll fourth-graders, a sampling frame that listed only *public* schools with fourth-grade classes would define a markedly different population. One task that often faces the survey researcher is to decide what to do about differences between his or her target population and the available sampling frame(s). Options include redefining the target population to match the sampling frame(s), augmenting the available sampling frame(s) to more nearly match the target population, or striking a compromise. In many cases, available sampling frames will be somewhat out-of-date, and it will be necessary to update current lists. An example from social research is the use of computerized lists of registered voters just prior to an upcoming election. Such lists often fail to include recently registered voters and must be updated using newly acquired forms if they are to be current.

We discuss the choice of a sampling method in a later section of this chapter and in the balance of this book. You will find that various sampling methods require different types of sampling frames and that the availability of a sampling frame of a particular type might strongly influence the choice of a sampling method. For example, a procedure that involved selection of public schools with fourth-grade classes and sampling of fourth-grade students only within sampled schools could be conducted using sampling frames that are readily available in most states. In contrast, a sampling method that required direct selection of fourth-graders from throughout a state would be impossible without the construction of an entirely new frame.

CONSTRUCT SURVEY INSTRUMENTS

When used in an interview survey, whether face-to-face or by telephone, the instrument with which information is collected from respondents is called an *interview protocol*. The instrument used to collect information in a mail survey is called a *questionnaire*. Doing a good job of constructing either type of instrument requires careful planning and meticulous attention to detail.

The desire to have each respondent answer exactly the same questions (but not necessarily give exactly the same answers) is fundamental in survey research. To achieve this goal, each respondent must understand and interpret the questions in precisely the same way. Clarity and precision in formulating and stating questions are obviously required. For example, suppose you were doing a survey of youth with the objective of understanding their transition from dependence on their parents to the independence of young adulthood. One question might concern the conditions of the youths' leaving the confines of their parents' homes to set up housekeeping on their own. The question, "How did you first leave your parents' home?", although a bit open-ended, might appear to be sensible and straightforward. What if you received responses ranging from "In my boyfriend's car" to "With great difficulty" to "Only after I had obtained a job and saved enough money to rent my own apartment"? Given those responses, it would be clear that the question meant very different things to the three respondents. The first thought you were asking about a means of conveyance. The second thought you were seeking information on the emotional components of the transition from dependence to independence. The third thought you wanted to know about economic feasibility. Although these responses tell you something about the individual respondents, that is *not* the objective of survey research. On the assumption that each respondent is answering the *same* question, the information collected in surveys is aggregated so as to infer the characteristics of some target population. Answers to different questions cannot be aggregated in any useful or meaningful way.

Beyond issues of clarity and concreteness, those who construct interview protocols and mail survey questionnaires must face the inevitable issue of what to include and what to leave out. It is at this point that a set of detailed research questions, such as those described earlier in this chapter, will prove invaluable. The temptation to include a question just because it "might be interesting to find out" must be avoided. The simplest way to handle the problem is to include only those questionnaire items or protocol questions needed to answer one or more of the research questions that have been defined in the problem-definition phase of survey planning. The research questions can also be used as a basis for judging the adequacy of survey instruments: Have the data needed to answer the research questions been sought through the questionnaire or interview protocol?

In bringing attention to the issues of clarity and coverage, we have barely initiated an adequate discussion of survey instruments. Doing a thorough job would require a large portion of an entire volume on survey research methods. Fortunately, a number of excellent sources exist, including Babbie (1973), Berdie and Anderson (1974), Moser and Kalton (1972), and Warwick and Lininger (1975). We will have more to say about these sources, as well as others, in a later section of this chapter.

DEFINE A SAMPLING PLAN

A plan for sampling potential respondents is a crucial element in survey design. The alternative possibilities are far more wide-ranging than you might imagine and, beyond this introductory chapter, constitute the subject of this entire book. At this point, we will only suggest that collection of data is the most costly component of a survey, whether conducted through face-to-face interviews, over the telephone, or through the mail. A soundly designed sampling plan that provides for collection of sufficient data to realize the research objectives of a survey, but no more, is an extremely wise investment. In later chapters of this book, you will see that alternative sampling designs that meet the same survey objectives can vary in their associated sample size requirements by a factor of 50 to 1. So selection of the most efficient sampling design has very practical, dollars-and-cents implications.

For the sake of clarity, we have discussed the steps in planning a survey in the form of an ordered list. In doing so, we have run the risk of making it appear that these steps should occur sequentially and independently. Such is *not* the case. Virtually all aspects of survey design and development depend on each other, and changing one component is likely to have serious effects on almost all others. For example, the range of potential sampling plans is limited by the availability of sampling frames. In the educational survey example used earlier, we indicated that a sampling plan that required selection of a sample of fourth-graders from an entire state could not be used without a statewide frame of all fourth-graders. Even the choice of a data-collection method limits the sampling possibilities in a very practical way. In a nationwide survey of adults, the travel costs associated with an interview survey might require that respondents be chosen from a few randomly selected contiguous geographical areas (such as a few large cities and their surrounding areas), rather than at random from throughout the nation. So the development of a sampling plan must be attentive to earlier survey design decisions, budget limitations, and specific research objectives. It is not at all unusual to modify earlier decisions on the research questions to be addressed or even the choice of a survey method once the realities of cost and complexity have been identified through a detailed sampling plan.

DESIGN FIELD PROCEDURES

Activities that take place in the actual collection and receipt of survey data are termed *field procedures*. In an interview survey, field procedures include activities such as recruitment, training, and deployment of interviewers, distribution of protocol materials to interviewers, assignment of potential respondents to interviewers, collection of completed protocols from interviewers, rules for handling repeated calls to potential respondents who cannot be reached through initial attempts (for example, because they are not at home), supervision of interviewers, quality control procedures (such as validation of initially secured data), and recordkeeping on interviews attempted and completed. In a mail survey, field procedures include control of and preparation of records on the distribution of mailed questionnaires, control of and preparation of records on the receipt of returned questionnaires, follow-up of nonrespondents, and preparation of records on follow-up attempts. Field procedures can become quite complex in a survey that involves multiple stages of questionnaire distribution and receipt. For example, in an educational survey that sought data from public school teachers during their working hours, it is likely that many school systems would not allow questionnaires to be sent directly to teachers or to be returned directly to the survey research office. Typically, questionnaires would have to be distributed to the central offices of school systems, then sent to the principals of sampled schools, and finally distributed within schools to sampled teachers. The same procedure might have to be followed in reverse when collecting the completed questionnaires. At each step in the procedure, the possibilities for delay or loss of survey materials are obvious. To avoid (or at least reduce) these problems, field procedures might include the appointment of a survey coordinator in each school system's central office and in each school. These persons would be responsible for one stage of questionnaire distribution and receipt and would notify the survey office when they had completed each of their tasks. In this way, any delayed or lost materials could be traced immediately to a particular step in the distribution and receipt process, and the likelihood of success in obtaining useful data would be increased considerably.

PLANS FOR REDUCTION AND EDITING OF DATA

Data collected in a survey are rarely in a form that allows immediate analysis either by hand or by using a computer. Despite the best efforts of survey planners, field supervisors, interviewers, and coordinators, the resulting data are inevitably incomplete, partially contradictory, and (hopefully only to a limited degree) outlandish or impossible. In addition, the form of the data—a sequence of circled options on survey questionnaires or the words and marks of interviewers on completed interview protocols—does not permit immediate analysis. Two prior steps, data reduction and data editing, must take place before analysis can begin.

Data reduction is a procedure for transforming the data from the form in which they are received (either on questionnaires or interview protocols) to a form that can be processed by hand or by computer. When so-called closed option questions are used (questions that permit the choice of one or more of a number of fixed responses, such as those that appear in a multiple-choice test), data reduction might consist of keypunching numbers corresponding to the selected options onto computer-readable punched cards, entering the data directly into a computer through a video terminal, or using an optical mark reader to automatically transform the data from the original response documents to a magnetic tape that can be read by a computer. You have probably completed standardized tests by marking circles or boxes on a separate answer sheet with a number two pencil. It is possible to print survey questionnaires that are handled in exactly the same way. Respondents fill in circles to indicate their answer choices, and these choices are detected and recorded automatically when the questionnaires are passed through an optical scanning machine. The procedure is very fast and reliable and avoids the inevitable clerical errors that result from keypunching.

Editing of data involves the detection and resolution of errors. Errors can include omission of requested data or inclusion of incorrect data (either through carelessness or intent). Omitted data can sometimes be retrieved through additional interviews or estimated by using other data provided by a respondent. Plans for editing must include rules for detecting omitted data and plans for action when omissions have been found. Incorrect data can sometimes be detected through checks on the internal consistency of a respondent's answers. For example, a city manager who indicated a city population of 5000 persons and employment of 1000 full-time workers in his or her city government would undoubtedly be mistaken in one or the other of these figures. The error could be resolved through a telephone follow-up or by accepting one or the other figure as correct on the basis of additional information (for example, a list of names of the workers employed by that city) and adjustment of the other figure (for example, estimating that 500 citizens were served by each city worker and adjusting the total employment roll to 10).

In most surveys, some editing takes place before the original response documents are processed by computer. Interview protocols might be reviewed by interviewers in the field, with on-the-spot resolution of errors through additional questioning. Mail survey questionnaires might be scanned quickly by hand to ensure that several crucial questions have been answered completely and that reasonable responses have been obtained. Telephone calls might be used to resolve any omissions or unrealistic responses, as soon as the questionnaires have been received.

An additional editing phase usually takes place after the survey data have been reduced to computer-readable form. It is easy to program a computer to flag all responses that are not within an acceptable range (for example, negative numbers or numbers greater than 10 in response to a question on the number of children in the family) or to flag all responses that are not consistent with each other (for example, a respondent claims never to have married but indicates that he or she has two grandchildren). Errors detected by computer can be handled either by having the computer identify (flag) the offending record and then resolving the error by hand or by applying rules for the resolution of errors (for example, if the respondent's age is missing, estimate it by adding the age at which formal schooling ended and the number of years of postgraduation work experience). All the rules for detecting and resolving errors must be specified in a plan for data editing.

When "open-ended" questions are used in a survey instrument (questions that allow a respondent to construct, rather than select, a response), responses will be in narrative form. If survey data are to be analyzed quantitatively (by tabulating frequencies of response and constructing tables that display statistics on aggregate responses), these narratives must be reduced to a form that permits categorization and counting. Narrative responses are usually reduced to analyzable form by developing a set of mutually exclusive codes and then applying the coding scheme to the narratives so as to place a respondent's answer into one or more categories. For example, a survey question might ask, "How often do you and your family go out in the evening for entertainment?" A response might be, "Well, we eat at a restaurant about twice a week, and I go bowling once a week. My wife and I both play bridge every other Thursday, and we sing in a community choir once a month." To obtain consistent data across respondents, an operational definition of "entertainment" would have to be developed. Would eating at a restaurant count as entertainment? The question as stated asked about "you and your family." If a husband goes bowling without any other family member, does that count as family entertainment? Should types of entertainment (for example, active participation versus spectator status) be coded differently? All these rules have to be specified clearly and unequivocally in a data-reduction plan, so as to avoid their arbitrary imposition, in a myriad of inconsistent ways, when different response documents are analyzed. Development of coding rules and training of coders are critical components of a data-reduction plan whenever open-ended questions are used.

PLANS FOR ANALYSIS OF DATA

One of the best ways to ensure that the data collected through survey instruments are consistent with the research questions specified in initial survey planning is to develop detailed plans for data analysis. These plans should be developed prior to data collection. We have found a detailed analysis plan, together with a document that links each analysis to one or more items in a survey instrument and one or more research questions (called a *crosswalk*), to be an invaluable aid in sound survey design.

Traditionally, the statistical methods used to analyze survey data have been simple and direct. Because the data that result from most closed-option questions are categorical in nature, simple frequency tables and cross-tabulations of the sort illustrated on the following page are most often used.

The numbers in this hypothetical table are *cell percents*. They represent the percent of total youth crimes committed during 1983 that fall into each cell defined by the intersection of a type of crime and a section of the city. Other types of cross-tabulations often include *row percents*—the percent of instances of the type defined by a row of the table that fall into each cell of that row (for example, the percent of Commercial property crimes that occurred in each section of the city)—and *column percents*—the percent of instances of the type defined by a column of the table that fall into each cell of that column (for example, the percent of crimes in the Northwest section of the city that fall into each of the type-of-crime categories).

In addition to tabulations of data, survey results are often reported through graphs and figures and, increasingly, through multivariate statistical procedures that allow consideration of a number of survey variables in the same analysis. Weisberg and Bowen (1977) provide extensive information on sophisticated methods for analyzing survey data.

TABLE 1.1
Percent of Crimes Committed by Youth in 1983

Type of Crime	Section of the City			
	Northwest	Northeast	Southwest	Southeast
Street crime against persons	6.2	0.4	5.4	3.1
Commercial property crime	10.5	4.2	12.6	0.2
Residential property crime	4.2	5.7	0.6	19.8
Residential crime against persons	2.9	3.5	0.1	12.6

Regardless of the methods to be used with survey data, it is important to specify the analytic procedures in detail prior to collecting the data. If tabulations are to be used, it is helpful to construct *table shells*—blank tables with all titles, labels, row headings, and column headings completed. If simple statistics are to be computed, it is useful to enumerate the types of statistics that are desired (for example, the average age of respondents or the correlation between respondents' numbers of years of formal schooling and current annual income). Once a complete enumeration of desired statistical results has been specified, it is possible (and recommended) to construct a crosswalk showing which questionnaire or interview protocol items will provide the data needed to complete the desired tables or compute the desired statistics and which research questions will be answered by producing the tables and computing the statistics. A three-part crosswalk document that is indexed by research questions in one part, by questionnaire or interview protocol items in a second part, and by table shells and statistical analyses in a third part provides a complete check on the internal consistency of the planned survey: Every research question has associated data sources in the survey instruments and associated statistical analyses that are to provide an answer. Every questionnaire item or interview protocol item has at least one associated research question (that is, a reason for being in the survey instruments) and an associated statistical analysis (that is, plans for using the responses to the item). Every statistical analysis is being conducted so as to answer at least one research question and has associated sources of data that will permit the analysis to be completed.

Construction of detailed data analysis plans and elaborate crosswalk documents is a very time-consuming task. However, in our experience, the benefits more than outweigh the costs.

PLANS FOR REPORTING SURVEY RESULTS

In the preceding section, we have tried to convey our belief that a survey researcher should specify quite clearly, prior to actual data collection, what will be done with the data to be collected. A logical step in this process is the preparation of plans for reporting survey results. A reporting plan should include a detailed outline of all reports that are to result from the survey and a listing of the intended audiences for the reports. Although specifying the content of reports prior to collecting and analyzing the survey data might appear difficult, the process is made much less tedious by the list of research questions prepared during the initial survey planning steps and by the detailed plans for analysis of data prepared in the preceding step. Often, reports can be organized around major research questions, and a set of table shells can be used to structure a report within major sections. Again, this planning step might appear to be more trouble than it is worth, but it is a great comfort to know that, barring unforeseen difficulties, the survey *as designed* will yield the information necessary to produce all the desired reports.

PILOT SURVEY

A pilot survey is an essential part of every survey research study. A pilot survey is a miniature "dry run" of the main survey. It serves many important purposes: A pilot survey can be used to determine the response rate or cooperation rate likely to be encountered in a main survey, and this, in turn, can be used to formulate a judgment on the feasibility of conducting the main survey as planned. For example, a mail survey of convicted youthful offenders would be totally infeasible if a pilot survey suggested a likely response rate of 25 percent. A pilot survey can be used to verify the availability and condition of intended sampling frames. A pilot survey can be used to examine the clarity and adequacy of survey instruments. Recall that in a mail survey, it is essential that respondents understand each question and the directions for responding to it without any personal assistance. If any part of the survey is unclear or ambiguous, a well-designed pilot survey will allow the researcher to detect the problem. In an interview survey, it is essential that interviewers follow directions precisely and interpret a given protocol item in exactly the same way. By analyzing the response distributions produced by different interviewers in a pilot survey, it is often possible to detect lack of clarity in protocol items and directions. Finally, a pilot survey can be used to validate the feasibility of planned field procedures: Do the forms distribution plans work? Do the right people receive questionnaires? Are the interviewers able to locate the respondents assigned to them? Do the procedures for controlling receipt of documents work? Can omission of crucial information be remedied by using supplementary data-collection procedures? Do the computer programs used to scan questionnaires and flag errors work as planned? Can a range of responses to an open-ended question be used to suggest appropriate options for a closed-option question in the main survey? This list, although extensive, is incomplete. The function of a well-designed pilot survey is to allow as many difficulties as possible to be detected and corrected, before it is too late.

Regardless of the experience and expertise of a survey designer, it is impossible to foresee everything that might go wrong in an operational survey. No professional survey researcher would presume that he or she is capable of such a feat. Only a novice would undertake a major survey research study without at least one pilot survey.

REVISION OF SURVEY COMPONENTS

A pilot survey helps in determining what's wrong; it does not solve the problems. After a pilot survey has been conducted and the results have been analyzed, the next step is to revise those survey components that appear to be problematic. Since every component is related to every other, changes in one part usually require changes in all other parts. For example, if a questionnaire or interview protocol were revised, the new instrument might not provide the data needed to answer every desired research question. Therefore, the research questions would need revision. The plans for analysis of data, including the table shells, presume that specific questionnaire or protocol items will be used. They, too, would have to be modified if the survey instruments were changed. If the sampling frames were found to be deficient (either because they did not contain data thought to be available for classifying sampling units or because they were out-of-date, or for some other reasons), the sampling plans might have to be changed. Or it might be necessary to redefine the target population, thus affecting some research questions and reporting plans.

This need for revision should not be discouraging. If the pilot survey has done its job, problems will have been detected in advance of the main survey, and few unfortunate surprises will await the survey researcher after the fact.

CONDUCTING THE SURVEY

The steps in survey research that we have discussed to this point have involved planning and field testing. The tasks of conducting the main survey, reducing and editing the data, analyzing the data, and reporting the results still remain. These tasks are rarely easy and never go exactly as planned despite the survey researcher's best efforts. However, careful planning and use of planning results are the keys to success.

The response rate in a main survey will almost always differ somewhat from that suggested by a pilot survey. Some questionnaire or protocol items will produce a higher proportion of omitted responses than had been anticipated or, perhaps, multiple responses where only one had been expected. It will not be possible to apply a detailed data-analysis plan slavishly, since some desired data will not be available and initial analyses will suggest others that illuminate research questions in ways that could not be anticipated without poking through the data. Nonetheless, detailed planning will stand a survey researcher in good stead. He or she will not be faced with a mound of data, having no idea what to do with it. Those constant regrets of survey researchers who have not planned, for example, "If only I had asked X and Y, I would have been able to . . . ," will be minimal, if present at all. Short of problems that no one could anticipate in advance, the researcher will be able to find answers to the myriad of questions that gave birth to his or her study in the first place. Taking time to plan a survey carefully yields ample rewards.

The Objectives of Survey Research—A Little Theory

GENERALIZATION IN EMPIRICAL RESEARCH

Survey research shares many characteristics and objectives with other forms of quantitative, empirical research. Among these characteristics is the near universal quest for generalization. Generalization is at the core of empirical research, including research in education and the social sciences. In its broadest sense, we generalize whenever we go beyond observed cases. Even when we restrict ourselves to observed cases, we generalize when we imply anything about those cases that has not been observed directly.

Generalization can be informal—as in the suggestion that it is probably going to rain today, since there are dark clouds in the sky—or formal—as in the scientific statement that collection of data on all cases of a certain type would result in a specific value for the average of some characteristic of those cases. Inferential statistical procedures, including sampling methods, facilitate formal generalizations. In particular, they allow a researcher to calculate or estimate probabilities that quantitative generalizations will be correct or, in other cases, to construct probabilistic statements on the range of likely quantitative generalizations.

In later chapters, these statements on the use of inferential statistics in empirical research will be supported through detailed illustrations of sampling procedures. At this point we will merely attempt to convey the flavor of quantitative generalizations and the ways inferential statistics are used in such generalizations. For those with some background in statistics, our illustrations may seem superficial. At this point, however, we will forego formal definitions and the details of computational mechanics in favor of basic concepts. In later chapters we will provide at least enough detail.

Two types of statistical inference are used in quantitative, empirical research: hypothesis testing and estimation. Hypothesis testing is used most often in experimental research and less frequently when analyzing survey data. Estimation is used in virtually all analyses of survey data. Both procedures can be illustrated through simple, hypothetical examples. We will consider hypothesis testing first.

Suppose that you wanted to know the average IQ score of all 9-year-old students who were regularly enrolled in the public schools of the Midville school system (a mythical school system somewhere in the middle of the country). You could calculate the desired figure directly, by administering an IQ test to each of Midville's 10,000 9-year-old students and then averaging their scores. But if your sole interest was in the students' average IQ and you did not care about the IQ scores of individuals, testing all 10,000 students would be very wasteful of time and money. It would make far greater sense to select a sample of students from the 10,000 and use the average IQ score of the sampled students to infer the average you would have found had you tested all 10,000.

It is time for some new vocabulary: The 10,000 9-year-old students who are regularly enrolled in Midville's public schools compose a *population*. The average IQ score of these 10,000 students is an example of what is called a *population parameter*. The average IQ score of a sample of students that is selected from the population of 10,000 is called a *sample statistic*. In quantitative, empirical research, statistical inference takes place when a sample statistic is used to infer the value of a population parameter. Hypothesis testing is just one form of statistical inference, which we will now consider in greater detail.

Suppose that, in the absence of any information to the contrary, you believed that the average IQ of Midville's 10,000 9-year-old students was exactly the same as the average IQ of 9-year-olds throughout the nation: 100 IQ points. Without data that would allow you to compute the actual value of the 10,000 students' average IQ, this statement of belief is an hypothesis. In statistical parlance, it is called a *null hypothesis*. The converse of a null hypothesis, in this case a statement that the average IQ of the 10,000 9-year-olds equals some value other than 100, is called an *alternative hypothesis*. The result of an hypothesis test is a decision as to whether a null hypothesis, or an alternative hypothesis, is true. The decision is based on evidence gleaned from the data in a sample.

Now suppose that you were to select a sample of 200 students from Midville's population of 10,000 by using one of the scientific sampling procedures presented in a later chapter of this book. You would then administer an IQ test to the 200 sampled students and then compute their average IQ. You would use the value of this sample average as a basis for deciding whether you still believed that the entire population of students had an average IQ equal to 100 (in which case you would be "retaining your null hypothesis") or whether you now believed that the population average was different from 100 (in which case you would be "rejecting your null hypothesis"). The logic is as follows: Suppose that the average IQ of the 200 students was 120 points. Isn't it highly unlikely that you would find a well-selected sample of 200 students, chosen from a population with an average IQ of 100, that had an average IQ of 120? Of course it is! In the face of this sample average, the only sensible action would be to give up your initial belief that the population of 10,000 students had an average IQ of 100. Now we did not say that a population average of 100 was impossible; only that it is so unlikely as to be unbelievable in the face of the sample data.

But what if the average IQ of the sample of 200 students had been some value very close to 100, say 101? It is very plausible that a well-selected sample of 200 students would have an average IQ of 101 when chosen from a population with an average IQ of 100. With a sample average IQ of 101, there would be no good reason to give up your initial belief that the population average equaled 100. In other words, you would be wise to retain your null hypothesis.

Whether you decide to retain or reject a null hypothesis, you run the risk of making the wrong decision. When a null hypothesis is rejected on the basis of sample data, despite the fact that it is true, the mistake is called a *Type I error*. So if the population really had an average IQ of 100 but you decided to reject the null hypothesis because the average IQ of your sample of students was 120, you would be making a Type I error. The other kind of error you could make would be to retain your belief in the truth of the null hypothesis, despite the fact that it was false. For example, it might be the case that the average IQ of the population of 10,000 students was really 105, but the sample you chose had an average IQ of 101. You would therefore take the reasonable action of retaining your null hypothesis that the population average equaled 100. You would then make a mistake called a *Type II error*.

There is no way to eliminate the risk of making Type I or Type II errors, short of collecting data on the entire population of 10,000 students. However, it is possible to specify in advance the risks you are willing to take of making these types of errors and to control those risks by choosing a sufficiently large sample. How the risks of making Type I and Type II errors are related to sample size need not concern us here since the details are provided in later chapters. However, it is important to realize that statistical hypothesis testing allows you to limit the risks of either falsely rejecting a null hypothesis that is true or falsely retaining a null hypothesis that is not true. No less formal method of making inferences provides such advantages.

We noted above that another form of statistical inference, called *estimation,* is used quite frequently in survey research. Our example involving the average IQ of Midville's 9-year-olds will also serve to illustrate statistical estimation.

Suppose that you selected a sample of 200 students from Midville's population of 9-year-olds, just as in the hypothesis testing example. When you tested the 200 students, suppose you found a sample average IQ of 102 points. You could use the sample average of 102 as a *point estimate* of the population average: the average you would have found had you tested all 10,000 students. In doing so, you would be applying a totally acceptable and frequently used statistical procedure.

Although point estimates are ubiquitous, they often lead to the question, "How close is my estimate likely to be to the true value of the population parameter?" Of course, there is no way to tell for sure without measuring the entire population and computing the value of the population parameter directly. However, there is another form of estimation, called *interval estimation,* that can shed some light on the question.

Were you to use interval estimation in investigating the average IQ of Midville's 9-year-olds, you might make a statement such as the following: "I am 95 percent confident that the average IQ of Midville's 9-year-old public school students is between 97 and 107." Notice that in this statement you are not giving a single estimate of the value of the population average. Instead, you are giving a range of values that you believe includes the true population average. Before we conclude this example, some additional vocabulary is necessary. The statement in quotes, "I am 95 percent confident . . . ," is an example of a *confidence interval.* The value 97 is an example of a *lower confidence limit.* The value 107 is an example of an *upper confidence limit.* The percentage figure, 95, is an example of a *level of confidence.* All these figures are related to each other.

We hope the advantage of using an interval estimate of the population average rather than a point estimate is clear. With an interval estimate, you not only specify a range of values that you think will include the population average, you give an indication of how sure you are that you are right. A statement that you are "95 percent confident" means that you would give 20-to-1 odds were you to bet that the true population average falls between 97 and 107. Here again, we have tried to give the flavor of interval estimation without becoming mired in formalism and detail. A more complete exposition can be found in later chapters.

STATISTICAL AND SUBSTANTIVE GENERALIZATION

The data collected in survey research consist of answers to relatively few specific questions asked of a relatively small sample of respondents. The survey researcher generalizes from this set of data by making statements regarding the facts, judgments, opinions, and conditions of some larger population of persons, institutions, or objects. The researcher generalizes on both statistical and substantive grounds. At this point, we will describe the types of generalizations that typically take place when survey data are interpreted. In doing so, we hope to alert you to some pitfalls and to make you more keenly aware of the inferential leaps you will propose when you write your own survey reports and which others will ask you to accept when they invite you to agree with their interpretations.

Two types of error plague valid statistical generalizations: bias error and random error. Statistical generalization of survey results assumes that both types of error are sufficiently small. Bias error occurs when a sample of respondents fails to represent the target population. This can occur for a variety of reasons, including faulty sample design, a high nonresponse rate, or errors in conducting the survey. An inadequate sampling frame can also lead to substantial bias error. For example, if the target population includes all registered voters in the state but the sampling frame consists of a listing of registered Democrats, obvious bias will result from any interpretation of voting patterns, positions on social issues, etc. Nonresponse in survey research is seldom random. The people who respond to surveys are rarely like those who do not, on any issue of interest. If nonresponse is high enough, serious distortions can occur when sample data are generalized to a target population. In later chapters, we treat the question of adequate response rates more explicitly and try to answer the question of how much nonresponse is too much. Even though a sampling plan is well-designed, bias error can occur if the plan is not carried out correctly. Inadequate randomization when drawing a sample, usually arising from the use of some physical process (such as drawing names from a box) rather than use of a random number table as described in later chapters, is a frequent problem in small surveys conducted by nonprofessionals.

Some random error is inevitable in survey research whenever sampling is used. However, the magnitude of random errors can be controlled by drawing a sample that is sufficiently large. How large is large enough depends on a number of factors, including the purposes of the survey and the characteristics of the population that is being sampled. Once again, we must ask you to withhold questions on details until we deal with sample size requirements explicitly in later chapters.

Statistical generalization presumes that an adequate sampling plan has been developed and that the survey has been conducted in complete accord with this plan, so as to minimize both bias and random errors. To the extent that this is true, sample data can be generalized to the target population on statistical grounds. Any flaws in sampling design or in sampling practice threaten the validity of statistical generalizations.

Valid substantive generalization of survey results is even more difficult than statistical generalization, since it involves a multiplicity of assumptions. These are illustrated schematically in Figure 1.1. The first assumption listed is that the questions asked on the survey instruments were "construct valid." Construct validity is a fundamental and, unfortunately, complex issue. It is discussed in full detail in Cronbach (1971) and in Cronbach and Meehl (1955). Our treatment here will be brief and incomplete. A construct is an unobservable, constructed variable. It is a label that is attached to a consistent set of behaviors. Researchers use constructs in virtually all social science interpretations, often without explicit recognition of their actions. For example, an opinion survey of voters might be interpreted by saying that "voters have a positive attitude toward the party in power." But what is an "attitude?" It is not directly observable. It is a summarizing label that is attached to a set of agreements with positive statements about the

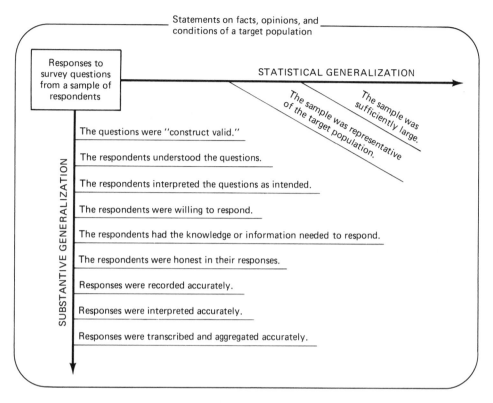

FIGURE 1.1. **Statistical generalization and substantive generalization in survey research.**

party in power. It is an inference *from* the "yes" responses to a series of questions *to* a conclusion about the meaning of those responses. In short, it is a construct. Is it valid to assume that a voter who marks "yes" in response to instructions to agree or disagree with each of a series of statements on the party in power really has a "positive attitude" toward that party? Do statements such as "The party has acted responsibly on economic issues," "The party has the interests of the wage earner at heart," and "The party has a number of strong leaders" provide an adequate sample of potential judgments? What if the statement, "The party exemplifies honesty in government," had been added? Would the "yes" responses still be in the majority? This example barely initiates an adequate discussion of the construct validity question. Constructs are defined in the context of a body of theory. Appropriate demonstrations of construct validity show that the implications of that theory are supported by, and not contravened by, research findings.

Given that the questions asked of respondents tapped the constructs of interest, a number of other issues must be faced before substantive generalizability of survey results is assured. Did the respondents understand the questions that were asked? Just because they provided answers is no assurance that respondents understood the questions, particularly with closed-option questions. Even if questions were understood at some level, respondents might not have interpreted them as the survey researcher intended. The seemingly straightforward question, "How do you feel?" admits a number interpretations that are perfectly valid in different contexts, as exemplified by the responses, "With my hands" and "Fine, thank you." Substantive generalization of survey results assumes that the respondents understood the survey questions and interpreted them as the researcher intended.

Respondents must have been willing and able to respond to the survey questions if generalizable results are to be obtained. Willingness is suggested by cooperation and is easily assessed. Ability presumes that the respondents had available to them the knowledge or information needed to respond. This is often a critical factor when a survey seeks facts, but it can also play an important part in securing valid judgments and opinions. If you were asked your opinion on animal vivisection for use in cancer research, would you know how you felt about the issue right now? What if you were given only 15 seconds to think of an answer and respond, as often happens in interview surveys? Interpretation of responses to opinion surveys assumes that respondents know their opinions. Interpretation of responses to surveys that seek factual information assumes that respondents were willing to provide the facts and had the facts available in a form that would permit accurate response. In both cases, honesty is a critical assumption. If respondents are dishonest in their replies, no valid generalizations can be made.

Recording errors are an added threat to substantive generalization. Valid interpretation of survey results presumes that respondents accurately recorded their responses in mail surveys, and that interviewers accurately recorded responses in interview surveys. Random recording errors that are rare in their occurrence pose no serious threat to validity, but systematic errors, such as omitting "no opinion" responses, can seriously bias interpretations.

In interview surveys that incorporate open-ended questions, interviewers often must interpret what a respondent says and then record that interpretation or select one of a number of response codes as a representation of the response. Valid generalization presumes that such interpretations have been made accurately and consistently.

Finally, responses must be accurately transcribed to the medium in which they will be analyzed (usually a computer tape or disk) and then aggregated accurately if valid generalizations are to be made.

This list of requirements for valid statistical and substantive generalization may appear forbidding, but it is not overstated. We hope that the need for care in survey planning and practice is now abundantly clear. In addition, we hope that you will keep these issues in mind as you interpret the results of surveys that you conduct as well as those conducted by others. Look carefully at the questions that were asked and think about them as you write or read the interpretations. Examine the sampling plan and the rate of response or cooperation that was achieved in the survey. Think about the likely availability of information to respondents whenever the survey concerns facts and conditions. Think about the opportunity and willingness of respondents to form and divulge their judgments whenever opinions are involved. Be inquisitive and withhold final judgment until you are convinced that generalization of findings is warranted and appropriate.

A Guide to Further Reading

In this brief overview, we have attempted to introduce some of the most important practical and theoretical ideas surrounding survey research. We hope our discussion of practice will help you avoid some pitfalls, and the theoretical matter will start you thinking about survey research in some new and useful ways. But our treatment of theory and practice is far from complete. To gain more than a surface view of survey research, you will have to spend time in the rich literature on the method. Here are some suggestions for additional reading.

SOURCES ON METHOD

Two books by E. R. Babbie, *The Practice of Social Research,* and *Practicing Social Research,* both published by Wadsworth (Belmont, Calif.: 1975a,b), discuss survey research in the context of other social research methods. The comparisons Babbie makes provide illuminating contrasts between survey research and other approaches to social science.

An earlier book by E. R. Babbie, entitled *Survey Research Methods* (Belmont, Calif.: 1973), is an elementary text that covers all phases of survey research, from initial survey planning through analysis of data and reporting of results. It is easy to read and does not presume any prior background in survey methods.

Two more general texts on survey method, *The Sample Survey: Theory and Practice* by D. P. Warwick and C. A. Lininger (New York: McGraw-Hill, 1975) and *Survey Method in Social Investigation* by C. A. Moser and G. Kalton, 2d ed. (New York: Basic Books, 1972), also cover all phases of survey research. Each is more complete than Babbie's introductory book but requires somewhat more sophistication on the part of the reader. Moser and Kalton use examples from Great Britain, and some of their terms and writing constructions reflect British language customs, but the richness of detail in their work more than compensates for the inconvenience of occasionally novel language. They also include a concise review of sampling methods.

The mechanics of questionnaire design, primarily for mail surveys, is well-presented in D. Berdie and J. Anderson's *Questionnaire Design and Use* (Metuchen, N.J.: Scarecrow Press, 1974). They offer practical hints on increasing the clarity of questionnaires, holding the respondent's interest, designing introductory letters, etc.

Years of research on the psychology of interviewing are reflected in *The Dynamics of Interviewing* by R. L. Kahn and C. F. Cannell (New York: Wiley, 1957). The book offers practical guidance on various interviewing techniques and methods, with careful consideration of situational variables, all soundly based on solid research evidence.

Methods of analyzing survey data are discussed in J. A. Davis's *Elementary Survey Analysis* (1971) and in J. A. Sondquist and W. C. Dunkelberg's *Survey and Opinion Research: Procedures for Processing and Analysis* (1977), both published by Prentice-Hall, Englewood Cliffs, N.J. Davis's book is, as the title suggests, a short introduction to survey analysis. In contrast, the Sondquist and Dunkelberg book is more complete and more demanding. Another more sophisticated book that treats the analysis of survey data is H. F. Weisberg and B. D. Bowen's *An Intro-*

duction to Survey Research and Data Analysis (San Francisco, Calif.: W. H. Freeman, 1977). All these books provide good ideas on how to analyze survey data so as to maximize their usefulness.

REPORTS ON IMPORTANT EDUCATIONAL AND SOCIAL SURVEYS

One of the most effective ways to learn about a research method is to study the results of exemplary practice. In the last 15 years, a number of impressive nationwide surveys have been conducted to examine various social institutions and the nature and status of our educational system. Although they are not without fault, these surveys illustrate the application of sound research practice, often on a massive scale. The reports of the surveys described below are often available in federal government document repositories at college and university libraries throughout the United States.

EQUALITY OF EDUCATION IN THE UNITED STATES

This survey was conducted in 1965–1966 as a result of the Civil Rights Act of 1964. That act directed the Secretary of Health, Education, and Welfare to report to the Congress, no later than July 4, 1966, on the status of racial isolation in the public schools of the United States. The survey was directed very broadly at the issue of school segregation and included an attempt to investigate the effect of segregation on the school achievement and attitudes of public school students from the early elementary grades through the last year of high school. A massive mail survey resulted in data from school superintendents, school principals, teachers, and students in every state. Nearly half a million students were surveyed and tested in the study.

One of the objectives of the survey was to provide separate estimates of the degree of school segregation for urban and rural populations in various geographic regions of the nation. In addition, the quality of schooling and the effects of schooling were to be compared for members of various racial groups.

These objectives dictated a complex sampling design that would yield estimates of a number of parameters for a number of subpopulations. The design used was two-stage sampling, with counties and metropolitan areas as sampling units in the first stage. Secondary schools were sampled within selected counties and metropolitan areas at the second stage of sampling. All elementary schools that sent a sizable percentage of their students to selected secondary schools were then added to the sample. In all, data were collected from 1170 schools, of which 349 were in metropolitan areas and 821 were rural. Both counties and schools were stratified, so as to ensure a large minority representation in the sample.

The "Coleman Survey," as the study has come to be known, has been criticized for a variety of reasons, including a response rate of 67 to 70 percent. Nonetheless, it is one of the landmark surveys of education in the United States and warrants careful review by any serious student of survey methods. The results of the survey are contained in a large report entitled *Equality of Educational Opportunity* by J. S. Coleman et al. (1966). It is available from the U.S. Government Printing Office, Washington, D.C., and can also be obtained from the ERIC Document Reproduction Service under number ED012 275.

NATIONAL ASSESSMENT

The purpose of this continuing survey is to assess changes over time in the educational achievement of four age groups throughout the United States and to produce comparative statistics on these changes for four geographic regions (northeast, southeast, central, and west), two racial groups (white and black), and seven categories of size and type of community, ranging from "main section of a big city" to "extreme rural."

Achievement test exercises in 10 subject areas have been administered for over a decade now to children and young adults, both in and out of schools, in four age groups: 9-year-olds, 13-year-olds, 17-year-olds, and young adults (age 26–35).

The National Assessment of Educational Progress (NAEP) is so designed that no examinee is required to answer every test question or to complete every exercise. This use of multiple test booklets, coupled with a design for sampling examinees, is known as *multiple matrix sampling*.

The design for sampling NAEP examinees in schools includes three stages of sampling. At the first stage, counties or groups of contiguous counties are sampled. At the second stage, both public and nonpublic schools are sampled within selected first-stage sampling units (counties or groups of counties). Finally, at the third stage of sampling, random samples of pupils are selected within each sampled school and pupils are randomly assigned to one of the test booklets. The design for sampling 17-year-olds and young adults who are not in school is equally complex.

You can secure additional information on this study from a report entitled *National Assessment of Educational Progress. General Information Yearbook* (1974).

THE GALLUP POLL

The Gallup organization conducts a large number of surveys for a wide variety of purposes. One such survey provides data for a monthly report on social, political, and economic issues, based on opinions expressed by members of the adult population of the United States living in private households. This survey has been conducted regularly since 1935.

The Gallup Poll is an interview survey. The respondents are selected by using a multistage sampling design in which the nation is divided into geographic regions and categories of community size. Census Bureau data are used to create these strata. A random sample of 320 locations is selected from the geographic and size-of-community strata, so as to ensure balanced coverage of the entire nation. Interviewers in each location are given maps with a randomly selected starting point and detailed instructions on sampling of respondents from households in their assigned area. The selection procedure is designed to yield a representative sample of the adult (18 years old and older) population living in private households in the United States.

You can find more information on this survey in *The Gallup Report,* a monthly periodical published by the Gallup Poll in Princeton, N.J.

Another private polling organization that conducts many surveys of the general population is the Harris Poll. The *Sourcebook of Harris National Surveys* by

E. Martin, D. McDuffee, and S. Pressler, published by the Institute for Research in Social Science, University of North Carolina at Chapel Hill in 1981, contains information on these surveys.

CONSUMER SURVEYS

The Survey Research Center, which is part of the Institute for Social Research at the University of Michigan, is one of the leading centers of survey research in the nation. Among the center's projects is a periodic survey of households in the United States, conducted since 1946, to obtain data on finance-related issues. Members of households are interviewed to obtain information on their attitudes toward personal finance, the economy, and market conditions or prices.

A sample of dwelling units that is representative of the entire nation is selected through a four-stage design. Dwellings are selected from 74 areas in 37 states and the District of Columbia. The United States is first divided into four large geographic regions (northeast, northcentral, south, and west). Next, counties and large metropolitan areas are assigned to one of 74 categories (strata) that are balanced on total population. Then these primary categories are divided and subdivided into successively smaller sampling units containing city blocks or clusters of addresses. At the fourth stage of sampling, 5 to 10 small segments are selected from each third-stage sampling unit. Each segment contains four housing units, from which one adult respondent is randomly selected to be interviewed. About 1500 interviews are conducted for each survey.

Additional information on these surveys can be found in *Surveys of Consumers: Contributions to Behavioral Economics,* edited by R. T. Curtin and published by the Institute for Social Research at the University of Michigan in Ann Arbor (1979).

The National Longitudinal Study of the High School Class of 1972 by W. B. Fetters, published by the National Center for Educational Statistics, U.S. Department of Health, Education, and Welfare in 1974, contains a report on the design and results of a massive survey of high school students in the United States.

The survey was designed to secure data on the flow of students through the nation's secondary schools and into postsecondary education or part-time and full-time employment. The data were intended to assist in identifying the major points at which adolescents make decisions that affect their educational and occupational choices during the 6- to 8-year period following their departure from high school (as graduates or nongraduates).

Data were collected from over 21,000 seniors in a stratified sample of 1200 high schools, as well as from counselors and principals in those schools. Students reported on their high school experiences, their attitudes and opinions about high school and careers, their plans for the future, and a variety of biographical information. A school questionnaire, completed by principals or their designates, was used to secure data on each school's grade structure, curriculum, absence rate and dropout rate, racial-ethnic composition, age of buildings, etc. A counselor questionnaire was administered to two randomly selected counselors in each sampled school. It was used to secure data on counselors' experience, education, assignments, activities, methods, workloads, and resources.

The initial survey was conducted late in the 1972 school year. Only 87 percent of the initially selected schools agreed to participate, although repeated follow-up surveys resulted in a 94 percent participation rate by students.

One important feature of the survey is its longitudinal design. Students who supplied data during the 1972 survey have been resurveyed periodically to gain information on their later educational and career choices and experiences. At present, three follow-up surveys have been completed. Various reports, available from the National Center for Education Statistics, U.S. Department of Education, Washington, D.C., provide information on the results of these follow-up surveys.

2

Basic Concepts of Sampling Theory

As is true in virtually every specialized field of study, sampling theory has its own technical vocabulary. In this chapter, we introduce a portion of that vocabulary by describing the basic structural elements that form the underpinnings of sampling theory. Some of these elements relate to the collection of data (sampling) and others to the analysis of data once collected (estimation). Several of the topics discussed in detail in this chapter were introduced informally in Chapter 1. Our discussion here differs from that in Chapter 1 mainly in its precision and its attention to statistical concerns.

Some of the topics we discuss in this chapter (such as populations and samples) may be familiar to you from general courses in applied statistics. Others (such as sampling frames and without-replacement sampling) may be new, since they have special significance in problems that involve sampling from finite populations.

Defining Populations and Constructing Sampling Frames

POPULATIONS

In any sampling study, there is a definable group or aggregation of elements to which the results of the study are to be generalized. This aggregation of elements is called the *target population* of the study. There is also a definable group or aggregation of elements from which samples are selected. This aggregation of elements is called the *operational population*. In the best of all worlds, the target population and the operational population are identical. In reality, these two populations often differ—a situation that will be discussed in the section on sampling frames.

26

In educational research studies, populations are often composed of students or educational institutions. In social surveys, populations can consist of many types of elements, including persons, institutions, geographic units, or political units. In behavioral science studies, the populations of interest are almost always animate but need not be human.

Precise definition of the target population is essential in studies involving sampling. The researcher must be able to determine, unequivocally, whether any given element falls within or outside the population. As we pointed out in Chapter 1, defining the target population precisely is not as simple as it might at first appear. For example, in an educational research study, the researcher might want to generalize to the population of fourth-grade students in a selected school system. This definition of the target population is, unfortunately, fraught with equivocal cases. Does it include fourth-grade students who attend nonpublic schools as well as those who attend public schools? Does it include students who attend nongraded schools but who would, because of their age, be placed in the fourth grade were they to attend a graded school? Does it include fourth-grade students who reside within the boundaries of the school system but attend school outside the system? Does it include the converse of these students, that is, fourth-graders who reside outside the boundaries of the school system but attend school within the system? Does it include or exclude fourth-grade students who attend "special" schools, such as schools for the physically handicapped or mentally or emotionally impaired? Answers to these questions are critically important in defining the target population. In fact, different answers define a different target population and will lead to different procedures for sampling and generalization of survey results.

The sampling and estimation procedures discussed in this book assume that populations consist of a finite number of elements. Classical statistical procedures of the sort encountered in most applied statistics courses assume that populations from which samples are drawn are infinitely large. Sampling from a finite population sometimes results in greater statistical efficiency and in other cases makes little practical difference. Because samples are assumed to be drawn from finite populations, you will find that some statistical formulas in this book differ from those you have encountered in earlier statistics courses.

SAMPLING UNITS

Populations are made up of elements termed *sampling units*. The sampling units into which a population is divided must be unique, in the sense that they do not overlap, and must, when aggregated, define the whole of the population of interest. Because of the way public schools are organized in the United States, sampling units arise naturally in many educational research studies; they often consist of students, classrooms, schools, or school systems. These elements are easily identified as unique units and are readily counted. In social science research, it may be more difficult to define appropriate sampling units, since the populations of interest might not be subdivided into standard administrative categories. Also, social science studies often admit a larger number of choices of sampling units. For example, what sampling units would be best in a study of the sociology of mental illness? Should one sample patients? Mental institutions? Public health clinics? Clinical psychologists and psychiatrists in private practice?

Each choice is likely to lead to a decidedly different study and to different conclusions.

When "selecting a sample," one is in fact selecting some number of sampling units from the aggregation that forms the operational population. For a unit to be selected, it must be identifiable. A list that uniquely identifies all the units in a finite population, in a particular order, is termed a *sampling frame*. The sampling frame is equivalent to the operational population and, in fact, defines the operational population.

We discussed sampling frames briefly in Chapter 1. We pointed out that desired sampling frames are not always readily available and that one of the first tasks of a survey research study might be to assemble a sampling frame from existing archival data or to update an existing sampling frame that is out of date.

If a sampling frame consists of all the elements in the target population and additional elements as well, we say that it contains *overregistration*. A sampling frame that contains fewer elements than the target population is said to suffer from *underregistration*. Neither condition is necessarily fatal in most survey research studies, although detection of the conditions prior to sampling is essential. If overregistration can be detected and if the sampling units that fall outside the target population can be identified, it might be possible to eliminate the effects of overregistration by modifying an intended sampling plan. If underregistration can be identified prior to sampling, it might be possible to modify the sampling frame through updating or some other procedure that adds omitted units.

In some studies, underregistration in one sampling frame is handled by adding a separate frame that contains the omitted units. For example, in political surveys, a list of newly registered voters could be added to a frame of voters who had maintained their registration by voting in the previous election. In surveys of public school teachers, it might not be possible to obtain a statewide sampling frame early in a school year. The sampling frame from the previous school year would undoubtedly contain both underregistration and overregistration, since some teachers employed during the previous school year would have left the profession or the state and other teachers would have joined the teaching ranks for the first time during the current school year. This problem could be handled by selecting a sample of school systems, and then having the sampled school systems update last year's sampling frame of teachers by eliminating those no longer teaching and adding the names of new teachers.

Desirable Properties of Sampling Procedures

The theory of sampling provides a variety of methods for selecting units from a finite population. These methods, called *probability sampling procedures,* have several characteristics in common. First, the methods are always applied to populations in which the units that compose the population and the units that are excluded from the population are explicitly defined. Second, every potential sample of a given size that could be drawn from the population, together with the probability of selecting each sample, must be specifiable prior to sampling taking place. It is not necessary that every potential sample have the same probability of selection, just that each selection probability can be specified. Third, every sampling unit in the population must have a probability of selection that is greater than zero; thus every sampling unit must have *some* chance of being sampled. Samples selected through probability sampling procedures are said to *represent* the population from which they were drawn.

The concept of probability sampling might be better understood by considering the converse—sampling methods that violate at least one of the rules of probability sampling procedures. We will consider an example from education first and then a classic case from the field of political science. Assume that the mean reading achievement of sixth-grade pupils in a city school system is to be estimated by testing pupils in half of its elementary schools. The sampling problem is to select schools in which sixth-grade pupils will be tested. Several simple procedures are available. The schools in which testing is conducted could be selected from a specified section of the city that contains half the city's elementary schools, or for convenience, schools closest to the city's school research office could be selected. With the second method, the school closest to the research office would be selected first, the second closest selected next, and so forth, until half the elementary schools in the city had been "sampled."

We can show that these two sampling methods are not probability sampling procedures by considering each ot the three defining characteristics discussed above. The first characteristic is satisfied: the units that compose the population and those excluded from the population are explicitly defined; all elementary schools with sixth-grade classes are in the population, and all other schools are not. The second characteristic is also satisfied: all samples of schools of a given size can be enumerated, and the probability of selecting any potential sample of schools can be specified. In the first sampling procedure, samples composed only of schools in a contiguous area of the city would be selected with a probability of one, and samples containing any schools outside this area would have selection probabilities equal to zero. In the "convenience" sampling procedure, using the city's school research office as a starting point, the "sample" of schools closest to the school research office would be chosen with a probability of one, and all other samples would have selection probabilities of zero. These methods both fail to possess the third characteristic of a probability sampling procedure. In each method, half the sampling units in the population have selection probabilities equal to zero. In the first method, schools outside the contiguous area of the city chosen for sampling could not enter the sample. In the "convenience" method, schools that were more than half way down a list arranged by distance from the city's school research office could not enter the sample.

Intuitively, one might suspect that these sampling methods would not yield samples that were representative of the citywide population of elementary schools with sixth-grade classes. Populations in cities are grouped into neighborhoods in which family incomes and occupational levels of heads of households are relatively homogeneous. Since these factors are closely related to school achievement (Coleman et al., 1966; Burkhead, 1967), selection of schools from a contiguous area for estimation of the citywide sixth-grade reading achievement mean would probably result in an overestimate or an underestimate, depending on which "side of the tracks" was selected. The sampling procedure based on convenience would suffer from the same deficiencies. The city's school research office would be a part of the city government and would likely be located somewhere near the heart of the city. Since neighborhood socioeconomic status generally increases with distance from the center of the city, the convenience procedure would probably lead to an underestimate of the citywide sixth-grade reading achievement mean. Neither of these methods of school selection could be expected to result in a sample that was representative of all the city's schools with sixth-grade classes. The population that *was* represented by the samples resulting from these methods could not be easily defined and would not be of general interest.

A classic case of bias in sampling that arose in the field of political science could be interpreted as having resulted from the use of a non-probability sampling procedure. In 1936 the *Literary Digest* conducted a massive nationwide poll of potential voters in an attempt to predict the outcome of that year's U.S. presidential election. The *Digest's* sample size was in excess of 10 million persons, although the survey achieved a response rate of only 20 percent. The *Digest* predicted that Herbert Hoover would win the election by a 20 percent margin based on the poll. When the election was over, the *Digest* found it had predicted the right winning margin, but the wrong winning candidate. Apart from its low response rate, the principal problem with the *Literary Digest* survey was its use of a sampling procedure in which potential voters who did not have a telephone or own an automobile had no chance of entering the sample. The sampling frame for the poll was constructed from telephone directories, automobile registration lists, and similar sources. Although automobiles and telephones seem to be necessities of life in the United States today, they were luxuries in 1936 and were far more likely to be possessed by supporters of Mr. Hoover than by supporters of Mr. Roosevelt. The *Literary Digest's* sampling procedure was not a probability sampling procedure for the same reason that selection of schools nearest a city's school research office is not a probability sampling procedure: a large number of sampling units were assigned selection probabilities of zero. The resulting sample did not represent the population of registered voters in the United States who voted in the 1936 presidential election.

Desirable Properties of Estimation Formulas

SOME DEFINITIONS

In addition to providing procedures for collecting data, sampling theory provides formulas for estimating such population characteristics as means, totals, and proportions. Statistics computed from sample data are termed *estimates,* and the values they seek to estimate are termed *population parameters.* Formulas for computing sample estimates are termed *estimators.* Some examples might help to clarify these terms.

Suppose one wanted to know the mean welfare payment received by persons assisted through the Los Angeles County Welfare Office during the month of November, 1983. Assume that the mean payment is to be estimated by reviewing only a sample of cases from the county's records. All persons assisted through the Los Angeles County Welfare Office during the month of November, 1983 would constitute the target population. The mean welfare payment received by all persons in this population is an example of a population parameter. An estimate of this parameter could be formed by computing the mean welfare payment received by persons whose files were sampled from the county's records. The formula used to compute the sample mean is an example of an estimator of the population mean. The usual formula for the sample mean (sum the welfare payments of sampled individuals and divide the sum by the sample size) is not the only estimator that could be used in this problem. Whether a better estimator could be found depends on the procedure used to sample welfare recipients. For example, separate sampling frames could be constructed for male recipients and female recipients, and the sample could be composed of specified numbers of persons from each list. With this sort of sampling scheme, a weighted average of the sample mean for males and the sample mean for females would likely be a better estimator of the population mean than would the sample mean of the two groups combined.

Several different estimators are discussed in conjunction with the probability sampling procedures presented in Chapters 3 through 7. When used with their respective sampling procedures, these estimators often have several desirable properties. In keeping with the literature of sampling theory, we will term these features *properties of estimators,* even though they actually depend on the sampling methods used as well as the estimation formulas.

ESTIMATOR BIAS

When a population is finite, the number of different samples that can be selected from it is also finite. Furthermore, each sample can be defined in terms of the sampling units of which it is composed.

For example, suppose a social service agency has five interviewers, and a researcher wishes to sample two of the five. All the potential samples can readily be identified. The sampling frame in this problem consists of a list of the interviewers, labeled, say, from 1 to 5. Ten different samples of two interviewers each can be constructed:

Sample	Sampling Units (Interviewers) in Sample
1	1, 2
2	1, 3
3	1, 4
4	1, 5
5	2, 3
6	2, 4
7	2, 5
8	3, 4
9	3, 5
10	4, 5

TABLE 2.1
Alternative Estimators of the Mean Caseload of a Population of Five Interviewers

Interviewer	Caseload
1	14
2	18
3	18
4	16
5	24
Population mean:	90/5 = 18

Sample	Interviewers	Average Caseload	Largest Caseload
1	1, 2	16	18
2	1, 3	16	18
3	1, 4	15	16
4	1, 5	19	24
5	2, 3	18	18
6	2, 4	17	18
7	2, 5	21	24
8	3, 4	17	18
9	3, 5	21	24
10	4, 5	20	24
Average of estimates:		180/10 = 18 (unbiased)	202/10 = 20.2 (biased)

Bias = 20.2 − 18 = 2.2

Suppose that the average monthly caseload of the entire population of five interviewers was of research interest. If the average monthly caseloads of the interviewers in each two-person sample were to be computed, these averages would vary. It is likely that none of the 10 sample averages (each a sample estimate) would coincide with the average monthly caseload of all five interviewers (the population parameter).

Assume that each potential sample had an equal chance of being selected and that once an interviewer had been sampled, he or she could not be chosen again. With this sampling plan, the average of the 10 sample estimates would necessarily equal the average for all five interviewers (the population parameter). Because this is true, the estimator of the average for all five interviewers would be unbiased.

More generally, if the average value of sample estimates is computed for all samples of a given size and, regardless of the sample size, that average is equal to the population parameter being estimated, the estimator is said to be *unbiased*. An estimator is said to be *biased* if, for all samples of a given size, the average of all sample estimates is either larger or smaller than the population parameter. The *amount of bias* is defined as the difference between the average of all sample estimates and the value of the population parameter.

Suppose that the five interviewers in the previous example had the caseloads shown in Table 2.1. If we wanted to estimate the mean caseload of the five interviewers from those of a sample of two interviewers, we could use any of several estimators. The most obvious (and the best) choice would be the average caseload of the sampled interviewers. This estimator is unbiased, as illustrated by the data in Table 2.1. Notice that the average of the sample estimates is 18, which is equal to the true mean caseload of the population of five interviewers. Another possible estimator, considered here for purposes of illustration only, would be the largest caseload of the two sampled estimators. This estimator is biased, as the data in Table 2.1 show. The average of the largest caseloads is 20.2, which is 2.2 points higher than the true population mean of 18.0. The bias of this estimator thus equals 2.2.

VARIANCE, MEAN SQUARE ERROR, AND EFFICIENCY

An *error of estimation* is defined as the difference between the value of an estimate and the value of a population parameter. The *variance* of an unbiased estimator is defined as the average of squared errors of estimation. The average is taken over all potential samples of a given size.

For a given sampling procedure and sample size, the most desirable unbiased estimator is the one with the smallest variance. The smaller the variance of an unbiased estimator, the lower the probability that large estimation errors will occur; in the extreme case of zero variance and no bias, every sample would yield an error-free estimate since every estimate would equal the population parameter.

Computation of the variance of an unbiased estimator is illustrated in Table 2.2. In Table 2.1, the average caseloads of samples of two interviewers were used to estimate the average caseload of a population of five interviewers. The variance of this estimator is computed in Table 2.2 and is found to equal 4.2.

The variance of a biased estimator is also defined as the average of squared differences. Here, each difference equals an estimate minus the average of

TABLE 2.2
Variance of an Unbiased Estimator of the Mean Caseload of a Population of Five Interviewers

Sample	Interviewers	Average Caseload	Estimate − Population Mean	Squared Difference
1	1, 2	16	−2	4
2	1, 3	16	−2	4
3	1, 4	15	−3	9
4	1, 5	19	1	1
5	2, 3	18	0	0
6	2, 4	17	−1	1
7	2, 5	21	3	9
8	3, 4	17	−1	1
9	3, 5	21	3	9
10	4, 5	20	2	4
	Total:	180	0	42

Variance = 42/10 = 4.2

estimates over all potential samples of a given size. The average of the squared differences is also taken over all potential samples of a given size.

Although it is desirable that estimators be unbiased, a biased estimator will usually be preferred when two conditions hold: first, the degree of bias is small; second, the variance of the biased estimator is considerably smaller than that of alternative unbiased estimators.

A measure of average estimation error that can be used to choose among biased and unbiased estimators is the *mean square error* (MSE). The mean square error of an estimator equals the sum of the estimator variance and the square of the estimator bias,

$$\text{Mean square error} = \text{variance} + (\text{bias})^2$$

Computation of the variance and mean square error of a biased estimator is illustrated in Table 2.3. You may recall that we used the largest caseload of a sample of two interviewers as an estimator of the average caseload of a population of five interviewers. The variance of this estimator is equal to 14.8, and with a bias of 2.2, the mean square error equals 19.64, which is more than four times the size of the variance of the unbiased estimator considered earlier. Intuitively, one would not think that a population mean could be estimated accurately or efficiently by using the largest observation from a sample. The data in Table 2.3 confirm this intuition.

For an unbiased estimator, the mean square error reduces to the variance, since the bias is zero. For a given sample size, an estimator that has a smaller mean square error than another is said to be more *efficient* than the other. When

TABLE 2.3
Variance of a Biased Estimator of the Mean Caseload of a Population of Five Interviewers

Sample	Interviewers	Largest Caseload	Estimate- Population Mean	Squared Difference
1	1, 2	18	0	0
2	1, 3	18	0	0
3	1, 4	16	−2	4
4	1, 5	24	6	36
5	2, 3	18	0	0
6	2, 4	18	0	0
7	2, 5	24	6	36
8	3, 4	18	0	0
9	3, 5	24	6	36
10	4, 5	24	6	36
	Total:	202	22	148

$$\text{Variance} = \frac{148}{10} = 14.8$$

$$\begin{aligned}\text{Mean square error} &= \text{variance} + (\text{bias})^2 \\ &= 14.8 + 2.2^2 = 19.64\end{aligned}$$

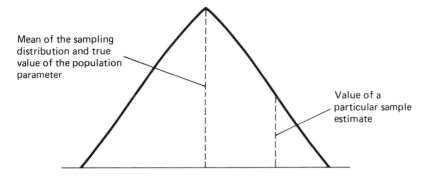

FIGURE 2.1. **Sampling distribution of an unbiased estimator.**

different sampling procedures are employed, a less efficient estimator might be preferred if its sampling procedure is more convenient or less costly, even though the sample size required to achieve a given mean square error is larger.

Figures 2.1 and 2.2 are intended to illustrate the concepts of bias, variance, and mean square error. Figure 2.1 shows the *sampling distribution* of an unbiased estimator. A sampling distribution of an estimator is the frequency distribution that would result if estimates of a population parameter were computed for all potential samples of a given size, and then all of these estimates were tabulated and plotted. Notice that the distribution in Figure 2.1 is centered around the true value of the population parameter and that even though some samples provide estimates that differ markedly from the population parameter, the average of all sample estimates (the central value of the sampling distribution) equals the population parameter.

In Figure 2.2, the sampling distribution is centered around an average that differs from the true value of the population parameter. The difference between this average and the value of the population parameter is equal to the bias.

A single sample estimate is noted in each of Figures 2.1 and 2.2. In Figure 2.1, because the estimator is unbiased, the variance of the estimator would be calculated by subtracting the value of the population parameter (assuming it were known) from each sample estimate, squaring the differences thus found, and averaging these squared differences across all sample estimates. This procedure was illustrated in Table 2.2. In Figure 2.2, with a biased estimator, the

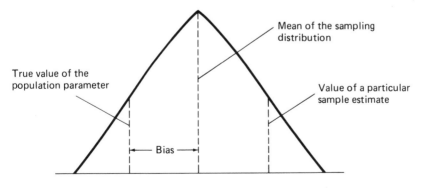

FIGURE 2.2. **Sampling distribution of a biased estimator.**

variance would be calculated by subtracting the average of all sample estimates (the mean of the sampling distribution) from each sample estimate, squaring these differences, and then averaging the squares across all sample estimates. The mean square error would equal this variance plus the square of the bias. This procedure is illustrated in Table 2.3.

CONSISTENCY

Errors of estimation are almost inevitable when population parameters are estimated from sample data. However, the magnitude of probable errors can and should be controlled. With some sampling and estimation procedures, the magnitude of probable estimation errors can be reduced by drawing a larger sample until, finally, estimation errors are eliminated completely when the sample size is equal to the population size. Such procedures are said to provide *consistent* estimation. In this book, a procedure is said to provide *inconsistent* estimation if one of two conditions hold: (1) the mean square error of the estimator does not get steadily smaller as the sample size is increased, or (2) estimation errors can occur even when the sample size equals the population size. The definition of a consistent estimator given here is unique, but it is one we have found to have practical value. It is closer to the definition provided by Cochran (1977, p. 21) than to the more rigorous statement found in Hansen et al. (1953, pp. 72ff). In the infinite-population case, Hansen et al. define a sequence of estimators as consistent if the sequence converges in probability to the value of the population parameter. Additional assumptions are then made to extend this definition to the finite-population case (Hansen et al., 1953, p. 74).

Figure 2.3 illustrates the mean square error of a consistent estimator when the sample size is increased to the point that it equals the size of the population. Notice that the mean square error associated with a larger sample size is never larger than the mean square error associated with a smaller sample size. Also, when the sample size (n) is equal to the population size (N), the mean square error is equal to zero, indicating that the sample estimate is free of error.

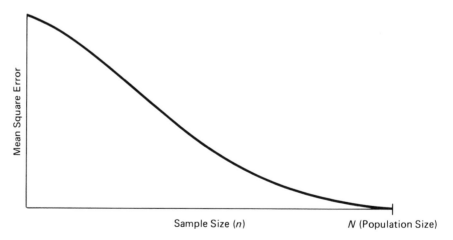

FIGURE 2.3. **Mean square error of a consistent estimator as sample size (n) is increased to equal population size (N).**

Although consistency is usually discussed as a property of estimators, it depends a great deal on the sampling plan as well. In many sampling procedures, units are sampled *without replacement*. That is, once a unit is sampled, it is removed from the population so that it cannot be selected a second time. When sampling is done without replacement, an unbiased estimator is usually, but not always, consistent. In contrast, when units are sampled *with replacement* (that is, an element can enter the sample more than once), estimators are often inconsistent even when they are unbiased. In sampling with replacement, there are many different samples that are the same size as the population. Each of these samples can lead to a different estimate of the population parameter, so the mean square error is larger than zero. In sampling without replacement, there is only one sample that is the same size as the population.

Consistency is a desirable property of estimators. However, an inconsistent estimator can still be useful, particularly when the required sample size is only a small fraction of the population size. In Chapter 7, we will discuss several with-replacement sampling and estimation procedures that provide efficient and useful estimates, even though they are inconsistent.

A Note on Succeeding Chapters

Chapters 3 through 7 present a variety of sampling and estimation procedures. Each procedure is illustrated with concrete, numerical examples from education or the social sciences. The situations portrayed in these examples are fictitious but are intended to approximate reality.

The examples from education are unique in that all of them employ the same set of real achievement test data. These examples resemble situations that have occurred in school administration and therefore have the potential of modeling realistic practice. In addition, use of the same data set with a variety of sampling and estimation procedures affords the opportunity of comparing results.

Rational selection among sampling and estimation procedures depends on a number of factors including the population to be sampled, the shapes of sampling distributions, the magnitudes of population variances, and the magnitudes of intercorrelations among sampling-related variables. Thus a procedure that is best when applied to one population might not be very efficient when applied to another. Keeping this in mind, the specific results presented in this book should be generalized with caution. Data for most of the examples from education were provided by a single school system. The sampling and estimation procedures that worked best with these data might not be optimal when applied to data from another school system. However, we feel reasonably confident that the general patterns of efficiency found with these data are likely to apply to reading achievement scores for sixth-graders from a number of school systems.

The examples from the social sciences are more likely to make use of fictitious data. They are included primarily to illustrate the great variety of practical situations in which sampling and estimation methods can facilitate social science research.

Numerical results are shown in detail in examples presented early in each chapter. We suggest that you take the time to verify each numerical value, so as to ensure your understanding of the underlying formulas. In addition, exercises are provided at the end of each chapter, including this one. We strongly suggest that you complete each exercise and then check your results with the answers provided in Appendix E. As is true of virtually all quantitative subjects, understanding of sampling and estimation methods comes only through application. Unless you attempt to solve problems, you will never know that you can actually apply sampling and estimation methods in useful ways.

Exercises

1. What kinds of decisions would have to be made in order to develop an unequivocal, operational definition of a target population of "health care facilities" in a medium-sized city?

2. If you wanted to conduct a survey of parents with children enrolled in the public schools of a small school system, how might you obtain or develop a sampling frame? Cite at least three possibilities, and list some of the advantages and disadvantages of each.

3. Which of the following are probability sampling procedures?

   A. The population consists of all paid-up members of a country club. Each member is assigned a unique number. Twenty percent of the members are sampled by drawing numbered slips of paper from a well-stirred bowl.
   B. The population is the same as in part A. Members are listed alphabetically, and then the first twenty percent of the members are selected from the alphabetically arranged list.
   C. In a public opinion survey, a television station assigns a reporter to a busy downtown street corner and instructs her to interview the first 25 people who pass by.

4. An estimator has a bias of -3.0 and a variance of 16.0. What is the mean square error of the estimator?

5. Two estimators of the same population parameter have the following bias and variance values:

Estimator	Bias	Variance
A	0.0	8.0
B	0.5	6.0

Which estimator is more efficient, A or B?

3

Simple Random Sampling: Procedures, Estimators, and Applications

The theory of sampling from finite populations provides statistical models for a number of sampling procedures that are useful in educational and behavioral science research. This chapter describes one such procedure and presents estimators of parameters that can be used to describe a variety of social, behavioral, and educational characteristics.

Definition of Simple Random Sampling

Of the procedures considered in most reviews of the theory of sampling from finite populations, simple random sampling is based on the most elementary probabilistic model. Although simple random sampling is generally not the most efficient procedure available, it is reviewed here for two reasons. First, the simplicity of the model helps to clarify concepts common to all probability sampling procedures. Second, the efficiency of simple random sampling provides a convenient standard against which to appraise the efficiencies of more complex procedures.

In simple random sampling from a finite population, once the sample size has been determined, each of the finite number of samples that can be drawn has an equal probability of selection. The number of samples of size n that can be drawn without replacement from a population of size N is given by the binomial coefficient

$$b(n;N) = \frac{1 \cdot 2 \cdot \cdot \cdot N}{(1 \cdot 2 \cdot \cdot \cdot n)[1 \cdot 2 \cdot \cdot \cdot (N - n)]} \tag{3.1}$$

Equation (3.1) is not as mysterious as it might at first appear. In fact, its derivation is so simple that we will violate our general rule of omitting proofs in this book, so as to illustrate the flavor of proofs in finite-population sampling theory.

Think of the problem of drawing a sample of size n from a population of size N as one of filling n slots in a sequence:

$$\underline{1} \ \underline{2} \ \underline{3} \ \underline{4} \ \cdots \ \underline{n}$$

The first slot could be filled with any of the N elements in the population. After that slot had been filled, $N - 1$ elements would remain in the population, and any of them could be used to fill the second slot. After the first two slots had been filled, $N - 2$ elements would remain, and each would have the potential of filling the third slot. If you were to continue reasoning this way, once you reached the last slot to be filled, the nth, you would see that $n - 1$ elements had been used to fill preceding slots. Thus $N - (n - 1) = N - n + 1$ elements would remain in the population and would have the potential of filling the nth slot. By this logic, the number of ways in which n elements could be sampled from N is given by the product $N(N - 1)(N - 2) \cdots (N - n + 1)$. However, there is one flaw in this argument. The product we have found gives the number of ways n elements can be selected from a population of size N and arranged in all possible ways. But if a sample consists of a given set of elements, we define it as the *same* sample, regardless of the order in which the elements appear. For example, a committee consisting of Bob, Sue, and Dick is the same committee, regardless of the order of the names (for example Dick, Sue, and Bob or Sue, Bob, and Dick). So to determine the number of distinct samples of n elements that could be drawn from N, it is necessary to *divide* the product given above by the number of different ways the n elements could be arranged, once they had been selected. This number can be found by considering the n slots once again. Any of the n sampled elements could be placed in the first slot. Once that slot had been filled, any of the remaining $n - 1$ elements could be placed in the second slot. After the first two slots had been filled, any of the remaining $n - 2$ elements could be placed in the third slot, and so on. Finally, when all but the nth slot had been filled, only one element would remain to fill it. So the number of different ways a sample of n elements could be arranged is given by the product $n(n - 1)(n - 2) \cdots 2 \cdot 1$.

The number of distinct samples of size n that can be drawn from a population of size N is given by the number of different ways n elements can be selected and arranged when drawn from a population of size N, divided by the number of different arrangements of the n elements, after they have been selected:

$$\frac{N(N - 1)(N - 2) \cdots (N - n + 1)}{n(n - 1)(n - 2) \cdots 2 \cdot 1}$$

This fraction can be seen to be equivalent to Equation (3.1) by multiplying the numerator and denominator by the same product, $(N - n)(N - n - 1) \cdots 1$, and reversing the order of the numbers:

$$\frac{N(N-1)(N-2)\cdots(N-n+1)}{n(n-1)(n-2)\cdots 1}\frac{(N-n)(N-n-1)\cdots 1}{(N-n)(N-n-1)\cdots 1}$$

$$=\frac{1\cdot 2\cdots N}{(1\cdot 2\cdots n)[1\cdot 2\cdots (N-n)]}=b(n;N)$$

When a simple random sampling procedure is used, each of the $b(N;n)$ samples in the population has a probability of selection equal to $1/b(n;N)$. Selection of samples is accomplished using a table of random numbers, which is discussed in Appendix A.

Since the intent of this book is to illuminate the application rather than the theory of sampling from finite populations, we will omit all remaining proofs of statistical results. A number of excellent books on the theory of sampling and estimation are available for readers who require proofs of major theorems or discussion of sampling and estimation methods that are more complex than those reviewed in this book. We recommend Hansen et al. (1953, Vol. II), Sukhatme (1964), Kish (1965), Raj (1968), Som (1973), and Cochran (1977) as good sources for additional reading.

Estimating Population Means and Totals

Many problems in education and the behavioral sciences require knowledge or estimation of population means or totals. For example, a city planner might want to know the total number of families who will seek low-income housing units during the coming year, or a school superintendent might be interested in the average age of students at the time of their graduation from high schools in his or her school district. Both types of parameters can be estimated from data secured through simple random sampling. We will consider first the estimation of means and then the slight modification of estimators needed to accomodate population totals. Estimation of population percentages and proportions is considered in a later section of this chapter.

FORMING THE ESTIMATES

The arithmetic mean of data collected through simple random sampling without replacement is an unbiased and consistent estimator of the mean of the population from which the sample was drawn. If n units are selected from a population of size N, the sample mean is given by

$$\bar{y}=\frac{1}{n}\sum_{i=1}^{n} y_i \tag{3.2}$$

where y_i denotes the value of the dependent variable for the ith sampled unit. This provides an unbiased estimate of the population mean

$$\bar{Y}=\frac{1}{N}\sum_{i=1}^{N} y_i \tag{3.3}$$

Similarly, the expression

$$y = N\bar{y} = \frac{N}{n} \sum_{i=1}^{n} y_i \tag{3.4}$$

is an unbiased and consistent estimator of the population total

$$Y = \sum_{i=1}^{N} y_i \tag{3.5}$$

ESTIMATOR VARIANCE

The probability theory underlying simple random sampling provides a basis for sample selection and a formula for unbiased estimation. However, additional information is needed to use simple random sampling in practice. If a population mean or total is estimated through any probability sampling procedure, some error of estimation will necessarily result. For practical application of a sampling procedure, the tolerable level of estimation error must be specified. It is necessary to state the precision with which the population parameter must be estimated in order to determine how large a sample should be used.

A fundamental assumption employed throughout the theory of sampling from finite populations is that estimators of population parameters have a normal probability distribution. Since a normal distribution is completely determined once its mean and variance are known, this assumption allows statements on the precision of estimation to be formulated solely in terms of means and variances. Consequently, determination of the variances of sample estimators receives considerable attention in the sampling theory literature.

If samples are drawn using simple random sampling without replacement, the variance of the sample average is given by

$$V(\bar{y}) = \frac{S^2}{n} \frac{N - n}{N} \tag{3.6}$$

where

$$S^2 = \frac{\sum_{i=1}^{N} (y_i - \bar{Y})^2}{N - 1} \tag{3.7}$$

and all other terms have been defined previously.

The variance of the sample mean is often written in the form

$$V(\bar{y}) = \frac{S^2}{n} (1 - f) \tag{3.8}$$

where $f = n/N$, the fraction of the population that is sampled. The standard error of $\bar{y}$ is equal to the square root of this variance.

Determination of the variance of the sample mean from Equation (3.8) requires a score for every element in the population. This is true since S^2 depends on the squared deviation of the score for each population element y_i from the population mean $\bar{Y}$. Since the population variance is rarely known when the population mean is unknown, an estimator of $V(\bar{y})$ is required to determine the precision of the sample mean $\bar{y}$.

An unbiased estimator of the variance of the sample mean is given by

$$v(\bar{y}) = \frac{s^2}{n}(1 - f) \tag{3.9}$$

where

$$s^2 = \frac{\displaystyle\sum_{i=1}^{n}(y_i - \bar{y})^2}{n - 1} \tag{3.10}$$

and all other terms have been defined earlier.

Like $\bar{y}$, s^2 can be determined from the data provided by the n sampled elements that were selected from the population of size N using simple random sampling without replacement.

When a population total, Y, is estimated using Equation (3.4), the variance of the estimator is given by

$$V(y) = \frac{N^2 S^2}{n}(1 - f) \tag{3.11}$$

Once again, all terms in this equation have already been defined.

As was true for the estimation of a population mean, when a population total is not known, the variance of the estimator is not likely to be known. Therefore, an estimator of the variance is needed to compute an indicator of estimation precision. An unbiased estimator of the variance of the estimated population total is given by

$$v(y) = \frac{N^2 s^2}{n}(1 - f) \tag{3.12}$$

where all terms in this equation have already been defined.

This variance and its square root can be calculated using the data in a single sample of size n. The square root of the estimated variance is the estimated standard error, a value that is needed to compute confidence limits.

CONFIDENCE LIMITS

If it is assumed that sample means are normally distributed, the estimators $\bar{y}$ and s can be used to form confidence limits on the population mean. These confidence limits take the form

$$\hat{\bar{Y}}_U = \bar{y} + \frac{ts}{\sqrt{n}} \sqrt{1-f}$$

$$(3.13)$$

$$\hat{\bar{Y}}_L = \bar{y} - \frac{ts}{\sqrt{n}} \sqrt{1-f}$$

where $\hat{\bar{Y}}_U$ is an upper confidence limit and $\hat{\bar{Y}}_L$ is a lower confidence limit.

With the exception of t, we have already defined all of the terms in Equation (3.13). If the sample size is large, say more than 30, t can be interpreted as the value on the ordinate of a standard normal distribution that corresponds to the desired level of confidence. Table 3.1 contains values of t for some frequently used confidence levels. If the sample size is smaller than 30, the value of t should be read from a table of Student's t-distribution for $n-1$ degrees of freedom. Use of the t-distribution when the sample size is small is an approximation that becomes increasingly poor as the distribution of individual scores (y_i) departs from a normal distribution.

Confidence limits on a population total, Y, can be calculated from the sample total y and the sample standard deviation of element scores, s, as follows:

$$\hat{Y}_U = N\bar{y} + \frac{Nts}{\sqrt{n}} \sqrt{1-f}$$

$$(3.14)$$

$$\hat{Y}_L = N\bar{y} - \frac{Nts}{\sqrt{n}} \sqrt{1-f}$$

where $\hat{Y}_U$ denotes an upper confidence limit on Y and $\hat{Y}_L$ denotes a lower confidence limit on Y. The value t is treated in exactly the same way, whether confidence limits are calculated for a population mean or a population total.

DETERMINATION OF REQUIRED SAMPLE SIZE

The size of the sample needed to estimate a population mean or a population total with a specified level of precision is one of the first issues that arise in any practical sampling problem involving these parameters. If the sample means or sample totals calculated for successive samples of elements are assumed to be normally distributed, sample size requirements can be determined from the standard error of $\bar{y}$ or the standard error of y, respectively. We will consider the population mean first.

The statement that the population mean $\bar{Y}$ should be estimated with a specified level of precision can be expressed mathematically by requiring that any

TABLE 3.1

Values of a Standard Normal Variable (t)

Confidence level (%)	80	90	95	99
Value of t	1.28	1.64	1.96	2.58

sample mean and the population mean differ by no more than a specified amount, say ε, with a probability of $1 - \alpha$ or more. The difference between the mean of a specific sample and the population mean, $\bar{y} - \bar{Y}$, is called the *error of estimation* for that sample.

The theory of sampling allows a researcher to specify a sample size that is sufficiently large to ensure a high probability that errors of estimation can be kept within desired limits. In mathematical terms, it is possible to specify a sample size that will ensure that the

$$\text{Prob}\,\{-\varepsilon \le \bar{y} - \bar{Y} \le \varepsilon\} \ge 1 - \alpha \tag{3.15}$$

where α is a specified probability value, frequently set to one of the values 0.10, 0.05, or 0.01. The smaller the values of ε and α, the larger the required sample size. For a population of size N with a variance equal to S^2, the sample size necessary to satisfy the probability statement given in Equation (3.15) is equal to

$$n = \frac{(tS/\varepsilon)^2}{1 + (1/N)(tS/\varepsilon)^2} \tag{3.16}$$

In this equation, t denotes a value on the abscissa of a standard normal distribution corresponding to the desired level of probability α; that is, α is the probability that a standard normal random variable takes on a value that is either larger than t or smaller than $-t$. If n is small, it may be desirable to read the value of t from a table of Student's t-distribution with $n - 1$ degrees of freedom. If n is large, say at least 30, a table of the standard normal distribution will suffice.

Equation (3.16) is an approximation that is based on the assumption of normally distributed scores in the population. If this assumption holds to a reasonable approximation, the sample size determined from Equation (3.16) will be adequate. If the distribution of individual scores is highly skewed, however, the sample size specified by Equation (3.16) may allow errors of estimation larger than ε to occur with probability higher than α.

Table 3.2 contains values of the expression $(tS/\varepsilon)^2$ for values of t and ε likely to be used in practice, and a wide range of population variances. Table 3.3 contains sample sizes found by evaluating Equation (3.16) over the range of values of $(tS/\varepsilon)^2$ contained in Table 3.2 and a range of population sizes from 10 to 30,000. These tables are included so that readers can easily check the reasonableness of their computations. For example, with a t of 1.64, an ε of 0.5, and a population variance S^2 of 200, the value of $(tS/\varepsilon)^2$ should fall between the tabulated values of 1076 and 2421. If the size of the population from which samples are drawn is 1200, the required sample size should fall between the tabulated values 428 and 827.

The sample size needed to estimate a population total with a probability $1 - \alpha$ of making an error no larger than ε is given by the following slight modification of Equation (3.16). Here again, it is assumed that the estimator of the population total varies from sample to sample in accordance with a normal probability distribution.

$$n = \frac{(NtS/\varepsilon)^2}{1 + (1/N)(NtS/\varepsilon)^2} \tag{3.17}$$

TABLE 3.2
$(tS/\epsilon)^2$ as a Function of Standard Normal Deviates t, Allowable Estimation Errors ϵ, and Population Variances S^2

			Population Variance S^2							
t	ϵ	$\dfrac{t}{\epsilon}$	1	4	9	16	100	225	400	625
1.28	2.0	0.64	0.410	1.638	3.686	6.554	40.96	92.16	164	256
1.96	2.0	0.98	0.960	3.842	8.644	15.37	96.04	216	384	600
1.28	1.0	1.28	1.638	6.554	14.75	26.21	164	369	655	1024
1.64	1.0	1.64	2.690	10.76	24.21	43.03	269	605	1076	1681
1.96	1.0	1.96	3.842	15.37	34.57	61.47	384	864	1537	2401
1.28	0.5	2.56	6.554	26.21	58.98	105	655	1475	2621	4096
1.64	0.5	3.28	10.76	43.03	96.33	172	1076	2421	4303	6724
1.96	0.5	3.92	15.37	61.47	138	246	1537	3457	6147	9604
2.58	0.5	5.16	26.63	107	240	426	2663	5991	10,650	16,641
1.28	0.2	6.40	40.96	164	369	655	4096	9216	16,384	25,600
1.64	0.2	8.20	67.24	269	605	1076	6724	15,129	26,896	42,025
1.96	0.2	9.80	96.04	384	864	1537	9604	21,609	38,416	60,025
2.58	0.2	12.90	166	666	1498	2663	16,641	37,442	66,564	104,006

TABLE 3.3
Sample Sizes Required for Simple Random Sampling as a Function of $(tS/\epsilon)^2$ and Population Size N

	Population Size N						
$(tS/\varepsilon)^2$	10	30	100	300	1200	10,000	30,000
0.410	0.394[a]	0.404	0.408	0.409	0.410	0.410	0.410
1.638	1.41	1.55	1.61	1.63	1.64	1.64	1.64
3.842	2.78	3.41	3.70	3.79	3.83	3.84	3.84
26.63	7.27	14.1	21.0	24.5	26.1	26.6	26.6
67.24	8.71	20.7	40.2	54.9	63.7	66.7	67.1
166	9.40	25.4	62.4	107	146	163	165
384	9.70	27.8	79.3	168	291	370	379
666	9.90	28.7	86.9	207	428	624	652
2663	10.0	29.7	96.0	269	827	2103	2445
6147	10.0	29.9	98.0	286	1004	3807	5101
16,641	10.0	29.9	99.0	295	1119	6247	10,702
66,564	10.0	30.0	100	299	1179	8694	20,671

[a] In practice, the fractional sample sizes listed would be rounded up to the next integer. Fractional values are shown so that computations may be verified.

Estimation of a population mean using simple random sampling without replacement is illustrated in the following example. The example includes determination of a required sample size, computation of confidence limits, and since data for the entire population were available, assessment of the actual error of estimation and the accuracy of the estimated confidence interval. The example employs real data on students' reading achievement from a medium-sized school district.

Example 3.1. Estimating Average Achievement in a School District

This hypothetical example illustrates the kind of testing problems that arise in school districts and the way in which sampling theory can be used to solve such problems.

CONTEXT AND PROBLEM

Midcity is a predominantly upper-middle-class community with a population of about 60,000. Though a majority of the heads of households in Midcity are professionals, some areas of the community have concentrations of low-income families. In past years, the average achievement of pupils in Midcity was higher than that of most national norm populations and considerably higher than that of pupils in most large cities. In 1983, the mean raw score for sixth-graders on the reading subtest of the Stanford Achievement Series was 68. Since Midcity is exceptionally homogeneous racially and socioeconomically, one might expect the variance of achievement scores to be smaller than usual. However, in 1983, the variance of sixth-grade achievement was of about the same magnitude as in larger cities.

Until 1984, the Office of Research of Midcity had regularly administered achievement tests in reading and arithmetic to every child in grades 2 through 6. In the spring of 1983, budget cuts forced drastic curtailment of the testing program. The state education department required that each school district provide an average reading achievement score for sixth-graders each year. Midcity's Office of Research therefore investigated the possibility of using probability sampling to secure districtwide achievement statistics. It was decided to use simple random sampling to estimate this average for 1984.

The decision to use sampling led to some immediate questions:

1. How precisely should the average reading achievement of sixth-graders be estimated?
2. How many pupils need be sampled in order to estimate the average with desired precision?
3. How should pupils be chosen for testing?

To answer these questions, representatives of the school district and the state education department met to discuss their use of test results and their mutual needs for data.

SOLUTION AND RESULTS

ESTIMATION PRECISION

It was decided that an estimate of mean achievement that did not differ from the actual mean by more than 0.2 grade equivalent units would be acceptable, but an estimate differing by no more than 0.1 grade equivalent units would be more satisfactory. The choice between the two levels of precision was to be made after the number of pupils required to reach each level had been determined. To avoid future disagreement, it was decided that the chances of exceeding the allowable error should be no more than 1 in 20.*

CALCULATION OF SAMPLE SIZES

The first question having been reduced to two alternatives, required sample sizes could be determined through use of Equation (3.16)

$$n = \frac{(tS/\varepsilon)^2}{1 + (1/N)(tS/\varepsilon)^2} \tag{3.16}$$

The allowable error ε was specified as either 0.1 or 0.2 grade equivalent units. Since α was specified to be 0.05, the appropriate value of t, read from a table of the normal distribution, was 1.96. The number of sixth-grade pupils in the district, N, was 1180. The standard deviation S of reading achievement scores for sixth-graders was still unknown. At first glance, the situation appeared hopeless: to determine the sample size necessary to estimate one characteristic of the population (namely, the mean), another characteristic, the standard deviation, had to be known. However, reading achievement tests had been regularly administered to sixth-graders in Midcity every year through 1983. There having been no marked changes in school district boundaries or population in Midcity since 1983, the unknown standard deviation for 1984 was not expected to differ greatly from the standard deviation for 1983. From the 1983 data, it was found that $S = 21.3$ raw score points.

In evaluating Equation (3.7), y_i was the 1983 reading achievement raw score for the ith pupil, $\bar{Y}$ was the average y_i for all pupils in the district, and N was the number of sixth-graders in the district in 1983.

From Equation (3.16), it was found that a sample of 319 pupils would be required if 0.1 grade equivalent units was the acceptable error limit for estimating mean achievement in the district and that 100 pupils should be sampled if an error of 0.2 grade equivalent units could be tolerated.

Sample sizes were calculated as follows. Acceptable error limits ε had been specified in grade equivalent units. Since the variance of pupil achievement in

* The acceptable error in estimating an institutional test parameter and the level of probability with which an error should not be exceeded is an administrative concern rather than a statistical matter. The administrative decision will depend on the intended use of the estimate, the costs of achieving various levels of precision, and the costs or losses associated with errors in estimation. Since institutional test data have traditionally been used for purposes of public information and not as guides to specific school administrative action, the losses associated with errors in estimation are difficult to ascertain. A technical discussion of decision theory and finite-population sampling theory may be found in Cochran (1977, pp. 83ff.).

1983 was available in raw score points, it was necessary to convert the error limits to raw score units. In 1983, the mean reading achievement was 68 raw score points. It was expected that the 1984 value would be similar. For Form W of the Stanford Achievement Test, a raw score of 68 corresponded to 6.6 grade equivalent units. A score 0.1 grade equivalent units higher, 6.7, corresponded to raw scores of 70 and 71, and a score 0.1 grade equivalent units lower, 6.5, corresponded to a raw score of 67. Thus error limits of ±0.1 grade equivalent units corresponded to limits of +3 and −1 raw score points. These error limits were averaged, and an ε of 0.1 grade equivalent units was assumed equal to an ε of 2 raw score points. By a similar argument, an ε of 0.2 grade equivalent units was assumed equal to an ε of 4 raw score points.

With $\varepsilon = 2$ and $t = 1.96$, $t/\varepsilon = 0.98$; with $S = 21.3$ and $N = 1,180$ pupils, $[(t/\varepsilon)(S)]^2 = [0.98(21.3)]^2 = 435.7$. These values were used in Equation (3.16),

$$n = \frac{435.7}{1 + \frac{1}{1180}(435.7)} = \frac{435.7}{1.369} = 318.2$$

Upon rounding up to the next integer, it was found that 319 pupils should be sampled.

For an allowable estimation error $\varepsilon = 0.2$ grade equivalent units, or 4 raw score points, $t/\varepsilon = 0.49$. The other factors used in Equation (3.16) were the same as above. In this case, Equation (3.16) gave a value for n equal to 99.7. Rounding to the nearest integer, it was found that a sample of 100 pupils was required.

Since 219 additional pupils were required if the allowable error in estimating the mean was 0.1 rather than 0.2 grade equivalent units, the increase in precision was regarded as being too costly. It was decided that 100 pupils would be tested and an error of estimation equal to 0.2 grade equivalent units would be accepted.

Tables 3.2 and 3.3 can be used to verify the approximate magnitude of the sample sizes given above. Table 3.2 contains values of $(tS/\varepsilon)^2$ for a range of values of t, ε, and S^2. Values of the expression are shown for $t = 1.96$ and $\varepsilon = 2$, as used in this example. The population variance in this example ($S^2 = 454.8$) lies between the values 400 and 625 contained in the table. For $S^2 = 400$, $(tS/\varepsilon)^2 = 384$, and for $S^2 = 625$, $(tS/\varepsilon)^2 = 600$. The value 435.7 computed above lies between the two values in the table, as expected. The population size in this example, 1180, is close to 1200. In Table 3.3, the required sample size for $N = 1200$ and $(tS/\varepsilon)^2 = 384$ is 291. For $N = 1200$ and $(tS/\varepsilon)^2 = 666$, the required sample size is 428. The sample size 319, computed in this example, lies between 291 and 428, providing a check on the calculation.

SELECTION OF PUPILS

The third question noted above has to do with the selection of pupils. Each sixth-grade pupil in the city should have an equal chance of being sampled. Simple random sampling thus embodies a certain fairness.

A selection procedure that afforded each sixth-grader an equal chance of selection was easily developed. The research office of Midcity had a complete list of the sixth-graders enrolled in the district, which they numbered from 1 to 1180. It was necessary to select 100 of these numbers in such a way that each

TABLE 3.4
Estimation of Variance of 1984 Reading Achievement Scores for Sixth-Grade Pupils in Midcity

Pupil Number	Reading Achievement Raw Score (y_i)	$y_i - \bar{y}$	$(y_i - \bar{y})^2$
1	49	−18.2	331.24
2	60	−7.2	51.84
3	81	13.8	190.44
⋮	⋮	⋮	⋮
100	74	6.8	46.24
		Sum =	46,110.24

number had an equal chance of selection. Tables of random numbers have been created for just such sampling problems. These are found in many elementary statistics texts, and elaborate tables are available as separate volumes. A list of available tables and procedures for using such tables are provided in Appendix A.

Having sampled 100 of the 1180 sixth-graders, the research office administered the Stanford Achievement Test to the selected pupils. The mean reading achievement for sixth-graders in the district was then estimated using Equation (3.2). It was found, let us say, that $\bar{y} = 67.2$ raw score points or 6.5 grade equivalent units. This was reported to the state education department and to the public.

CALCULATION OF CONFIDENCE LIMITS

The research office of Midcity was interested in the error of estimation. Since the actual mean achievement of sixth-graders in 1984 was unknown, the error of estimation could not be determined directly. It was possible, however, to form confidence limits for the mean by using Equations (3.13). Equation (3.10) was first evaluated to estimate the variance of 1984 reading achievement scores. Table 3.4 illustrates the use of Equation (3.10).

From Equation (3.10),

$$s^2 = \frac{\sum_{i=1}^{n} (y_i - \bar{y})^2}{n - 1} = \frac{46,110.24}{100 - 1} = 465.76$$

The estimated standard deviation s was found to equal 21.58. Note that s for 1984 was found to be very close to the 1983 value, 21.3.

It was decided to compute 95 percent confidence limits on the actual mean achievement in the district [Equation (3.13)]. The following values were known or computed: $n = 100$, $f = \frac{100}{1180}$, $\bar{y} = 67.2$, and $s = 21.6$. Upper and lower 95 percent confidence limits were found from Equation (3.13) to equal

$$\hat{Y}_U = 67.2 + \frac{1.96(21.6)}{\sqrt{100}} \sqrt{1 - \frac{100}{1180}} = 67.2 + 4.05 = 71.25$$

and

$$\hat{Y}_L = 67.2 - \frac{1.96(21.6)}{\sqrt{100}} \sqrt{1 - \frac{100}{1180}} = 67.2 - 4.05 = 63.15$$

It might be helpful to review the definition of confidence limits. If one states that the actual mean achievement of sixth-graders falls somewhere between $\hat{Y}_U$ and $\hat{Y}_L$, the statement will be true for some samples and false for others. If many different samples are drawn and 95 percent confidence intervals are computed for each, 95 percent of the statements will be true. That is, in the course of a very large number of repeated samplings and computations of 95 percent confidence limits, the actual district mean will fall between the upper and lower confidence limits 95 percent of the time. In 5 percent of the samples, the actual mean will fall outside the limits.

Recall that though the situation portrayed in this example is fictitious, the data are actual values obtained from the testing records of Midcity. In fact, reading achievement scores were available for all of the 1180 sixth-graders in Midcity for 1984. Thus the sample estimate can be compared to the corresponding parameter for the population. Knowing the value of the parameter to be estimated, one can, in this case, determine the actual error of estimation.

The actual mean is 68.0 raw score points. Thus the estimate formed from data on 100 pupils is 0.8 raw score points less than the actual value. For the sample chosen, the actual mean is between the computed confidence limits (71.3 and 63.1). However, 5 percent of the potential samples would yield confidence intervals that would not contain the actual mean.

DISCUSSION

This example shows that the average reading achievement of sixth-graders in a medium-sized city is well-estimated using simple random sampling with a sampling fraction of 9 percent. Using this sampling fraction, 95 percent confidence intervals on mean achievement have an average width of ±4 raw score points or ±0.2 grade equivalent units. With a sampling fraction of 27 percent, 95 percent confidence intervals have an average width of ±2 raw score points or ±0.1 grade equivalent units.

The next example illustrates the use of simple random sampling without replacement to estimate a population total. In this example, we select a sample size arbitrarily and then use sample data to estimate a population total, the variance of the estimated population total, and a 90 percent confidence interval on a population total. We then estimate the sample size that would be required to estimate a population total within prescribed error limits, at a specified level of confidence. Both the example and the data used are fictitious.

Example 3.2 Estimating the Annual Client Load of a Mental Health Facility

This hypothetical example illustrates the estimation of a population total and some statistics associated with the precision of that estimate. The problem illustrated here typifies the administrative uses of sampling methods in a social welfare context.

CONTEXT AND PROBLEM

The director of a mental health agency was faced with the unpleasant task of preparing an annual report on the agency's activities. As is often the case, detailed records on client services had been compiled by counselors in the agency, and these records had been dutifully filed with little analysis of their contents. The agency director wanted to report the number of clients who had been served by the agency during the past year but did not want to find daily totals for each of the 260 days (Monday through Friday of each week, including holidays) the agency had been open during the year.

The director reasoned that he might select a sample of the days for which records were available and then estimate the annual client load from these sample data. Being naive in the ways of sampling, he arbitrarily decided to take a "10 percent sample." He assigned sequential numbers from 1 to 260 to each of the days the agency had been open during the previous year, and then selected 26 of these numbers at random by using an abbreviated random number table in

TABLE 3.5
Clients Served by a Mental Health Agency

Day Number	Number of Clients Served
1	72
2	93
3	90
4	81
5	96
6	79
7	72
8	82
9	99
10	96
11	83
12	88
13	92
14	95
15	97
16	82
17	76
18	81
19	91
20	83
21	73
22	84
23	99
24	96
25	89
26	71

the back of a graduate-level statistics book. Although the arbitrary decision to use 10 percent of the operating days as a sample size made little sense, the director's technique for sampling those days was quite reasonable.

SOLUTION AND RESULTS

ESTIMATION OF THE TOTAL

The numbers of clients served by the 10 counselors at the mental health agency on each of the 26 sampled days are shown in Table 3.5. The total number of clients served during the 26 sampled days was 2240, and the mean number served per day was found to equal

$$\bar{y} = \sum_{i=1}^{26} y_i = \tfrac{2240}{26} = 86.1538$$

This sample mean can be used to estimate the total number of clients served during the year by applying Equation (3.4):

$$y = N\bar{y} = 260(86.1538) = 22{,}400$$

VARIANCE OF THE ESTIMATE

Knowing a bit about statistics, the agency director reasoned that his estimate of the total number of clients served by the agency during the past year was likely to differ from the true population total. He knew that the magnitude of this estimation error could not be determined with certainty but that the standard error of the estimator would tell him something about the likely size of the estimation error. The director therefore used Equation (3.12) to compute an estimate of the variance of the estimated population total and then took the square root of this number to determine the estimated standard error.

The estimated variance is given by

$$v(y) = \frac{N^2 s^2}{n}(1 - f)$$

$$= \frac{260^2(8.983^2)}{26}\left(1 - \frac{26}{260}\right)$$

$$= 209{,}808(0.90) = 188{,}827.2$$

The one value in the formula for the estimated variance that might appear puzzling is the estimated standard deviation s among elements (daily totals in this case). The value 8.983 was found by using Equation (3.10) with the data in Table 3.5 and then taking the square root of the resulting element variance:

$$s^2 = \frac{\sum_{i=1}^{26}(y_i - \bar{y})^2}{26 - 1}$$

$$= \frac{2017.3846}{25} = 80.6946$$

The standard deviation among daily totals is given by

$$s = \sqrt{80.695} = 8.983$$

the value used in computing the estimated variance of the number of clients served during the past year.

The estimated standard error of the estimated total number of clients served by the agency during the year is equal to the square root of the estimated variance:

$$s_y = \sqrt{188,827.2} = 434.543$$

This standard error is a very large number, suggesting that the population total has not been estimated very precisely by using the data from a sample of 26 days.

CONFIDENCE INTERVAL

To determine just how precise his estimate was, the agency director decided to compute a 90 percent confidence interval on the total number of clients served during the year and used Equations (3.14) as follows:

$$\hat{Y}_U = N\bar{y} + \frac{Nts}{\sqrt{n}} \sqrt{(1 - f)}$$

$$= 22,400 + \frac{260(1.708)(8.983)}{\sqrt{26}} \sqrt{(1 - 0.1)}$$

$$= 22,400 + 782.341(0.9487)$$

$$= 22,400 + 742.19 = 23,142.19$$

which rounds up to an upper confidence limit of 23,143.

$$\hat{Y}_L = N\bar{y} - \frac{Nts}{\sqrt{n}} \sqrt{(1 - f)}$$

$$= 22,400 - 742.19 = 21,657.81$$

which rounds down to a lower confidence limit of 21,657.

These results confirm that the director was right to suspect poor estimation precision after seeing the estimated standard error of the estimated population total. By using data from only 26 days, he only knows, with 90 percent confidence, that the agency served somewhere between 21,657 and 23,143 clients during the past year. The difference between these two values is 1486, or a range of ±743 clients around the estimated population total of 22,400. Thus the agency director is far from certain about the number of clients served during the year.

One of the numbers used in computing confidence limits, the value of t, requires some explanation. From Table 3.1, the value of t that is associated with a 90 percent confidence level is 1.64. This value comes from the standard normal distribution. Since the sample size of 26 was smaller than 30, we chose to use a critical value from Student's t-distribution rather than the corresponding value from the normal distribution. With a sample size of 26, the appropriate number of degrees of freedom for the t-distribution is 25, and the corresponding critical value is 1.708.

REQUIRED SAMPLE SIZE

With a sample of 26 days chosen from the population of 260 days using simple random sampling without replacement, we have seen that the agency director can estimate the total number of clients served within a range of ±743, with 90 percent confidence. Suppose the agency director decides that this level of precision is unacceptable and wants a far more precise estimate of the total number of clients served, say within ±100, with 90 percent confidence. How large a sample would be required to achieve this precise an estimate?

The answer to this question can be found by assuming that our sample standard deviation of the number of clients served each day, s, is equal to the actual population value S and then using this value in Equation (3.17), together with the other numbers we have already discussed. The value of ε in this equation is the director's upper limit on estimation error, 100 clients:

$$
\begin{aligned}
n &= \frac{(NtS/\varepsilon)^2}{1 + (1/N)(NtS/\varepsilon)^2} \\
&= \frac{[260(1.64)(8.983)/100]^2}{1 + \frac{1}{260}[260(1.64)(8.983)/100]^2} \\
&= \frac{38.420^2}{1 + \frac{1}{260}(38.420^2)} \\
&= 1476.12/6.677 = 221.06
\end{aligned}
$$

Rounding this value up to the next whole number results in the astounding conclusion that the agency director would have to sample 222 days out of the 260 the agency was open during the year in order to estimate the total number of clients served annually within ±100, with 90 percent confidence. If this level of precision were really necessary, the director might logically conclude that sampling was not helpful in solving this problem. By calculating daily totals for the remaining 38 days, he would know exactly how many clients the agency had served!

Estimating Population Proportions

A large number of research problems in education and the social and behavioral sciences involve one or more population proportions. For example, a cognitive psychologist might want to know the proportion of adult males in the United States with right cerebral dominance, or an educational administrator might want to know the proportion of students in a school system who need remedial instruction in mathematics. In both of these examples, the parameter of interest is the proportion of elements in a population that possess a certain attribute. In other problems, the parameter of interest is the proportion of elements in the population for which some variable falls within a specified interval. For example, the chief academic officer of a college might want to know the proportion of students enrolled in the college who have a grade point average between 3.0 and 3.5. The proportions of interest in each of these examples can be estimated using simple random sampling without replacement.

FORMING THE ESTIMATES

Cochran (1977, pp. 50–54) and Som (1973, pp. 40–43), among others, show that the formulas for estimating population means through simple random sampling can be used to estimate population proportions as well, with just a slight reinterpretation of observed values. Suppose it is desired to estimate the proportion of elements in a population for which some measurement is within a specified interval, say between C_1 and C_2. The population proportion and the proportion calculated from a simple random sample can be treated as a population mean and a sample mean, respectively, by assigning the value 1.0 to all measurements that fall between C_1 and C_2 and the value zero to all measurements that fall outside this interval. The mean of such recoded measurements in the sample equals the sample proportion, and the mean of measurements in the population equals the population proportion.

If y_i denotes the measurement for the ith element, transformed to either 1.0 or zero by the rules given above, Equations (3.2) and (3.3) can be used to calculate sample and population proportions, respectively. The numerator of Equation (3.2) equals the number of elements in the sample with measurements that fall between C_1 and C_2. When this numerator is divided by the sample size, the resulting ratio equals the proportion of elements in the sample with measurements between C_1 and C_2. This value is denoted by p. The numerator of Equation (3.3) equals the number of elements in the population with measurements between C_1 and C_2. When this numerator is divided by the population size, the result equals the population proportion of elements with measurements in the specified interval. The population proportion is denoted by P. Clearly, all these arguments also hold when the objective is to estimate the proportion of elements in the population that have some common attribute, such as the proportion of kittens in Key West, Florida, that have six toes (this proportion is likely to be higher than in other locales, since Ernest Hemingway's cats were prone to being superdigital and prolific).

All the statistical results for estimating population means through simple random sampling apply directly to the estimation of proportions. The sample proportion p is an unbiased and consistent estimator of the population proportion P. Formulas for calculating and estimating estimator variance apply as well.

ESTIMATOR VARIANCE

Equation (3.6) for the variance of the sample mean can be simplified when all scores are zeros and ones. For sample proportions, the variance formula reduces to

$$V(p) = \frac{P(1-P)}{n} \frac{N-n}{N-1} \tag{3.18}$$

Equation (3.18) follows directly from Equation (3.6) with p treated as a sample mean:

$$V(p) = \frac{S^2}{n}(1 - f) \tag{3.19}$$

since for proportions,

$$S^2 = \frac{N}{N-1}P(1 - P)$$

Note that the variance of the sample proportion p depends on P, the population proportion. This is analogous to the relationship between the variance of the sample mean $\bar{y}$ and the population mean $\bar{Y}$, which is expressed in Equations (3.6) and (3.7).

An unbiased estimator of the variance of the sample proportion can be formed by transforming the observed measurements to zeros and ones, as described above, and applying Equations (3.9) and (3.10) for estimating the variance of the sample mean. The analogous equations are as follows:

$$v(p) = \frac{s^2}{n}\frac{N-n}{N} \tag{3.20}$$

where

$$s^2 = \frac{n}{n-1}p(1 - p) \tag{3.21}$$

Combining Equations (3.20) and (3.21), the estimated variance of the sample proportion can be written as

$$v(p) = \frac{N-n}{(n-1)N}p(1 - p) \tag{3.22}$$

CONFIDENCE LIMITS

As was true for the population mean, it is possible to find confidence limits on the proportion of elements in the population that have some measurement value or some attribute. Most books on sampling from finite populations, including Cochran (1977, pp. 57–60) and Jessen (1978, pp. 64–66), discuss the derivation of exact confidence limits, and several approximations.

Only the normal approximation, which is most similar to the method used for finding confidence limits on the population mean, will be discussed here. The accuracy of the normal approximation depends on the population size N, the sample size n, the population proportion P, and the sample proportion p. The error of approximation is most sensitive to the quantity np, the observed number of elements that have some attribute or have measurements in a specified interval. A table containing the smallest values of np for which the normal approximation affords acceptable accuracy is provided by Cochran (1977, p. 58).

TABLE 3.6
Smallest Values of *np* (and Sample Sizes *n*) for Use of the Normal Approximation When Finding Confidence Limits on a Population Proportion *P*

Sample Proportion p	Observed Number of Elements with a Specified Attribute or Measurements within a Specified Interval (np)	Sample Size n
0.5	15	30
0.4	20	50
0.3	24	80
0.2	40	200
0.1	60	600
0.05	70	1400

It is reproduced here as Table 3.6. In this table, p is chosen so that np is the smaller of the number of sample elements with the specified attribute or the number of sample elements without the specified attribute. The table is constructed so that a probability error of no more than 1 percent is made at either end of a 95 percent confidence interval on P.

Upper and lower confidence limits on the proportion of elements with a specified attribute or with measurements between C_1 and C_2 are approximated by

$$\hat{P}_U = p + \left(t\sqrt{1-f}\,\sqrt{\frac{p(1-p)}{n-1}} + \frac{1}{2n} \right)$$

$$\hat{P}_L = p - \left(t\sqrt{1-f}\,\sqrt{\frac{p(1-p)}{n-1}} + \frac{1}{2n} \right)$$

(3.23)

where

N = population size

n = sample size

f = sampling fraction, $\dfrac{n}{N}$

t = a value from the normal distribution corresponding to the desired level of confidence (Table 3.1)

DETERMINATION OF REQUIRED SAMPLE SIZE

It is often necessary to determine the size of the sample required to estimate a population proportion with desired precision. The allowable error in estimating a population proportion can be expressed as follows:

$$\text{Prob}\{-\varepsilon \le p - P \le \varepsilon\} \ge 1 - \alpha$$

(3.24)

This states that the sample proportion and the population proportion differ by no more than a specified value ε with probability greater than some large value $1 - \alpha$. Ordinarily, $1 - \alpha$ is set to one of the values 0.99, 0.95, or 0.90, which means that α equals 0.01, 0.05, or 0.10.

Using the normal approximation to the distribution of the sample proportion p, an equation for required sample size that is analogous to Equation (3.16) is as follows:

$$n = \frac{(t/\varepsilon)^2 P(1 - P)}{1 + (1/N)[(t/\varepsilon)^2 P(1 - P) - 1]} \tag{3.25}$$

This equation is derived from Equation (3.23). For simplicity, the $1/2n$ term, which improves the normal approximation to confidence limits on P, has been ignored. The effect of this simplification is negligible in all practical cases.

Note that the sample size required for estimation of P depends on the value of P itself. In practical applications, either a conjectured value for P or a P value of 0.5 is used in Equation (3.25). The latter procedure makes use of the fact that the product $P(1 - P)$ is largest (equal to 0.25) when P equals 0.5. Thus the maximum sample size is required when the population proportion equals 0.5, and required sample size diminishes symmetrically as the population proportion

TABLE 3.7
Sample Sizes Required for Estimation of Proportions Using Simple Random Sampling, as a Function of Standard Normal Deviates t, Allowable Estimation Errors ε, Population Proportions P, and Population Sizes N

t	ε	P	Population Size N			
			100	500	1000	10,000
1.28	0.05	0.1	37.3	52.9	55.7	58.6
1.96	0.05	0.1	58.3	108.5	121.6	136.4
1.28	0.10	0.1	13.0	14.4	14.5	14.7
1.96	0.10	0.1	25.9	32.4	33.5	34.5
1.28	0.20	0.1	3.6	3.7	3.7	3.7
1.96	0.20	0.1	8.0	8.5	8.6	8.6
1.28	0.05	0.3	58.2	108.1	121.1	135.8
1.96	0.05	0.3	76.5	196.4	244.1	312.6
1.28	0.10	0.3	25.8	32.3	33.3	34.3
1.96	0.10	0.3	44.9	69.6	74.7	80.0
1.28	0.20	0.3	8.0	8.5	8.5	8.6
1.96	0.20	0.3	16.9	19.4	19.8	20.1
1.28	0.05	0.5	62.3	123.6	140.9	161.2
1.96	0.05	0.5	79.5	217.5	277.7	370.0
1.28	0.10	0.5	29.3	37.9	39.4	40.8
1.96	0.10	0.5	49.2	80.7	87.7	95.1
1.28	0.20	0.5	9.4	10.1	10.1	10.2
1.96	0.20	0.5	19.5	23.0	23.5	24.0

is either larger or smaller than 0.5. If P is thought to lie within a specified range of values, one conservative procedure for determination of sample size would assume P equals the point closest to 0.5 in the specified range.

The relationship between the probability level α and the value of t is the same in Equation (3.25) as in Equation (3.16). Here again, α is the probability that a normal random variable with mean zero and variance one assumes an absolute value exceeding t. Table 3.7 contains illustrative sample sizes for various values of t, ε, P, and N. The table can be used to verify computations of required sample sizes.

The example that follows illustrates the use of simple random sampling without replacement to estimate the proportion of pupils in a school district who have achievement test scores below a specified level. The data used in this example are the same as those used in Example 3.1. The example includes the determination of a required sample size, the estimation of a population proportion, and the determination of the precision of that estimate.

Example 3.3 Estimating the Proportion of Low Achievers Using Simple Random Sampling

CONTEXT AND PROBLEM

The context of this example is the same as that of Example 3.1. Until 1983, Midcity maintained annual records on the proportion of sixth-grade pupils with reading scores at least 1 grade equivalent unit below norm. This proportion had long been viewed as an indicator of the quality of the city's reading curriculum. A curtailment of funds in 1983 required that the proportion of low achievers be estimated by testing sampled pupils rather than testing every sixth-grader. Simple random sampling was proposed. The usual questions of required sample size and desired estimation precision had to be resolved.

SOLUTION AND RESULTS

DETERMINING SAMPLE SIZE

There were 1180 sixth-graders in Midcity at the beginning of 1983–1984. Data for previous school years indicated that the proportion of sixth-graders with achievement scores at least 1 grade equivalent unit below norm varied somewhat but had never been less than 0.15 and never more than 0.22. Since neither the composition of the city population nor the instructional program had changed drastically, the current proportion of low achievers was expected to be close to those for earlier years. However, no one was certain that this was the case.

To determine the sample size required to estimate the citywide proportion of low achievers, an acceptable estimation error had to be specified.

After some consideration, it was decided that the proportion of low achievers should be estimated to within ± 0.02 at 95 percent confidence if the sample size required did not exceed the city's testing budget but that an estimate within ± 0.05 at 95 percent confidence could be tolerated.

From the discussion following Equation (3.25), it was clear that the sample size associated with a specified error of estimation is largest when the proportion of low-achieving pupils is closest to 0.50. The most conservative strategy that could have been adopted in determining sample size requirements would have been to ignore data from previous years and assume that the proportion of low achievers was 0.50. To be a bit less conservative, earlier information would have been used, and a population proportion slightly larger than the largest proportion previously observed would have been assumed. For the data from Midcity, an assumed proportion of low achievers equal to 0.25 appeared reasonable and conservative.

Using Equation (3.25), four values of required sample size were determined. They corresponded to assumed proportions of low-achieving sixth-graders equal to 0.50 and 0.25 and errors of estimation equal to 0.05 and 0.02. For an assumed population proportion of 0.50 and error limits of 0.05, Equation (3.25) was evaluated as follows,

$$n = \frac{(\frac{1.96}{0.05})^2(0.50^2)}{1 + \frac{1}{1180}[(\frac{1.96}{0.05})^2(0.50^2) - 1]} = \frac{384.16}{1 + \frac{383.16}{1180}} = 290 \text{ pupils}$$

For other combinations of assumed proportions of low achievers and specified errors of estimation, Equation (3.25) provided the following values for required sample sizes:

Assumed Proportion of Low-Achieving Sixth-Grade Pupils	Acceptable Error of Estimation	Required Sample Size
0.50	0.02	792
0.25	0.05	232
0.25	0.02	714

Within its available budget, Midcity could not afford the sample size required to achieve an error of estimation of 0.02 with 95 percent confidence. In fact, the budget barely allowed testing the 290 pupils required for a 0.05 error of estimation and an assumed population proportion of 0.50. It was decided to test 290 sixth-grade pupils, selected through simple random sampling.

SELECTING THE SAMPLE AND FORMING THE ESTIMATE

The selection process employed was quite similar to that described in Example 3.1. A list of all sixth-grade pupils in the city was assembled, and pupils on the list were numbered from 1 to 1180. A table of random numbers was then used to select 290 pupils in such a way that each potential sample had an equal probability of selection.

The norms for the Stanford Achievement Series show that pupils achieving a raw score of 46 or less for the combined Word Meaning and Paragraph Meaning subtests of Form W will have grade equivalent scores of 5.0 or lower. Since the testing in Midcity took place during the first month of the school year, pupils with raw scores of 46 or less would have scores at least 1 grade equivalent unit below norm.

Equation (3.2) was used to estimate the proportion of low achievers. Fifty-seven pupils, 19.7 percent of the sample of 290, had reading achievement scores of 46 or less.

FORMING CONFIDENCE LIMITS

To estimate the variance of the sample proportion of low achievers and to find confidence limits on the population proportion, Equations (3.20) and (3.21) were used. From Equation (3.20), the estimated variance of the sample proportion of low achievers was found to be

$$v(p) = \frac{1180 - 290}{(289)1180} (0.197)(0.803) = 0.00041$$

The corresponding standard error was found to equal 0.02, or 2 percent.

To compute 95 percent confidence limits on the population proportion P, Equations (3.23) were used with $t = 1.96$. The sampling fraction was $\frac{290}{1180}$ or 0.246. An upper 95 percent confidence limit on P was calculated as

$$\hat{P}_U = 0.197 + \left(1.96\sqrt{1 - 0.246}\sqrt{\frac{0.197(0.803)}{290 - 1} + \frac{1}{2(290)}}\right)$$

$$= 0.197 + 0.042 = 0.239$$

A lower 95 percent confidence limit on P was found to equal

$$\hat{P}_L = 0.197 - 0.042 = 0.155$$

Thus approximate 95 percent confidence limits on the citywide percentage of pupils with achievement scores at least 1 grade equivalent unit below the national norm were found to be 15 and 24 percent.

As in Example 3.1, the situation portrayed in this example is fictitious. However, the data used in Example 3.3 are values obtained from the test records of Midcity. Actually, the reading subtest of the Stanford Achievement Series was administered to all sixth-grade pupils in Midcity in September of 1984. This procedure was consistent with regular testing practice in Midcity, and in accordance with requirements imposed by the state Department of Education. The results of citywide testing showed that 229 of 1180 sixth-graders earned reading achievement scores of 46 or less. Thus 19.4 percent of the sixth-graders were low achievers, by the criterion used in this example. The estimated value of 19.7 percent is very close to the actual percentage. In this case, the population proportion is well within the confidence limits computed from the sample. However, in 5 percent of a large number of samplings, the confidence intervals would not include the population proportion.

DISCUSSION

Sampling fractions required to estimate the proportion of low achievers depend on the allowable error of estimation and the availability of information on the population proportion. When no prior information on the population proportion was considered, and 95 percent confidence intervals with an average width of 0.04 (± 0.02) were desired, the required sampling fraction was found, for the data in this example, to be 67 percent. Using data from previous years to set an upper bound on the population proportion, the required sampling fraction was found to be 61 percent. For 95 percent confidence intervals with an average width of 0.10 (± 0.05), required sampling fractions were found to be 20 or 25 percent of the total population, with and without use of earlier data to set an upper bound on the population proportion, respectively.

Summary

Simple random sampling is the most elementary of the finite-population sampling procedures discussed in this book. While it is not likely to be the most efficient procedure available in most applications, it provides a convenient standard for comparison of the efficiencies of more complex procedures. It also serves to motivate important assumptions and concepts that apply throughout the theory of sampling from finite populations.

The examples discussed in this chapter illustrate the use of simple random sampling in both educational and social welfare contexts. The data used in the examples from education were real, whereas the social welfare example made use of fictitious data. Therefore, any temptation to reify the results of the social welfare example should be avoided.

When applied to data from a medium-sized school district, simple random sampling afforded adequate precision for estimating the mean reading achievement of sixth-graders by using a sampling fraction of 9 percent. Ninety-five percent confidence intervals on the proportion of low achievers required sampling over 61 percent of the population in order to realize an average confidence interval width of 0.04 and over 20 percent for an average interval width of 0.10. Depending on precision requirements in specific situations, simple random sampling might be practical for estimation of mean achievement in a school district but might require too large a sampling fraction for estimation of proportions of low achievers. Generalization of the results shown in Examples 3.1 and 3.3 to other school districts should be undertaken with a good bit of caution. As is always true in sampling applications, there is no substitute for use of data from the local context to estimate likely precision.

Exercises

1. The local director of the U.S. Department of Agriculture Food Stamp Program wants to estimate the average monthly food stamp allotment per family in her community. She knows that 10,000 families in the community receive some food stamps each month. She asks her staff to select a simple random sample of 100 recipient families from the office's records, and to record their food stamp allotments for the most recent month. The data reported to her are as follows: (1) The total dollar amount of food stamps provided to the 100 sampled families in the most recent month was $12,000. (2) The sample sum of squared deviations of monthly food stamp allotments to these 100 families was equal to 61,875 in the most recent month; that is,

$$\sum_{i=1}^{100} (y_i - \bar{y})^2 = 61{,}875$$

Use this information to answer the following questions:

A. Estimate the mean monthly food stamp allotment per family for the entire population of 10,000 families.
B. Estimate the standard error of the sample mean monthly food stamp allotment per family.
C. Calculate a 90 percent confidence interval on the population mean monthly food stamp allotment per family.
D. Assume that the sample standard deviation of individual family food stamp allotments is equal to the true population value of the standard deviation; that is, assume $S = s$. Determine the number of families that would have to be sampled in order to estimate the mean monthly food stamp allotment per family within $\pm\$1$, with 90 percent confidence.

2. In a sample of 50 white rats, drawn randomly and without replacement from a laboratory population of 1,000, 28 were found to salivate at the sight of pellet food. Use these data to answer the following:

A. Using an unbiased estimator, compute an estimate of the proportion of white rats in the laboratory population that would be expected to salivate at the sight of pellet food.
B. Taking the most conservative possible value for the population proportion of white rats that would salivate at the sight of pellet food, how many rats would have to be sampled in order to estimate the population proportion within ± 0.10 with 95 percent confidence?

3. The vice president for business affairs at a small private college wants to estimate the total income the college will receive during the coming semester from students' tuition payments. The college charges $100 per academic credit hour. A sample of students is chosen randomly from the population of enrollees, and each sampled student is asked how many credit hours he or she intends to take during the coming semester. The results of this survey are as follows:

Class	Sample Size	Population Size	Average Number of Credit Hours to be Taken by *Sampled* Students
Freshmen	20	500	14.3
Sophomores	25	300	12.5
Juniors	30	200	12.6
Seniors	25	200	14.8
Totals:	100	1200	

Except when calculating the average number of credit hours all sampled students intend to take during the coming semester, ignore class distinctions when using these data to answer the following questions. Also note that the standard deviation of the number of credit hours sampled students intend to take during the coming semester is equal to 2.5 (this figure was calculated across all classes).

A. Estimate the total amount of income the college can anticipate earning from student tuition during the coming semester.

B. Calculate a 90 percent confidence interval on the total income the college will earn from students' tuition payments during the coming semester.

C. How many students would have to be sampled to estimate the total dollar amount the college will earn from students' tuition payments during the coming semester, within ±$25,000, with 95 percent confidence?

4

Stratified Sampling: Procedures, Estimators, and Applications

The Nature of Stratified Sampling

Simple random sampling requires the measurement of a single variable to form estimates of population parameters. Stratified sampling involves two kinds of variables: a sampling variable and a set of classification variables. A *sampling variable* is one for which estimates of population parameters are desired. *Classification variables* are used to classify sampling units into categories or strata.

A stratified sampling procedure is a two-stage process. Each sampling unit in the population is first assigned to one of a number of nonoverlapping strata according to its value on one or more classification variables. In the second stage of the process, independent samples are selected from each stratum, statistics are computed from sample observations, and these statistics are combined across strata to form estimates of population parameters.

Stratified sampling should be quite useful in a number of research applications in education and in the social and behavioral sciences. When estimating the percentage of voters who are likely to vote for a particular candidate in a senatorial election, for example, Candidate X, it would make sense to stratify potential voters by political party affiliation, by region of the state involved, and possibly by sex and race as well. Independent samples of potential voters would be drawn from each stratum and used to form separate estimates of the percentage of voters in each stratum who are likely to vote for Candidate X. These estimates would then be combined across strata to form an overall estimate of the percentage of the entire electorate that is likely to vote for Candidate X. In another situation, one might want to estimate the average score all fifth-graders in a large school system would earn on an arithmetic test. In this case, one might first classify students into strata by sex. Independent samples of boys and girls would be drawn to form separate estimates of average achievement—one for each stratum. These estimates would then be combined, using a formula given below, to estimate the mean arithmetic achievement of the entire fifth-grade population.

In some situations, strata form domains of possible research interest. For example, it might be important to know the percentage of male and female voters likely to support Candidate X. More often, the subpopulations defined by strata are not of interest in themselves, but form a partitioning that leads to statistically efficient estimation. The theory of stratified sampling can be applied whenever strata are distinct, in the sense that each sampling unit can be classified into one and only one stratum, and the collection of strata covers the entire target population.

The Benefits of Stratified Sampling

Stratified sampling is more complex and possibly more costly than simple random sampling. In simple random sampling, all units in the population must be identified and listed, but stratified sampling requires additional effort. At least one classification variable must be observed for each sampling unit, and each unit must then be assigned to a stratum. To justify the added costs of observation, measurement, and classification, stratified sampling must provide some marginal benefits when compared to simple random sampling.

The primary benefit afforded by stratified sampling is increased statistical efficiency. This may be expressed in two ways. Wise stratification prior to sampling avoids selection of undesirable samples that might be selected through other procedures. For a given sample size, the variance of some estimators can be significantly reduced through the use of stratified sampling. These two statements resolve to the same point, since estimator variance is reduced by eliminating samples that would result in poor estimates.

The way in which stratified sampling eliminates undesirable samples can perhaps be seen more clearly through a specific example. Consider once again the problem described in Example 3.1. In that problem, it was desired to estimate the mean reading achievement of sixth-grade pupils in a school district. The required sample size was determined by specifying that estimates should be within 0.2 grade equivalent units of the actual mean with 95 percent confidence. Thus it was acknowledged that 5 percent of the potential samples would result in estimates that were more than 0.2 grade equivalent units from the actual mean. Such samples would consist of large numbers of pupils with reading scores that were inordinately high or inordinately low.

One way to reduce sample size while retaining the level of precision specified in Example 3.1 would be to eliminate potential samples composed almost entirely of high achievers or of low achievers. From previous studies of the correlates of achievement, one might infer several characteristics of high achievers and low achievers. Pupils who do well on achievement tests are likely to have high scores on ability tests. Such pupils are also likely to come from families with high socioeconomic status. Conversely, low achievers are likely to have low scores on ability tests and, more frequently than high achievers, come from families with low socioeconomic status. If pupils could be classified by ability test scores prior to selection for testing, potential samples could be restricted to those that contained a mixture of low-, average-, and high-ability pupils. The association between ability test scores and achievement test scores is sufficiently high that this stratification would eliminate samples consisting predominantly of high achievers or of low achievers. Thus many undesirable samples could be avoided. Classification of pupils by an indicator of socioeconomic status, such as family income, would have the same effect. If the only potential samples were those composed of pupils from a mixture of well-to-do, middle-income and poor families, many potential samples that would yield inaccurate estimates would be avoided.

Another benefit of stratified sampling results when strata are domains of interest in their own right. In that case, stratified sampling can provide efficient estimates of overall population parameters while at the same time yielding estimates for important subpopulations.

Finally, stratified sampling affords flexibility that is missing in some other sampling procedures. An important assumption of stratified sampling is independence of selection in each stratum. Selection procedures that are independent from stratum to stratum need not be identical for all strata. For example, simple random sampling of pupils could be used in a stratum composed of small schools, whereas classrooms, rather than pupils, could be sampled in a stratum consisting of large schools. Different sampling procedures could be used within each stratum to improve sampling efficiency and to accommodate local administrative requirements.

In summary, stratified sampling affords greater flexibility than does simple random sampling and may provide significant increases in precision of estimation.

Estimating Population Means

Estimation formulas for stratified sampling and the use of stratified sampling in general survey applications are discussed by Kish (1965, pp. 75–112) and Yates (1981, pp. 140–143), among other authors. Angoff (1970), Lord (1959), and Peaker (1953) consider the potential benefits of stratification in a behavioral science application when they discuss the construction of test norms. They also present some applicable formulas.

In this chapter, we discuss two sets of formulas for estimating population means. The first set is very general, in that it is appropriate for use with any sampling and estimation procedure that leads to unbiased estimation of the mean of each stratum. However, the sampling of elements from any one stratum must be totally independent of the sampling of elements from any other stratum. The other set of formulas applies only when simple random sampling is used in each stratum.

ESTIMATION FORMULAS FOR THE GENERAL CASE

THE BASIC ESTIMATOR

Assume that a population of N elements has been divided into K strata in such a way that the kth stratum contains N_k elements. If from stratum k a sample of n_k elements is drawn to form an unbiased estimate of the mean of the sampling variable for that stratum, an unbiased estimator of the overall population mean is given by

$$\bar{y}_{st} = \left(\frac{1}{N}\right) \sum_{k=1}^{K} N_k \bar{y}_k = \sum_{k=1}^{K} W_k \bar{y}_k \tag{4.1}$$

where

$\bar{y}_k$ = unbiased estimator of the mean on the sampling variable in stratum k

$W_k = N_k/N$ = proportion of the overall population that is contained in stratum k

Equation (4.1) is based on three important assumptions. First, it is assumed that the number of elements N_k in each of the K strata is known. Second, it is assumed that sampling in each stratum is independent of sampling in all other strata. Finally, it is assumed that the estimator of the mean of the sampling variable in each stratum (denoted by $\bar{y}_k$ for the kth stratum) is unbiased. The estimator of the mean of the sampling variable for the entire population, $\bar{y}_{st}$, is unbiased only if all of these assumptions are satisfied. We have used the subscript st in our notation for this estimator of the population mean to indicate that it applies when data have been collected through stratified sampling.

The estimator given by Equation (4.1) is a weighted average of the subpopulation estimates computed from the data collected in each stratum. The terms W_k in the last part of the equation are frequently called *stratum weights*.

ESTIMATOR VARIANCE

If we again make the assumptions listed under Equation (4.1), we can compute the variance of the estimated population mean, $\bar{y}_{st}$, from the formula

$$V(\bar{y}_{st}) = \sum_{k=1}^{K} W_k^2 V(\bar{y}_k) \tag{4.2}$$

where $V(\bar{y}_k)$ is the variance of the estimated mean in the kth stratum. Like the estimator of the population mean, Equation (4.1), the formula for the variance of the estimator is a weighted average of terms that apply to each of the K strata. The weights in Equation (4.2) are the squares of the stratum weights used in Equation (4.1).

The potential increase in statistical precision afforded by stratified sampling can be discovered through a careful analysis of Equation (4.2). The variance of the estimated population mean depends only on the variances among units within each of the strata and not on the variation among the stratum means. This can be seen by noting that the terms $V(\bar{y}_k)$ become larger or smaller as the variance among units in the kth stratum becomes larger or smaller. Differences among the means of the various strata do not contribute to the variance defined by Equation (4.2), since these means do not appear anywhere in the formula.

If we could stratify a population so that all the units that fell into a given stratum had the same value on the sampling variable, the variance within that stratum would be zero. Since all estimates of the mean of that stratum would then have the same value, the variance of the estimated stratum mean, for example, for the kth stratum, the value $V(\bar{y}_k)$ would also equal zero. If all terms $V(\bar{y}_k)$ were zero, the variance of the estimated population mean, $V(\bar{y}_{st})$, would equal zero, and since the estimator $\bar{y}_{st}$ is unbiased, we would have perfect estimation with no estimation error at all. In reality, this ideal is never realized. However, when stratified sampling is effective, the variance among units in the same stratum will be noticeably smaller than the variance among units in the entire population. This is true because like units will be assigned to the same stratum. As a result, the variances of the estimated stratum means will be relatively small, as will the variance of the estimated population mean, $V(\bar{y}_{st})$.

ESTIMATION FORMULAS FOR RANDOM SAMPLING WITHIN STRATA

THE BASIC ESTIMATOR

If simple random sampling is used independently within each stratum and the estimation formulas of Chapter 3 are applied to the data collected from each stratum, all the assumptions of stratified sampling will be satisfied. From Equation (3.2), an unbiased estimator of the mean in the kth stratum $\bar{y}_k$ is given by the sample average

$$\bar{y}_k = \frac{1}{n_k} \sum_{i=1}^{n_k} y_{ik}$$

where y_{ik} denotes the value of the sampling variable for the ith unit sampled from the kth stratum. Using this formula in Equation (4.1), an unbiased estimator of the population mean is given by

$$\bar{y}_{st} = \frac{1}{N} \sum_{k=1}^{K} \frac{N_k}{n_k} \sum_{i=1}^{n_k} y_{ik} \tag{4.3}$$

ESTIMATOR VARIANCE

From Equation (3.8), the variance of the sample mean in the kth stratum, $\bar{y}_k$, is given by

$$V(\bar{y}_k) = \frac{S_k^2}{n_k} (1 - f_k)$$

where $f_k = n_k/N_k$ denotes the sampling fraction in the kth stratum, the variance among elements in the kth stratum is denoted by S_k^2, and is given by the formula

$$S_k^2 = \frac{\sum_{i=1}^{N_k} (y_{ik} - \bar{Y}_k)^2}{N_k - 1} \tag{4.4}$$

and $\bar{Y}_k$ is the mean of the sampling variable for all units in the kth stratum. The variance S_k^2 also requires measurement of the sampling variable for all units in the kth stratum.

The variances of the sample means in each stratum can be combined through Equation (4.2) to yield the following formula for the variance of the estimator of the overall population mean, $\bar{y}_{st}$:

$$V(\bar{y}_{st}) = \sum_{k=1}^{K} W_k^2 \frac{S_k^2}{n_k} (1 - f_k) \tag{4.5}$$

Equation (4.5) is a special case of Equation (4.2) that applies only when independent simple random sampling is used in each stratum.

ESTIMATION OF ESTIMATOR VARIANCE

As in simple random sampling, the variance among units within each stratum will rarely be known when the means within the strata are unknown. Therefore, the actual variance of $\bar{y}_{st}$ cannot be computed, and a suitable estimate will be needed in any practical situation. By direct analogy to Equation (3.9), an unbiased estimator of the variance of the sample mean is given by

$$v(\bar{y}_{st}) = \sum_{k=1}^{K} W_k^2 \frac{s_k^2}{n_k} (1 - f_k) \tag{4.6}$$

where

$$s_k^2 = \frac{\sum_{i=1}^{n_k} (y_{ik} - \bar{y}_k)^2}{n_k - 1} \tag{4.7}$$

is an estimator of the variance among measurements on the sampling variable for units in the kth stratum.

CONFIDENCE LIMITS

If the estimator $\bar{y}_{st}$ is assumed to follow a normal distribution and the size of the sample upon which it is based is sufficiently large, Equations (4.3) and (4.6) can be used to derive approximate confidence limits on the population mean $\bar{Y}$. An approximate lower confidence limit can be computed by using the estimated mean $\bar{y}_{st}$ and the estimated variance $v(\bar{y}_{st})$ in the equation

$$\hat{Y}_{L_{st}} = \bar{y}_{st} - t\sqrt{v(\bar{y}_{st})} \tag{4.8a}$$

An approximate upper confidence limit is given by the corresponding formula

$$\hat{Y}_{U_{st}} = \bar{y}_{st} + t\sqrt{v(\bar{y}_{st})} \tag{4.8b}$$

where t is the value on the abscissa of a normal distribution that corresponds to the desired level of confidence. Values of t for several frequently used confidence levels are given in Table 3.1.

When stratified sampling is used in some educational and behavioral science applications, the sizes of samples within strata are quite small. When sample sizes are small, the estimates of variances within individual strata tend to be unstable.

Since the use of the normal distribution in computing confidence limits assumes that variances among units are known, the resulting confidence intervals tend to be optimistically narrow. If a variance is estimated by using a simple random sample of a normally distributed random variable, the estimator will have a chi-square distribution, and tables of Students' t-distribution can be used to compute more accurate confidence limits. These assumptions are not strictly met in stratified sampling, since the estimator is a weighted sum of chi-square random variables. However, theory developed by Satterthwaite (1946) can be used to form appropriate confidence limits on the population mean, even when sample sizes within strata are small. To use Satterthwaite's method, the value

$$g_k = \frac{N_k(N_k - n_k)}{n_k} = \frac{N_k - n_k}{f_k}$$

is computed for the kth stratum. The value of t in Equations (4.8) can then be read from a table of Student's t-distribution with degrees of freedom (d.f.) equal to

$$\text{d.f.} = \frac{\left(\sum\limits_{k=1}^{K} g_k s_k^2 \right)^2}{\sum\limits_{k=1}^{K} g_k^2 s_k^4 / (n_k - 1)} \tag{4.9}$$

Estimating Population Totals

Population totals can be estimated very readily using stratified sampling. The estimators are minor variations of corresponding formulas for estimating population means. The variances of the estimators can be calculated or estimated by using formulas that are minor variations of those used for calculating or estimating the variances of estimators of the population mean. We will give two sets of equations for estimating the population total and the variance of the estimator. As was the case for the population mean, one set will apply regardless of the sampling procedures used in each stratum, so long as the methods produce unbiased estimates of the within-stratum totals and sampling in one stratum is independent of sampling in all other strata. The other set of equations will apply when simple random sampling is used in each stratum.

THE BASIC ESTIMATOR

When stratified sampling is used with independent sampling in each stratum and unbiased estimation of the total Y_k in each stratum, an unbiased estimate of the overall population total can be found from the formula

$$y_{st} = N\bar{y}_{st} = \sum_{k=1}^{K} N_k \bar{y}_k \tag{4.10}$$

where $\bar{y}_k$ is an unbiased estimator of the mean in the kth stratum.

ESTIMATOR VARIANCE

The variance of the estimated population total is equal to the square of the population size multiplied by the variance of the estimated population mean:

$$V(y_{st}) = N^2 V(\bar{y}_{st})$$

which reduces to

$$V(y_{st}) = \sum_{k=1}^{K} N_k^2 V(\bar{y}_k) \qquad (4.11)$$

If $v(\bar{y}_k)$ is an unbiased estimator of the variance of the estimated mean in the kth stratum, an unbiased estimator of the variance of the estimated population total y_{st} is given by

$$v(y_{st}) = \sum_{k=1}^{K} N_k^2 v(\bar{y}_k) \qquad (4.12)$$

Our earlier observations on the interpretation of Equation (4.2) also apply to Equations (4.11) and (4.12). Estimating a population total using stratified sampling should result in more efficient estimation than would simple random sampling because with stratified sampling, the only sources of variance of the estimator fall within the various strata. To the extent that sampling units can be assigned to strata in such a way that they are homogeneous on the sampling variable, the within-stratum variances will be small, and the variance of the estimated population total will be reduced.

ESTIMATION FORMULAS FOR RANDOM SAMPLING WITHIN STRATA

THE BASIC ESTIMATOR

If simple random sampling is used within each stratum, an unbiased estimator of the population total is given by

$$y_{st} = \sum_{k=1}^{K} \frac{1}{f_k} \sum_{i=1}^{n_k} y_{ik} \qquad (4.13)$$

where $f_k = n_k/N_k$ is the sampling fraction within the kth stratum and, as defined earlier, y_{ik} is the value of the sampling variable for the ith unit selected from the kth stratum.

ESTIMATOR VARIANCE

The variance of the stratified sampling estimator of the population total can also be found by direct analogy to the variance of the stratified sampling estimator of the population mean. After a bit of algebraic manipulation to simplify the formula, the estimator variance can be determined by using the equation

$$V(y_{st}) = \sum_{k=1}^{K} \frac{1}{f_k} S_k^2 (N_k - n_k) \qquad (4.14)$$

In this equation, the term S_k^2 denotes the variance among units in the kth stratum. We have defined all other terms earlier.

ESTIMATION OF ESTIMATOR VARIANCE

Since in practical situations data for every unit in each stratum will not be available to compute the within-stratum variances, it will be necessary to estimate the variance of the estimated population total. An unbiased estimator of this variance is given by

$$v(y_{st}) = \sum_{k=1}^{K} \frac{1}{f_k} s_k^2 (N_k - n_k) \tag{4.15}$$

where s_k^2 denotes the estimated variance among units in the kth stratum and is defined explicitly by Equation (4.7).

CONFIDENCE LIMITS

Approximate confidence limits on the population total Y are given by the equations

$$\hat{Y}_{L_{st}} = y_{st} - t\sqrt{v(y_{st})}$$
$$\hat{Y}_{U_{st}} = y_{st} + t\sqrt{v(y_{st})} \tag{4.16}$$

These approximate confidence limits are subject to the same cautions noted above for the approximate confidence limits on the population mean (Equations 4.8). When sample sizes within strata are small, use of a value from the normal distribution for t will lead to optimistically narrow confidence intervals. Substitution of a value for t from Student's t-distribution is not strictly correct in this case either, because conventional practice will lead to an incorrect value for degrees of freedom. Equation (4.9) can be used to determine Satterthwaite's value for degrees of freedom, thereby improving the approximation of appropriate confidence limits.

Estimating Population Proportions

We discuss formulas for estimating population proportions only for the case of simple random sampling within strata. Formulas for the more general case where samples are selected independently within strata by any method that results in unbiased estimation, but not through simple random sampling, can be derived from analogous formulas for estimating population means.

FORMING THE ESTIMATE

When simple random sampling is used independently within each stratum, the methods for estimating population proportions described in Chapter 3 can be applied to data collected through stratified sampling. To estimate the proportion of elements in a population for which some measurement is in the interval

C_1 to C_2, values of the sampling variable for elements sampled from each stratum would be transformed to the values 1.0 and zero as follows:

$$y_{ik} = \begin{cases} 1.0 & \text{if the measurement for the } i\text{th element sampled from the } \\ & k\text{th stratum is between } C_1 \text{ and } C_2 \\ 0 & \text{if the measurement for the } i\text{th element sampled from the } \\ & k\text{th stratum is not between } C_1 \text{ and } C_2 \end{cases}$$

Transformed data can be used directly in Equation (4.3) to estimate the proportion of elements in the population with measurements between C_1 and C_2. The proportion of elements in the kth stratum with these measurement values is estimated by

$$p_k = \frac{\sum_{i=1}^{n_k} y_{ik}}{n_k}$$

Using these values in Equation (4.3), an unbiased estimator of the population proportion is

$$p_{st} = \sum_{k=1}^{K} \frac{N_k}{N} p_k = \sum_{k=1}^{K} W_k p_k \tag{4.17}$$

ESTIMATOR VARIANCE

From Equation (3.18), the variance of the estimated proportion for the kth stratum is

$$V(p_k) = \frac{P_k(1 - P_k)}{n_k} \frac{N_k - n_k}{N_k - 1} \tag{4.18}$$

Since sampling is independent in each stratum, these within-stratum variances can be substituted for the terms $V(\bar{y}_k)$ in Equation (4.2) to determine the variance of the estimated proportion for the entire population,

$$V(p_{st}) = \sum_{k=1}^{K} W_k^2 \frac{P_k(1 - P_k)}{n_k} \frac{N_k - n_k}{N_k - 1} \tag{4.19}$$

In simple random sampling, the variance of the estimated proportion depends on the actual proportion. In stratified sampling, the variance depends on the actual proportion within each stratum. An estimator of the variance is required for any practical application. By analogy to Equation (3.22), an unbiased estimator of the variance is given by

$$v(p_{st}) = \sum_{k=1}^{K} W_k^2 \frac{N_k - n_k}{(n_k - 1)N_k} p_k(1 - p_k) \tag{4.20}$$

When estimating means, the efficiency of stratified sampling is often large relative to that of simple random sampling. However, when proportions are

TABLE 4.1

Efficiency of Stratified Sampling Relative to Simple Random Sampling as a Function of Within-Stratum Proportions for Three Strata[a]

Assumed Actual Proportions within Strata	Relative Efficiency, %
0.4, 0.5, 0.6	102.7
0.3, 0.5, 0.7	111.9
0.2, 0.5, 0.8	131.6
0.1, 0.5, 0.9	174.4
0.05, 0.5, 0.95	217.4

[a] Population proportion = 0.5, strata are of equal size, and samples of equal size are drawn from each stratum.

estimated, stratified sampling usually affords only small gains in efficiency (Kish, 1965, pp. 88–89; Cochran, 1977, pp. 109–110). These small gains result from the relationship between the actual proportions within strata and the variances of these proportions. Recall that stratified sampling is efficient when within-stratum variances are small, relative to the variance among elements in the entire population. If proportions within strata are in the range 0.2 to 0.8, the corresponding variances, proportional to $P_k(1 - P_k)$, are almost equal and are not much smaller than the overall population variance.

Table 4.1 shows the relative efficiency of stratified sampling and simple random sampling for five hypothetical examples involving three strata. We have assumed that each stratum contains one-third of the population and that one-third of the total sample is drawn from each stratum using simple random sampling within strata. Actual proportions within each stratum are indicated in the table.

The relative efficiencies shown in Table 4.1 can be interpreted as percentage increases in sample sizes required to achieve the same estimator variance if simple random sampling were used rather than stratified sampling. It is clear that proportions within strata must vary widely before stratified sampling affords appreciable economies.

Applicational Issues in Stratified Sampling

This section treats several issues that arise when stratified sampling is applied to practical problems. The issues considered are (1) the number of strata to be used, (2) establishment of boundaries for strata when a continuous classification variable is used, (3) determination of the size of the sample to be selected from each stratum, and (4) determination of the overall sample size. We also relate some theoretical results on these topics to the use of stratified sampling in educational and behavioral science research.

Another major concern in the use of stratified sampling is the choice of stratification variables. We provide some theoretical results on this topic and then use a school testing example to illustrate the way in which empirical research findings can be used to guide the choice of stratification variables.

NUMBER OF STRATA

If sampling variables rather than classification variables (as defined on page 66) were available for stratification, it would seem reasonable to create as many strata as possible for a given sample size. Increasing the number of strata would provide increased precision of estimation, since estimator variance is determined by variation within strata and not by variation between strata. Cochran (1977, p. 132) shows that stratification of a uniformly distributed variable can reduce the variance of the estimated mean by a factor proportional to the square of the number of strata.

In many educational research applications, a number of classification variables that are highly correlated with sampling variables will be available for stratification. One might therefore assume that use of a large number of strata is desirable. A theoretical model developed by Dalenius and Gurney (1951) and empirical results obtained by Cochran (1961) show that such is not always the case. Even when the sampling variable and classification variable have a correlation as high as 0.95, use of more than six strata appears to afford little increase in sampling efficiency. If the sampling variable and classification variable have a correlation of 0.90 or less, four strata appear to be sufficient. Of course any increase in statistical efficiency is desirable, provided it is not accompanied by an excessive increase in costs.

Table 4.2 shows the efficiency of stratified random sampling relative to simple random sampling as a function of the number of strata used. The classification variable is assumed to be linearly related to the sampling variable, as in the Dalenius and Gurney model. Samples of equal size are assumed to be drawn from each stratum.

TABLE 4.2
Efficiency of Stratified Random Sampling Relative to Simple Random Sampling When Estimating a Mean

Number of Strata	Correlation between Sampling Variable and Classification Variable[a]		
	0.95	0.90	0.85
2	310[b]	255	218
3	505	357	279
4	649	415	310
5	746	450	327
6	813	472	336
Infinite	1020	526	361

[a] Classification variable and sampling variable are assumed to be linearly related with indicated correlations. Samples of equal size are assumed to be drawn from each stratum.
[b] In percentages.

The number of strata that should be used in any practical application depends on the variables available for stratification and their degree of relationship with the sampling variable. Also, a decision on the appropriate number of strata should consider the cost of defining the strata as well as the cost of allocating all population elements among the strata. In a later section we present correlations between achievement test scores and variables that might be available for stratification in a school testing program. Some recommendations on the number of strata that might be useful in this educational application are made from these correlations and the data in Table 4.2. This illustration is intended to model a rational approach to determining the number of strata that should be used in any practical problem involving only one continuous stratification variable.

STRATUM BOUNDARIES

The boundaries or endpoints of strata are of practical concern in stratified sampling. If K strata are used and a single continuous stratification variable is denoted by x, the problem to be solved is as follows. The $K - 1$ points x_1, x_2, . . . , x_{K-1} must be defined between the minimum and maximum values of x. The first point x_1 is the upper boundary of the first stratum and the lower boundary of the second, x_2 is the upper boundary of the second stratum and the lower boundary of the third, etc. The point x_{K-1} is the upper boundary of the $(K - 1)$st stratum and the lower boundary of the Kth. The lower boundary of the first stratum equals the minimum value of x, and the upper boundary of the Kth stratum equals the maximum value of x.

For a fixed number of strata, the problem of defining stratum boundaries has been solved by Dalenius (1957). Dalenius's solution is too complex to be applied in practice, but Dalenius and Hodges (1959) have developed an approximation that can be used in practical situations. Cochran (1961) has used the approximation with actual data and has found it to work well. Hess et al. (1966) also found that the method produces reasonably good results.

The Dalenius and Hodges approximation is applied as follows: The classification variable is divided into 15 to 25 intervals of equal width, between its minimum and maximum values. The number of units in the population that fall into each of the intervals is recorded. Let x_i denote the value of the classification variable at the upper limit of the ith interval, and let $f(x_i)$ denote the number of units in the population that fall in the interval from x_{i-1} to x_i. Thus $f(x_1)$ is the number of units in the first interval, $f(x_2)$ is the number of units in the second interval, etc. When the frequencies $f(x_i)$ have been recorded, the square root of the frequency in each interval is computed. This will result in an array of numbers

$$sq_i = \sqrt{f(x_i)}$$

The square roots of the frequencies, the sq_i, are then summed to form cumulative square roots for each interval. The cumulative square root of frequencies for the first interval is $Z_1 = \sqrt{f(x_1)}$, the value for the second interval is $Z_2 = \sqrt{f(x_1)} + \sqrt{f(x_2)}$, and so on for each interval.

The Dalenius and Hodges method defines as strata a new set of intervals containing equal portions of the cumulative square root frequency for the highest of the 15 to 25 originally constructed intervals. Assume that the classification variable was initially divided into L intervals and that the cumulative square root frequency values $Z_1, Z_2, \ldots, Z_L$, have been computed. The cumulative square root frequency for the highest interval Z_L is then divided by K, the number of strata to be created. The upper boundary of the first stratum equals the upper boundary of the interval with cumulative square root frequency Z_i closest to Z_L/K. The upper boundary of the second stratum equals the upper boundary of the interval with cumulative square root frequency Z_j closest to $2Z_L/K$, and so on until K strata have been defined. A numerical example should help to clarify the procedure.

Example 4.1. Boundaries for Stratification by Ability Test Scores

PROBLEM AND CONTEXT

Consider again the problem described in Example 3.1. In that example, it was desired to estimate the mean reading achievement of sixth-grade pupils in a medium-sized school district (Midcity). Sufficient funds were not available for testing every pupil, so the district mean was estimated from sample data. Simple random sampling was used in Example 3.1. Stratified sampling is assumed in the present example. The problem here is to determine the boundaries or endpoints of strata.

It is assumed that scores on the Lorge-Thorndike ability test are available for every sixth-grader in Midcity. Analysis of data from previous years shows that ability test scores have a correlation of about 0.90 with scores on the reading subtest of the Stanford Achievement Test. If ability test scores and reading achievement scores are linearly related, the efficiency of stratified sampling relative to simple random sampling would be 415 percent using four strata and 472 percent using six strata (Table 4.2). Because additional strata involve no additional expense and might reduce required sample size, six strata will be used in this example.

SOLUTION AND RESULTS

When raw scores on the Lorge-Thorndike ability test are tabulated for sixth-graders in Midcity, the minimum score is 10 and the maximum is 89. The Dalenius and Hodges procedure for determining stratum boundaries is applied by defining 20 intervals of equal width on the stratification variable. The intervals fall between the minimum Lorge-Thorndike score, 10, and the maximum score, 89. Frequencies, square roots of frequencies, and cumulative square root frequencies are shown for each interval in Table 4.3.

In this example, the number of intervals L is 20. The cumulative square root frequency for the twentieth interval is 138.05. Dividing by the number of strata to be formed ($K = 6$) yields a value for Z_L/K of 23.01. Multiples of Z_L/K are as follows:

$$\frac{2Z_L}{K} = 46.02 \qquad \frac{3Z_L}{K} = 69.03 \qquad \frac{4Z_L}{K} = 92.04 \qquad \frac{5Z_L}{K} = 115.05$$

TABLE 4.3

Distribution of Lorge-Thorndike Ability Test Scores for Sixth-Grade Pupils in Midcity

Interval Number (i)	Interval Limits (Lorge-Thorndike Ability Test Scores)	Frequency (Number of Pupils, f_i)	$\sqrt{f_i}$	Z_i = Cumulative $\sqrt{f_i}$
1	10.00–13.95	3	1.73	1.73
2	13.95–17.90	4	2.00	3.73
3	17.90–21.85	6	2.45	6.18
4	21.85–25.80	20	4.47	10.65
5	25.80–29.75	22	4.69	15.34
6	29.75–33.70	34	5.83	21.17
7	33.70–37.65	49	7.00	28.17
8	37.65–41.60	54	7.35	35.52
9	41.60–45.55	70	8.37	43.89
10	45.55–49.50	75	8.66	52.55
11	49.50–53.45	122	11.05	63.60
12	53.45–57.40	126	11.22	74.82
13	57.40–61.35	115	10.72	85.54
14	61.35–65.30	128	11.31	96.85
15	65.30–69.25	131	11.45	108.30
16	69.25–73.20	93	9.64	117.94
17	73.20–77.15	61	7.81	125.75
18	77.15–81.10	40	6.32	132.07
19	81.10–85.05	14	3.74	135.81
20	85.05–89.00	5	2.24	138.05

These are the values that divide the cumulative square root frequency scale into six equal intervals, the criterion proposed by Dalenius and Hodges.

To define stratum boundaries, cumulative square root frequency values for the intervals in Table 4.3 are compared to the multiples of Z_L/K computed above. The interval with upper limit 33.70 has a cumulative square root frequency of 21.17. It is the value closest to $Z_L/K = 23.01$. The upper boundary of the first stratum is therefore set at 33.70. The upper boundary of the second stratum equals the upper limit of the interval with cumulative square root frequency closest to $2Z_L/K = 46.02$. The interval with upper limit 45.55 has a cumulative square root frequency of 43.89. The second stratum therefore has boundaries 33.70 and 45.55. The value 33.70 defines not only the upper boundary of the first stratum but the lower boundary of the second. Three times Z_L/K equals 69.03. The interval with upper limit 53.45 has a cumulative square root frequency closest to 69.03 and defines the upper boundary of the third stratum. Boundaries for the remaining strata are found by comparing cumulative square root frequencies for the intervals in Table 4.3 to $4Z_L/K$ and $5Z_L/K$. When this process is completed, the six strata are defined as in Table 4.4.

The Dalenius and Hodges procedure for determining stratum boundaries can be applied using the STRASAMP-I computer program, described in Appendix B.*

ALLOCATION OF SAMPLES AMONG STRATA

A third consideration in applying stratified sampling is the number of units to be sampled from each stratum. Two procedures for allocation of an overall sample size among strata have been used extensively in sampling applications. One method, termed *proportional allocation,* guarantees that stratified sampling will be at least as efficient as simple random sampling. Proportional allocation has the additional advantage of simplicity. The other method, termed *optimal allocation,* minimizes estimator variance for a fixed sample size and a predetermined number of strata. The Dalenius and Hodges method for determining stratum boundaries assumes the use of optimal allocation. However, we will describe both methods of allocation and provide estimators of various population parameters for both methods.

PROPORTIONAL ALLOCATION

In proportional allocation, the number of units sampled from a stratum is directly proportional to the size of the population in that stratum. The sampling fraction is therefore constant for all strata and equals $f = n_k/N_k$ for all values of k. When proportional allocation is used, the size of the sample to be drawn from each stratum is easily calculated, and the general formulas given above for estimating means, totals, proportions, and variances can be simplified. An added advantage is estimation precision that is at least as high as that afforded by simple random sampling. Estimators of various population parameters under the assumption of proportional allocation with simple random sampling in each stratum are given below.

With proportional allocation and simple random sampling in each stratum,

TABLE 4.4
Stratum Boundaries in Lorge-Thorndike Ability Test Raw Scores for Sixth-Grade Pupils in Midcity

Stratum Number	Upper and Lower Limits of Stratum
1	10.00 and 33.70
2	33.70 and 45.55
3	45.55 and 53.45
4	53.45 and 65.30
5	65.30 and 73.20
6	73.20 and 89.00

---

* The STRASAMP-I Computer Program: Appendix B contains a FORTRAN computer program, entitled STRASAMP-I. This program allows a researcher to evaluate several alternative classification variables for which data are available for an entire population. When the desired number of strata is indicated and data for the population of interest have been provided, the program computes the Dalenius and Hodges stratum boundaries, the estimates of required sample sizes using both proportional allocation and optimal allocation, and the sample size for each stratum under proportional allocation. A complete description of the program, including required input values, program output, a program listing, and sample output, are contained in Appendix B.

the arithmetic average of the sampling variable for all sampled elements is an unbiased estimator of the population mean. Thus Equation (4.3) reduces to

$$\frac{\sum_{k=1}^{K} \sum_{i=1}^{n_k} y_{ik}}{n} \tag{4.21}$$

where y_{ik} denotes the value of the sampling variable for the ith element sampled from the kth stratum and n is the overall sample size.

Under these assumptions, the variance of the estimated population mean, Equation (4.5), simplifies to

$$V(\bar{y}_{st}) = \frac{1-f}{n} \sum_{k=1}^{K} W_k S_k^2 \tag{4.22}$$

and an unbiased estimator of the variance of $\bar{y}_{st}$ is given by

$$v(\bar{y}_{st}) = \frac{1-f}{n} \sum_{k=1}^{K} w_k s_k^2 \tag{4.23}$$

where all terms are as defined in the more general Equation (4.6).

Note that the sample stratum weights w_k have been substituted for the population stratum weights W_k in Equation (4.23), since a constant sampling fraction implies that $W_k = n_k/n = w_k$ for all strata.

The formulas used in estimating population totals and for calculating and estimating the variance of the estimator simplify as follows under the assumption of proportional allocation.

Equation (4.13) for estimating population totals reduces to

$$y_{st} = \frac{1}{f} \sum_{k=1}^{K} \sum_{i=1}^{n_k} y_{ik} \tag{4.24}$$

Equation (4.14) for the variance of the estimated population total y_{st} becomes

$$V(y_{st}) = \frac{N-n}{n} \sum_{k=1}^{K} N_k S_k^2 \tag{4.25}$$

Finally, Equation (4.15) for estimating the variance of the estimated population total reduces to

$$v(y_{st}) = \frac{N-n}{n} \sum_{k=1}^{K} N_k s_k^2 \tag{4.26}$$

When estimating population proportions with proportional allocation and simple random sampling within strata, Equation (4.21) is used, together with the usual transformation of the y_{ik} to 0 and 1.0. Equation (4.9) for the variance of the estimated proportion simplifies as follows:

$$V(p_{st}) = \frac{1-f}{nN} \sum_{k=1}^{K} \frac{N_k^2 P_k(1 - P_k)}{N_k - 1}$$ (4.27)

An estimator of this variance can be formed by substituting the sample proportions p_k for the population proportions P_k in Equation (4.27).

OPTIMAL ALLOCATION

The problem of allocating a fixed sample size among strata so as to minimize the variance of the estimated population mean, total, or proportion was solved by Neyman (1934). Cochran (1977) cites an earlier proof by Tschuprow (1923) that was discovered some time after Neyman's widely recognized publication. We will consider a special case of Neyman's solution. His general solution takes account of differences in the cost of sampling in different strata, whereas our special case assumes equal per-element sampling costs. This special case yields a simple solution and is likely to meet the needs of many applications in education and the social sciences.

Estimator variance is minimized for a fixed sample size n and a fixed number of strata K by setting the sample size for the kth stratum equal to

$$n_k = n \frac{N_k S_k}{\sum\limits_{k=1}^{K} N_k S_k}$$ (4.28)

Equation (4.28) assigns a sample size to each stratum in proportion to the product of the size of the stratum and the standard deviation in the stratum. The larger the stratum or the more variable the elements in the stratum, the larger the sample size assigned to that stratum.

The Dalenius and Hodges procedure for establishment of optimal stratum boundaries makes the product of the stratum size and the standard deviation in the stratum, $N_k S_k$, constant for all strata. Because this is true, optimal allocation is simplified when the Dalenius and Hodges procedure is used to define stratum boundaries. If $N_k S_k$ is equal to a constant, say b, Equation (4.28) reduces to

$$n_k = n \frac{b}{Kb} = \frac{n}{K}$$ (4.29)

Therefore, a total sample size of n is divided equally among the strata formed by the Dalenius and Hodges procedure.

When stratified sampling is used to estimate proportions, proportional allocation and optimal allocation provide nearly equal estimator variances. Moreover, stratified sampling often affords little advantage over simple random sampling for estimating population proportions, regardless of the method used to allocate sample sizes to strata. If, despite this disadvantage, stratified sampling is used to estimate population proportions, we recommend the use of proportional allocation. It provides somewhat simpler estimators than does optimal allocation, and when it is used, stratified sampling is at least as efficient as simple random sampling.

DETERMINATION OF SAMPLE SIZE

The last major decision that must be made before stratified sampling can be used in practice concerns the size of the overall sample. As was true in the case of simple random sampling, the between-element variance must be known if a required sample size is to be calculated precisely. In the case of stratified sampling, the situation is even worse, since between-element variances S_k^2 must be known for every stratum.

SAMPLE SIZE FOR ESTIMATING POPULATION MEANS

It is possible to estimate the sample size required to estimate a population mean within a specified error limit at a specified level of confidence by using estimates of the between-element variances within strata. The following general formula applies:

$$n = \frac{\sum_{k=1}^{K} W_k^2 s_k^2 / w_k}{(\varepsilon/t)^2 + (1/N) \sum_{k=1}^{K} W_k s_k^2} \tag{4.30}$$

where

$$s_k^2 = \frac{\sum_{i=1}^{n_k} (y_{ik} - \bar{y}_k)^2}{n_k - 1}$$

= estimated variance among elements in the kth stratum

ε = acceptable difference between the actual population mean and an estimate formed through stratified sampling

t = value on the abscissa of a standard normal distribution corresponding to the desired level of confidence. (See Table 3.1 for representative values of t.)

If either proportional allocation or optimal allocation is used to determine the sample sizes assigned to various strata, Equation (4.30) can be simplified. For proportional allocation, an estimate of the overall required sample size is given by

$$n = \frac{\left(\sum_{k=1}^{K} W_k s_k^2\right) \Big/ (\varepsilon/t)^2}{1 + (1/N)(t/\varepsilon)^2 \sum_{k=1}^{K} W_k s_k^2} \tag{4.31}$$

For optimal allocation of sample sizes to strata, an estimate of the required overall sample size is given by

$$n = \frac{\left(\sum_{k=1}^{K} W_k s_k\right)^2}{(\varepsilon/t)^2 + (1/N) \sum_{k=1}^{K} W_k s_k^2} \tag{4.32}$$

Use of Equations (4.30) through (4.32) requires estimates of the variances among elements within each stratum, s_k^2. Although we have given a formula for s_k^2, it should be realized that actual computation of these variances would require collection of data from each stratum. Usually, data are collected after the required sample size, stratum boundaries, etc., have been determined. As in simple random sampling, an advance estimate of the variance among elements can sometimes be obtained from earlier research. However, such data are far less likely to be available when stratified sampling is used because a variance estimate is needed for each stratum. This makes the problem of estimating required sample size far more difficult.

SAMPLE SIZE FOR ESTIMATING POPULATION TOTALS

When population totals are to be estimated, it is again possible to use estimates of within-stratum variances among elements to estimate required sample size. A general formula that is analogous to Equation (4.30) for population means is as follows:

$$n = \frac{\sum_{k=1}^{K} N_k s_k^2 / w_k}{(\varepsilon/t)^2 + \sum_{k=1}^{K} N_k s_k^2} \tag{4.33}$$

All terms in this equation are as defined in Equation (4.30) or in earlier sections of this chapter.

If either proportional allocation or optimal allocation is used to determine the distribution of sample sizes among strata, Equation (4.33) can be simplified. For proportional allocation, the appropriate expression for estimated sample size is

$$n = \frac{N(t/\varepsilon)^2 \sum_{k=1}^{K} N_k s_k^2}{1 + (t/\varepsilon)^2 \sum_{k=1}^{K} N_k s_k^2} \tag{4.34}$$

If our simplified version of Neyman's optimal allocation of sample sizes among strata is used, an estimate of required sample size is given by

$$n = \frac{\left(\sum_{k=1}^{K} N_k s_k\right)^2}{(\varepsilon/t)^2 + \sum_{k=1}^{K} N_k s_k^2} \tag{4.35}$$

All of the terms in this equation have been defined previously.

The following example illustrates the use of stratified sampling to estimate mean reading achievement in a school testing program. We also determine the number of strata to be used, calculate sample size requirements, and allocate sample sizes among strata.

Example 4.2. Estimating Mean Achievement in a School District

CONTEXT AND PROBLEM

To allow comparison of stratified sampling and simple random sampling, the problem described in Example 3.1 is considered once again. In that example, simple random sampling was used to estimate the average reading achievement of sixth-grade pupils in a medium-sized school district (Midcity). In this example, stratified sampling is used to estimate the same parameter.

As in Example 3.1, it is assumed that mean achievement in the district is to be estimated within 0.2 grade equivalent units at a 95 percent confidence level. As in Example 4.1, it is assumed that scores on the Lorge-Thorndike ability test are available for every sixth-grader in Midcity. Ability test scores are used for stratifying pupils into the strata defined in Example 4.1.

In addition to estimating mean achievement, the problems include determination of overall sample size requirements and allocation of sample sizes among strata. Both proportional allocation and optimal allocation are considered.

SOLUTION AND RESULTS

CALCULATION OF SAMPLE SIZE

To determine the sample size required to estimate a mean within specified error limits, estimates of element variances must be available for each stratum. For this example, estimates are required for the variance of reading achievement scores among pupils in each stratum. For simple random sampling, the variance among test scores in a previous school year was used as an estimate of variance for the current school year (Example 3.1). A similar procedure can be used for stratified sampling, but the assumptions involved are more complex. In simple random sampling, the overall variances among test scores were assumed to be approximately equal in two successive school years. To use data from a previous year in stratified sampling, the variances among scores within strata must be assumed approximately equal for two successive years. Equivalently, it must be assumed that the relationship between the sampling variable and the classification variable in a previous year is the same as the relationship between the variables in the current year.

If these assumptions are made, stratum boundaries are determined as in Example 4.1, using the classification variable for the current school year. The same classification variable for the previous year is then used to define strata having the computed boundaries. Pupils enrolled in the previous school year are assigned to their respective strata, and the variances for the sampling variable are computed for each stratum. These are used as variance estimates for the sampling variable for the present school year.

An alternative procedure for estimating within-stratum variances uses two variables for which data are available on every pupil currently enrolled. These variables are the classification variable and what could be termed a proxy sampling variable. Suppose mean reading achievement is to be estimated for 1984. The classification variable might then be ability test scores for 1984, and the proxy sampling variable might be language test scores for 1984. If it is assumed that the within-strata variances of language test scores are approximately the same as those of the reading test, the computed within-stratum variances for language could be used as estimates of those for reading. Both tests would be scored in the same units, for example, grade equivalent units. Before this procedure can be deemed practical, data on the comparability of within-stratum variances are needed for tests in different subjects.

For this example, the only data available are Lorge-Thorndike scores and reading achievement scores for a single school year. Reading achievement is

used as the sampling variable and Lorge-Thorndike scores are used for stratification. Actual within-stratum variances are computed for the sampling variable.
These variances are treated as if they were estimates when sample size requirements are determined.

To compute variances, pupils are first sorted into the strata defined in Table
4.4, using their scores on the Lorge-Thorndike ability test. The variance within
each stratum is then computed using Equation (4.5). The results are shown in
Table 4.5.

At this point, the standard deviations of reading achievement within strata
can be compared to the standard deviation for the entire population. Since the
variance of an estimate based on stratified sampling depends on the element
variation within strata, the comparison will indicate the efficiency of stratified
sampling relative to simple random sampling. From Example 3.1, the standard
deviation of reading achievement scores among sixth-grade pupils in Midcity is
21.3. The standard deviation within a stratum is therefore less than half that for
the entire population. The variance within a stratum is less than one-fourth that
for the entire population. Stratification by ability test scores is therefore quite
effective in separating pupils into groups that are homogeneous in reading
achievement. The efficiency of stratified sampling relative to simple random
sampling should therefore be high.

Equation (4.31) is used to compute required sample size using proportional
allocation. The sixth-grade enrollment in Midcity is 1172 pupils; the other necessary data are contained in Table 4.5. Computations are shown in Table 4.6.

As in Example 3.1, an error limit ε of 0.2 grade equivalent units is assumed
to equal an error limit of 4 raw score points. For a 95 percent confidence level,
$t = 1.96$, so $(\varepsilon/t)^2 = 4.16$ and $(t/\varepsilon)^2 = 0.240$. Using these values in Equation
(4.31),

$$n = \frac{\frac{108.92}{4.16}}{1 + \frac{1}{1172}(0.240)(108.92)} = \frac{26.183}{1.022} = 25.61$$

TABLE 4.5

**Statistics of Reading Achievement Scores within Strata Formed
from Lorge-Thorndike Ability Test Scores (Sixth-Grade Pupils
in Midcity)**[a]

Stratum	Number of Pupils	Variance of Reading Achievement	Standard Deviation of Reading Achievement	Stratum Weight
1	89	104.24	10.21	0.076
2	173	126.11	11.23	0.148
3	197	132.58	11.51	0.168
4	369	139.48	11.81	0.315
5	224	56.85	7.54	0.191
6	120	51.70	7.19	0.102

[a] Stratum boundaries are shown in Table 4.4.

TABLE 4.6

**Computation of Sample Size to Estimate Mean
Achievement of Sixth-Grade Pupils in Midcity, Using
Stratified Random Sampling with Proportional Allocation**

Stratum	Stratum Weight	Variance	Weight × Variance
1	$\frac{89}{1172} = 0.076$	104.24	7.92
2	$\frac{173}{1172} = 0.148$	126.11	18.66
3	$\frac{197}{1172} = 0.168$	132.58	22.27
4	$\frac{369}{1172} = 0.315$	139.48	43.94
5	$\frac{224}{1172} = 0.191$	56.85	10.86
6	$\frac{120}{1172} = 0.102$	51.70	5.27
			Sum = 108.92

Using optimal allocation of sample sizes to strata, required sample size is computed from Equation (4.32). In addition to the terms used in Equation (4.31), the sum of the products of stratum weights and standard deviations within strata is needed. Data needed to compute the sum are contained in Table 4.5. Computations are shown in Table 4.7.

Using these data in Equation (4.32), the required sample size is

$$n = \frac{10.265^2}{4.16 + \frac{1}{1172}(108.92)} = 24.75 \text{ pupils}$$

As expected, the indicated sample size using optimal allocation is slightly smaller than that required using proportional allocation. For practical purposes, the difference, 25 pupils versus 26 pupils, is negligible.

ALLOCATION OF SAMPLE SIZES

Using proportional allocation, the sizes of samples required from each stratum are determined by multiplying the stratum weights by the total sample size.

TABLE 4.7

**Computation of Sample Size to Estimate Mean
Achievement of Sixth-Grade Pupils in Midcity, Using
Stratified Random Sampling with Optimal Allocation**

Stratum	Stratum Weight	Standard Deviation	Weight × Standard Deviation
1	$\frac{89}{1172} = 0.076$	10.21	0.776
2	$\frac{173}{1172} = 0.148$	11.23	1.662
3	$\frac{197}{1172} = 0.168$	11.51	1.934
4	$\frac{369}{1172} = 0.315$	11.81	3.720
5	$\frac{224}{1172} = 0.191$	7.54	1.440
6	$\frac{120}{1172} = 0.102$	7.19	0.733
			Sum = 10.265

TABLE 4.8

Sample Sizes Required for Each Stratum, to Estimate Mean Achievement Using Stratified Random Sampling with Proportional Allocation for Sixth-Grade Pupils in Midcity

Stratum	Required Sample Size
1	1.95
2	3.79
3	4.30
4	8.07
5	4.89
6	2.61

Results are shown in Table 4.8. Fractional sample sizes are rounded up to the next highest integer; so two pupils are selected from the first stratum, four from the second, etc. The total sample size is thus 28 pupils.

When the Dalenius and Hodges procedure for constructing stratum boundaries is used with optimal allocation, samples of the same size are selected from each stratum. Dividing the sample size required under optimal allocation by the number of strata, the sample size per stratum is $24.75/6 = 4.13$. Rounding up to the next integer, five pupils would be selected from each stratum. In this case, then, 30 pupils would be selected using optimal allocation and 28 would be selected using proportional allocation. For practical purposes, this difference is insignificant.

ESTIMATION OF MEAN ACHIEVEMENT

To estimate mean achievement, strata were constructed in accordance with the Dalenius and Hodges boundaries shown in Table 4.4, and five pupils were selected from each stratum using simple random sampling. Basic data and computation of the estimated mean are shown in Table 4.9.

Using Equation (4.1), the estimated mean is computed by summing the products of within-stratum sample means and respective stratum weights. Stratum weights are shown in Table (4.7).

$$\bar{y}_{st} = 0.076(30.0) + 0.148(45.8) + 0.168(59.4)$$
$$+ 0.315(70.8) + 0.191(87.4) + 0.102(106.2)$$
$$= 68.87$$

The actual mean reading achievement of sixth-graders in Midcity is 68.0 raw score points. Thus the estimate is about 0.9 raw score points too high. In this case, the estimated mean is well within the prescribed error limits of ± 4 raw score points. However, 5 percent of the potential samples would result in estimates that were not within these limits.

DISCUSSION

Stratified random sampling was used to estimate the mean reading achievement of sixth-graders in a medium-sized school district. For 95 percent confi-

TABLE 4.9
Computation of Estimated Mean Reading Achievement for Sixth-Grade Pupils in Midcity, Using Stratified Random Sampling

Stratum Number	Sampled Pupil	Reading Achievement Score
1	117	22
1	1021	31
1	919	27
1	112	36
1	678	34
		$\bar{y}_1 = 30.0$
2	291	46
2	1164	54
2	708	37
2	791	44
2	422	48
		$\bar{y}_2 = 45.8$
3	923	51
3	716	66
3	23	62
3	357	71
3	572	47
		$\bar{y}_3 = 59.4$
4	149	69
4	818	76
4	72	69
4	1005	81
4	996	59
		$\bar{y}_4 = 70.8$
5	334	88
5	695	100
5	82	74
5	255	91
5	264	84
		$\bar{y}_5 = 87.4$
6	536	103
6	475	108
6	48	108
6	911	106
6	186	106
		$\bar{y}_6 = 106.2$

dence intervals with an average width of 4 raw score points, a sampling fraction of 3 percent was required. In this case, the method of allocating sample sizes among strata made little difference. Proportional allocation and optimal allocation resulted in nearly identical sample size requirements. Ability test scores were used quite effectively for stratification. The sample size required using stratified random sampling was 30 percent of that required using simple random sampling (see Example 3.1).

Choosing Variables for Stratification

DESIRABLE PROPERTIES OF CLASSIFICATION VARIABLES

Classification variables should be chosen so as to make values of the sampling variable as similar as possible for elements within the same stratum. With this objective in mind, Hansen et al. (1953, p. 229) suggest the use of classification variables that are highly correlated with the sampling variable:

> the most effective variable on which to stratify would be the characteristic to be measured; and since in practice this is not feasible, stratification on the most highly correlated data available will lead to the greatest reduction in variance.

In general, the greater the relationship between the classification variables and the sampling variable, the greater will be the effect of stratification. A strong linear relationship will be reflected in a high correlation. However, a classification variable having a strong nonlinear relationship with the sampling variable might also be useful for stratification, even though its correlation with the sampling variable is low.

Angoff (1970, p. 105) gives a formula for approximating the efficiency of stratified sampling relative to simple random sampling when estimating means:

$$V(\bar{y}_{st}) \cong V(\bar{y})(1 - R^2) \qquad (4.36)$$

where

$V(\bar{y}_{st})$ = variance of the estimated mean using stratified sampling
$V(\bar{y})$ = variance of the estimated mean using simple random sampling
R = coefficient of multiple correlation between the sampling variable and the classification variables

Equation (4.36) can also be used to approximate the efficiency of stratified sampling for estimation of population totals and proportions. For proportions, R is the coefficient of multiple correlation between the classification variables and the sampling variable transformed to zeros and ones, as described in Chapter 3 for simple random sampling.

Equation (4.36) helps to illustrate a second desirable property of classification variables. If two or more classification variables are used, each should have a strong relationship with the sampling variable. At the same time, the classification variables should not be highly related to each other. For linearly related variables, these two properties lead to large multiple correlation coefficients. From Equation (4.36), it is clear that the higher the value of R, the lower the value of $V(\bar{y}_{st})$ relative to $V(\bar{y})$, and the higher the efficiency of stratified sampling relative to simple random sampling.

USING RESEARCH RESULTS WHEN SELECTING CLASSIFICATION VARIABLES

Social scientists and educational researchers thrive on data and have a penchant for storing them in voluminous archives. These fields are therefore ripe for the application of stratified sampling. In many studies in these fields, a survey researcher might be faced with a plethora of potential classification variables. Knowing the general properties desired of classification variables, a survey researcher can therefore make good use of research on the relationships between potential classification variables and the sampling variable to be used in his or her current study. Reviews of literature should be conducted carefully and thoughtfully, so as to choose the best of available classification variables. In the following section, we illustrate such a review of literature for the purpose of choosing classification variables for use in estimating the parameters of achievement test distributions in school systems. This application is consistent with our earlier examples from the field of educational research.

CLASSIFICATION VARIABLES FOR SCHOOL TESTING PROGRAMS

School systems regularly collect large amounts of data on pupils and institutions. Many variables for which data are collected are highly correlated with achievement test scores. Stratified sampling should therefore be useful and efficient when parameters of achievement test distributions are to be estimated.

Variables for Stratifying Pupils. The extensive testing programs typically conducted in school systems provide several potential classification variables. Ability test scores collected on every pupil in a school system were shown in Example 4.2 to be quite useful for stratification. Achievement tests administered in some subject areas could be used to stratify pupils for testing in other subject areas. The tests used for stratification would be administered to all pupils, and those used as sampling variables would be given only to sampled pupils. Test scores from previous school years could also be used to stratify currently enrolled pupils. Bloom (1964, pp. 106–108) reports correlations between scores on achievement tests administered 2 to 4 years apart. Bloom's data, summarized from studies of 10 different investigators, show correlations between 0.60 and 0.92 without correction for attenuation due to unreliability. Generally, the shorter the interval between initial testing and retesting, the higher the correlations. Also, the higher the grade levels of the tested pupils, the higher the correlations. Some data summarized by Bloom are for pupils in a single school, and other studies included pupils enrolled in many schools. Ten of 14 correlations reported for tests administered no more than 3 years apart are higher than 0.85. These correlations are for tests given to pupils enrolled in the fifth grade or in higher grades. With a correlation between the classification variable and the sampling variable of 0.85 and with at least four strata, the efficiency of stratified sampling relative to simple random sampling is more than 300 percent (see Table 4.2). Thus initial test scores should be useful for stratifying pupils when estimating retest means.

A statewide study in Pennsylvania (Campbell, 1968) provides correlations between achievement test scores and several individual and institutional variables for pupils in grades 5 and 11. The institutional variables include the characteristics of the classrooms and schools in which the pupils were enrolled. In calculating the correlation between a classroom characteristic (such as the racial composition of a classroom) and the achievement of individual pupils, the classroom characteristic has the same value for every pupil in the class. Characteristics of schools (such as the urbanism of a school's location) are treated similarly. Variances among pupils will therefore be low for institutional variables, and these variables can be expected to have low correlations with pupils' test scores. The correlations found in the Pennsylvania study are reported in Table 4.10.

Of the variables investigated in the Pennsylvania study, ability test scores for previous school years are the most useful for stratification (see Table 4.10). Stratification with any of the other variables would afford little advantage over simple random sampling. For example, suppose the percentage of white enrollment in a pupil's classroom were to be used for stratification of fifth-graders.

TABLE 4.10
Correlations between Achievement Test Scores for Individual Pupils and Other Individual and Institutional Variables[a]

Variable	Correlation with Achievement Scores for Individual Pupils
Fifth-grade pupils	
Ability test scores for previous school year	0.702
Occupation of pupil's father	0.365
Experience of pupil's teacher in present school system (years)	0.218
Urbanism of location of pupil's school	0.227
Library books per pupil in pupil's school	0.268
Percent of white enrollment in pupil's classroom	0.376
Multiple correlation coefficient:	0.745
Eleventh-grade pupils	
Ability test scores for previous school year	0.792
Occupation of pupil's father	0.395
Percentage of white enrollment in pupil's classroom	0.404
Index of fiscal effort in pupil's school district	0.267
Staff/pupil ratio in pupil's school	0.214
Professional aspirations of pupil's teachers	−0.230
Multiple correlation coefficient:	0.814

[a] Data from a statewide study in Pennsylvania, after Campbell (1968).

This variable has a squared correlation (R^2) of 0.14 with achievement. Using an R^2 of 0.14 in Equation (4.36), the variance of the estimator of mean achievement for stratified sampling would be about 86 percent of that resulting from simple random sampling. The other variables for fifth-graders would result in even larger estimator variances. If all six variables shown in Table 4.10 for fifth-graders were to be used for stratification, the estimator variance would be almost as large as that resulting from the use of previous test scores alone. With a multiple correlation coefficient of 0.745, $R^2 = 0.55$. Estimator variance using stratified sampling would thus be about 45 percent of that provided by simple random sampling [Equation (4.36)]. Using previous test scores as the only classification variable, estimator variance would be reduced to 51 percent of that provided by simple random sampling. These figures are based on the assumption of linear relationships between achievement test scores and the classification variables. Strong nonlinear relationships might not be reflected in the multiple correlation coefficient, so Equation (4.36) might underestimate the effect of additional stratification variables.

The Pennsylvania results for eleventh-graders are similar to those for fifth-graders. Next to previous ability test scores, the percentage of white enrollment in classrooms has the highest correlation with current achievement test scores. With a squared correlation of 0.16, stratification by classroom racial composition would provide an estimator variance that was about 84 percent as large as that of simple random sampling. With achievement test scores as the dependent variable, the multiple correlation coefficient for the six eleventh-grade variables shown in Table 4.10 is not much larger than the zero-order correlation of previous test scores. Thus, use of more than one classification variable is probably not warranted in this application. The cost of recording and processing many classification variables would probably not be matched by savings from increased precision.

Variables for Stratifying Classrooms. In some sampling methods, classrooms or schools, rather than pupils, would be used as sampling units. Single-stage cluster sampling, discussed in Chapter 7, is one such method. Sampled classrooms could be selected through stratified sampling. If classrooms were used as sampling units, all pupils in sampled classrooms would be tested. Similarly, schools could be selected through stratified sampling, and all pupils enrolled in sampled schools would be tested. Classification variables useful for stratifying classrooms are test averages and indices such as average socioeconomic status or racial composition. The correlation between average achievement in classrooms and average ability test scores in classrooms would indicate the usefulness of ability test averages for classroom stratification. Such correlations are termed *ecological correlations,* since the correlated variables are averages. The averages are computed within sampling units, and the correlations are computed across sampling units. Ecological correlations are usually higher than correlations of the same variables for individuals, since attenuations due to unreliability of measurement are greatly reduced.

The Pennsylvania study provides correlations between average achievement test scores for pupils in the same classrooms and classroom averages for several pupil background variables. Correlations are also provided for classroom average achievements and some institutional variables for classrooms, schools, and school districts. These correlations are shown in Table 4.11.

Average scores on previously administered ability tests have the highest correlations with current achievement averages for fifth-graders and for eleventh-graders. Used as a classification variable, ability test averages for eleventh-graders should provide very efficient stratification of classrooms. With six strata, the efficiency of stratified sampling relative to simple random sampling would be at least 472 percent (see Table 4.2). Used individually, the other variables reported for eleventh-graders in Table 4.11 would result in considerably lower relative efficiencies.

For fifth-graders, the squared correlation of averages on previously administered ability tests and averages on currently administered achievement tests is 0.63. From Equation (4.36), the variance of the mean of classroom average achievements would be 37 percent as large using stratified sampling as using

TABLE 4.11

Correlations across Classrooms of Average Achievement and Other Variables for Pupils in the Same Classrooms[a]

Variable	Correlation with Average Achievement of Pupils in the Same Classroom
Fifth-grade pupils	
Classroom average of ability test scores for previous school year	0.793
Classroom average of occupations of pupils' fathers	0.622
Percentage of white enrollment in pupils' classrooms	0.658
Classroom average of education of pupils' fathers	0.585
Experience of pupils' teacher in present school system (years)	0.365
Eleventh-grade pupils	
Classroom average of ability test scores for previous school year	0.914
Classroom average of occupations of pupils' fathers	0.661
Per-pupil expenditure in pupils' school district	0.368
Percentage of white enrollment in pupils' classrooms	0.602
Classroom average of education of pupils' fathers	0.606

[a] Data from a statewide study in Pennsylvania, after Campbell (1968).

simple random sampling. The efficiency of stratified sampling relative to simple random sampling should thus be about 270 percent. If used individually, other classification variables for fifth-graders reported in Table 4.11 would result in far smaller relative efficiencies.

The variables in Table 4.11 need not be used individually. Two or more variables could be used for stratification simultaneously, either by creating a composite variable through linear regression or by using two-way or three-way stratification. The median correlations of averages over classrooms is 0.53 for the fifth-grade classification variables and 0.62 for the eleventh-grade classification variables. Averages of pupil background variables thus correlate as highly with each other as with average achievement. Their effectiveness as classification variables when used together is therefore unlikely to be much greater than their effectiveness when used individually. It should be recognized that the correlations reported in Table 4.11 are computed over classrooms throughout the state of Pennsylvania. The variability of classroom averages within a single school district would probably be somewhat lower, resulting in lower correlations with achievement test averages.

Variables for Stratifying High Schools. Correlations of variables that might be used for stratification of high schools were reported by Burkhead (1967). Variables considered include the ethnic and socioeconomic composition of high schools and statistics related to average achievement and ability test scores. Correlations of these characteristics of schools were computed over 39 high schools in Chicago. Correlations between test score statistics and the composition variables are high. Composition variables should therefore be useful for stratifying high schools. Simultaneous stratification on two or more variables would not be much more efficient than stratification on a single variable, since correlations between composition variables are also high. Some of Burkhead's data are shown in Table 4.12.

TABLE 4.12

Correlations over High Schools of Test Indices and Student Body Composition Variables[a]

	Grade 11 Ability Test	Grade 11 Reading Test	Median Family Income	Percent Nonwhite	Percent Substandard Housing
Grade 9 ability test index	0.98	0.94	0.92	−0.86	−0.79
Grade 11 ability test index		0.96	0.90	−0.84	−0.74
Grade 11 reading test index			0.91	−0.80	−0.73
Median family income				−0.79	−0.73
Percent nonwhite					0.57

[a] Data from the city of Chicago, after Burkhead (1967).

Table 4.13

Correlations over Schools of Median Test Scores and Student Body Composition Variables for Elementary Schools

City in the West	1967 Reading Achievement Median, Grade 6	1968 Reading Achievement Median, Grade 6	1968 Ability Test Median, Grade 6	1968–1969 Percent Minorities
1969 reading achievement median, grade 3	0.80	0.84	0.80	−0.75
1967 reading achievement median, grade 6		0.86	0.85	−0.68
1968 reading achievement median, grade 6			0.93	−0.78
1968 ability test median, grade 6				−0.73

City in the Midwest	1968 Arithmetic Achievement Median, Grade 6	1969 Verbal Aptitude Median, Grade 6	1968–1969 Percent White	1968–1969 Percent AFDC[a]
1969 reading achievement median, grade 6	0.84	0.93	0.66	−0.74
1969 arithmetic achievement median, grade 6		0.82	0.56	−0.68
1969 verbal aptitude median, grade 6			0.72	−0.73
1968–1969 percent white enrollment				−0.64

[a] Aid to Families with Dependent Children.

Variables for Stratifying Elementary Schools. Variables for stratifying elementary schools were investigated by Jaeger (1970). The parameter to be estimated was either mean achievement in a school district or the mean of the distribution of school achievement medians. Correlations between median ability test scores, median achievement test scores, and student body composition variables are shown in Table 4.13. Correlations were computed over schools in a large western city and in a midwestern city. Correlations are highest between median scores on tests administered in the same year to pupils in the same grade. Correlations between tests administered in different years or to pupils in different grades are above 0.80 but are lower than those for tests administered in the same year. Correlations between median test scores and student body composition variables are still lower (0.56–0.78). The best single variable for stratifying elementary schools is thus a median test score for the same year and grade as the sampling variable. Student body composition variables should provide stratified sampling with relative efficiencies between 140 and 250 percent, compared to simple random sampling if used individually for stratification of schools [Equation (4.36)].

In summary, individual test scores and test score averages for classrooms and schools appear to be the most useful variables for stratification of pupils, classrooms, and schools, respectively. The correlations between pupils' test scores and individual background variables are so low (Table 4.10) that background variables alone would not provide efficient stratification of pupils. Some classroom composition variables have reasonably high correlations with classroom mean achievement scores (Table 4.11). These composition variables should provide reasonable efficiency if used singly for stratification and might be useful for two-way or three-way stratification. Student body composition variables should be useful for stratifying elementary schools and high schools (Tables 4.12 and 4.13). When estimating mean achievement in a school district, stratification on these variables should provide relative efficiencies for stratified sampling of at least 140 percent, compared to simple random sampling.

The following example is totally fictitious and makes use of fictitious data. It is intended to illustrate the use of stratified sampling in estimating a population total.

Example 4.3 Estimating the Total Number of Crimes Committed by Youthful Offenders

CONTENT AND PROBLEM

The Department of Justice of a large state was interested in estimating the total number of crimes committed in that state by youthful offenders (defined as persons under the age of 18) during the calendar year 1982. In years past, this parameter had been calculated from data collected from every police department in the state. Historically, separate figures on the number of crimes committed by youthful offenders had been reported for police departments serving both incorporated and unincorporated areas in six different population categories:

Populations of 1,000,000 or more
Populations of at least 250,000 but less than 1,000,000
Populations of at least 100,000 but less than 250,000
Populations of at least 25,000 but less than 100,000
Populations of at least 10,000 but less than 25,000
Populations under 10,000

The numbers of police departments serving areas with populations in these categories were known, as was the total number of departments in the state.

In 1983 the Department of Justice did not have the funds necessary to conduct a census of police departments throughout the state. Nevertheless, the department wanted an estimate of the total number of crimes committed by youthful offenders during 1982. In addition, the department wanted to report crime figures for each of the six population categories listed above if, within budget limitations, sufficient data could be collected to compute reasonably precise estimates.

A policymaker in the Department of Justice was interviewed at length to determine how precise an estimate she wanted of the total number of crimes committed by youthful offenders. It was decided that this precision requirement would be used to calculate the number of police departments to be sampled and that estimates for the separate population categories would be computed only if resulting sample sizes within these strata provided sufficient precision. After considerable discussion, it was decided that the statewide estimate of the number of crimes by youthful offenders was to be within 100,000 of the true parameter, with 90 percent confidence.

SOLUTION AND RESULTS

Several of the stratified sampling design issues that face a survey researcher were addressed in the specifications of this problem. Both the number of strata to be used and the boundaries of strata were defined by the subpopulations of interest to the Department of Justice. It is therefore unnecessary to determine an appropriate stratification variable, to consult Table 4.2 on the relationship between relative precision and the number of strata used, or to invoke the Dalenius and Hodges procedure for calculating optimal stratum boundaries. However, it *is* necessary to estimate overall sample size requirements and then to allocate sample sizes among the six strata. We will consider these problems next.

ESTIMATION OF REQUIRED SAMPLE SIZE

Since a census of police departments had been conducted in years past, the mean and standard deviation of the number of crimes committed by youthful offenders were available for every year from 1941 through 1981 for all police departments serving areas in each of the six strata (population categories). It was decided to use data for the most recent year to estimate the required sample size.

The numbers of police departments serving areas in each population category and the standard deviations of the numbers of crimes by youthful offenders reported by these police departments in 1981 were as follows:

Population Category	Number of Police Departments	Standard Deviation
1. Over 1,000,000	2	7500
2. 250,000 to 1,000,000	5	6500
3. 100,000 to 250,000	50	1000
4. 25,000 to 100,000	100	400
5. 10,000 to 25,000	500	150
6. Under 10,000	1500	100

The total number of police departments in the state was 2157.

In order to simplify calculations, and because the number of strata and the stratum boundaries had been specified in the problem statement, we decided to use proportional allocation, at least initially, in distributing the overall sample size among strata. Therefore, we can use Equation (4.34) to estimate the number of police departments that should be surveyed. We can summarize or calculate the values needed to evaluate Equation (4.34) from the problem specifications and the data available for 1981:

Overall population size, N: 2157
Value on the normal distribution associated with a desired confidence level of 90 percent: 1.64
Maximum error limit (tolerable difference between the value of an estimate and the true value of the population total): 100,000
Size of each stratum, estimated variance in each stratum, and product of the two (from data tabulated above):

Stratum	Population	Variance (s_k^2)	Product ($N_k s_k^2$)
1	2	56,250,000	112,500,000
2	5	42,250,000	211,250,000
3	50	1,000,000	50,000,000
4	100	160,000	16,000,000
5	500	22,500	11,250,000
6	1500	10,000	15,000,000
		Total	416,000,000

An estimate of required sample size can now be found by using these data in Equation (4.34):

$$n = \frac{2157(1.64/100,000)^2(416,000,000)}{1 + (1.64/100,000)^2(416,000,000)}$$

$$= \frac{241.34}{1 + 0.1119} = 217.05$$

which rounds up to a required sample size of 218 police departments.

As is true in this example, when totals are estimated for large populations, the numbers involved often become very large. Computations with an ordinary hand calculator become very tedious (and we would hate to imagine a reader attempting to verify our computations or complete the exercises given at the end of each chapter with nothing more than a pencil and paper). One way around this problem is to use scientific notation, in which large numbers are expressed as a value multiplied by a power of 10. For example, 416,000,000 is equal to $416 \times 1,000,000$, and 1,000,000 is equal to 10 to the sixth power. Thus 416,000,000 can be written as 416×10^6. Similarly, 100,000 is equal to 10^5, and 1/100,000 is equal to $1/10^5 = 10^{-5}$. When numbers expressed in powers of 10 are multiplied, their exponents are simply added. When one number expressed in powers of 10 is divided by another, the exponent in the denominator is subtracted from the exponent in the numerator. Finally, when a number expressed in scientific notation is squared, its exponent is merely doubled. For example, $416,000,000/100,000^2$ is equal to

$$\frac{416 \times 10^6}{(10^5)^2} = \frac{416 \times 10^6}{10^{10}}$$

$$= 416 \times 10^{6-10}$$
$$= 416 \times 10^{-4} = 0.0416$$

Notice that multiplying by positive powers of 10 is equivalent to moving the decimal point to the right, and multiplying by negative powers of 10 is equivalent to moving the decimal point to the left. In each case, the decimal point is moved the same number of places as the expressed power of 10.

ALLOCATION OF SAMPLE SIZES

To allocate the overall sample size of 218 police departments among the six strata, we must multiply the total sample size by the six stratum weights. These computations are shown below.

Stratum	Stratum Weight	Product	Allocation
1	$\frac{2}{2157} = 0.00093$	0.2	2
2	$\frac{5}{2157} = 0.00232$	0.5	5
3	$\frac{50}{2157} = 0.02318$	4.9	5
4	$\frac{100}{2157} = 0.04636$	9.8	10
5	$\frac{500}{2157} = 0.23180$	48.9	49
6	$\frac{1500}{2157} = 0.69541$	146.7	147
Total	1.00000		218

The "Allocation" column requires some explanation. As is often done in actual survey designs, we have arbitrarily increased the sample sizes allocated to the smallest strata (in this case, Strata 1 and 2) to the point where a census would be conducted within these strata. We then adjusted the total sample size of 218 to account for these allocations ($218 - 7 = 211$) and allocated the adjusted sample size proportionally among the remaining strata. The advantage of this procedure is that it affords exact values of the subpopulation parameter (the total number of crimes committed by youth) for the two smallest strata, with very little reduction in the precision of estimates computed for the remaining strata.

ESTIMATION OF THE POPULATION TOTAL

Data on crimes committed by youthful offenders during 1982 were collected from the numbers of police departments indicated in the "Allocation" column above, for each stratum. Mean numbers of crimes committed by youthful offenders during 1982 were then computed and tabulated as follows:

Stratum	Mean Number of Crimes by Youthful Offenders
1	45,000
2	25,000
3	7,600
4	2,700
5	750
6	200

These data were used in Equation (4.10), together with the size of each stratum, to compute an estimate of the total number of crimes committed by youthful offenders throughout the state during 1982:

$$y_{st} = 2(45,000) + 5(25,000) + 50(7600)$$
$$+ 100(2700) + 500(750) + 1500(200)$$
$$= 1,540,000$$

This total would be truly frightening if it were based on actual data. However, note that the total population of our fictitious state is about 29,000,000 which is even larger than California's population. We can therefore hope that the crime rate has been exaggerated just as much in formulating this example.

CONSTRUCTION OF A CONFIDENCE INTERVAL

If we assume that the estimated variances within strata for 1981 apply to 1982 as well, we can construct a confidence interval on the statewide total of crimes committed by youth from Equation (4.16). Of course the first step is estimation of the variance of the estimated total, and for that, we use Equation (4.15).

To evaluate Equation (4.15), we need the estimated variances among elements in each stratum, s_k^2; the population size and sample size in each stratum, N_k and n_k; and the sampling fraction in each stratum, f_k. These values are tabulated below:

Stratum	Population	Sample	Variance	Sampling Fraction
1	2	2	56,250,000	1.000
2	5	5	42,250,000	1.000
3	50	5	1,000,000	0.100
4	100	10	160,000	0.100
5	500	49	22,500	0.098
6	1500	147	10,000	0.098

Equation (4.15) is evaluated by finding the inverse of the sampling fraction $1/f_k$ and multiplying this by the estimated variance among elements, s_k^2, and then by the difference between the population size and the sample size, $N_k - n_k$, in each stratum. These products are then summed across all strata. By inspecting Equation (4.15), it is easy to see the advantage of having done a complete census in strata 1 and 2. Since the population size and the sample size are identical in each of these strata, their difference is zero, and the first two strata do not contribute to the variance of the estimated total. The estimated variance of the estimated population total, $v(y_{st})$, is computed from the data for strata 3 through 6, as follows:

$$v(y_{st}) = \frac{1}{0.1}(1,000,000)(50 - 5)$$

$$+ \frac{1}{0.1}(160,000)(100 - 10)$$

$$+ \frac{1}{0.098}(22,500)(500 - 49)$$

$$+ \frac{1}{0.098}(10,000)(1500 - 147)$$

$$= 450,000,000 + 144,000,000 + 103,545,920 + 138,061,220$$

$$= 835,607,140$$

The standard error of the estimated population total is equal to the square root of the estimator variance. In this case, its value is 28,906.87.

A 90 percent confidence interval on the statewide total of crimes by youthful offenders can now be found from Equation (4.16):

$$\hat{Y}_{L_{st}} = 1,540,000 - 1.64(28,907) = 1,492,592$$

and

$$\hat{Y}_{U_{st}} = 1,540,000 + 1.64(28,907) = 1,587,408$$

These confidence limits suggest that a sample size of 218 police departments, a bit more than 10 percent of the population of 2157 departments, was certainly sufficient to provide the desired 90 percent confidence interval with a width of no more than ±100,000. Whether the confidence interval includes the true population total is, of course, unknown.

ESTIMATED TOTALS FOR INDIVIDUAL STRATA

The numbers of crimes committed by youthful offenders in areas served by police departments in the first two strata were computed from census data and therefore can be reported with complete confidence. The values are

For Stratum 1: 2(45,000) = 90,000
For Stratum 2: 5(25,000) = 125,000

In Strata 3 through 6, police departments were selected using simple random sampling. Equations (3.4) and (3.14) can be used to compute estimates of totals for these strata as well as confidence intervals around these estimates. Using a 90 percent confidence level, the resulting values are:

Stratum	Estimated Total	Upper Limit	Lower Limit
3	380,000	414,790	345,210
4	270,000	289,680	250,320
5	375,000	391,689	358,312
6	300,000	319,270	280,730

Detailed computations have not been shown for these values because they follow from equations that have been discussed in detail in Chapter 3. However, we strongly recommend that you compute estimated totals and associated confidence limits for Strata 3 through 6 to ensure that you can apply the appropriate equations confidently.

DISCUSSION

Using a sample of 218 police departments, a bit more than 10 percent of the entire population, the statewide total number of crimes committed by youthful offenders during 1982 has been estimated with a confidence interval that is well within the width specified in the problem design statement. Estimates for individual strata vary in their precision and might be judged too imprecise to be reported. The widest 90 percent confidence interval, for stratum 3, has a width of almost 70,000. Confidence intervals on totals in strata 4 through 6 are somewhat narrower, with widths that are less than 40,000. Whether these intervals are too wide to provide a basis for reasonable policy decisions is a matter of administrative judgment. If the Department of Justice report is being compiled solely to inform the public and/or state legislators on the incidence of crime in the state, the figures reported for individual strata might be sufficiently informative and precise.

Exercises

1. The director of institutional research at a large university wanted to estimate the mean score of the university's incoming freshman class on the verbal section of the Scholastic Aptitude Test (SAT) without having to administer the test to all students. Instead of selecting a simple random sample of students, the director decided to make use of students' scores on the regularly administered Spielberger State Anxiety Measure. Since this measure, which was administered to all incoming freshmen during registration week, had a correlation of 0.85 with the verbal section of the SAT, it was likely to be useful as a classification variable in a stratified sampling procedure. Based on the correlation between the classification variable and the sampling variable, the director of research decided to use four strata. He came to you with the following data, and asked you to derive stratum boundaries that would minimize the variance of his estimator of the mean SAT score. You decided to apply the Dalenius and Hodges procedure with 20 data intervals. Find optimal stratum boundaries.

Anxiety Score	Frequency	Anxiety Score	Frequency
11	4	31	28
12	8	32	30
13	10	33	29
14	18	34	28
15	19	35	26
16	26	36	24
17	26	37	27
18	28	38	22
19	37	39	19
20	43	40	20
21	50	41	19
22	46	42	16
23	52	43	12
24	50	44	15
25	43	45	14
26	49	46	10
27	37	47	7
28	41	48	9
29	32	49	5
30	30	50	4

2. A social scientist was studying the relationship between the socioeconomic status of adults in a small city and their attitudes toward pornography. She conducted an interview survey that included the question, "Should the city government pass a regulation outlawing 'adult' bookstores?" Recent census data were used to divide the adult residents of the city into five socioeconomic-status strata. Simple random samples of respondents were selected from each stratum. The population size, sample size, and proportion of respondents who answered "Yes" were as follows for each stratum:

Stratum	Population Size	Sample Size	Proportion "Yes"
Lower class	6,000	100	0.60
Working class	10,000	200	0.50
Lower middle class	5,000	100	0.40
Upper middle class	3,000	100	0.35
Upper class	1,000	100	0.25

a. Estimate the proportion of the citywide population of adults who, if interviewed, would answer "Yes."
b. Estimate the variance of your estimator.
c. Determine a 95 percent confidence interval on the population proportion of adults who would answer "Yes."

   d. Compute the efficiency of stratified sampling relative to simple random sampling in this application.

3. A researcher in religious studies developed a scale of "religiosity" and wanted to estimate the mean religiosity of the adult population in a neighboring city. He decided to select a stratified random sample of adults from four previously defined strata, using optimal allocation of sample sizes to strata. From a combination of earlier research in another city and the work of religious leaders in the city under study, the researcher knew the size of the population in each stratum and had estimates of the variance of scores on the religiosity scale in each stratum. These data are given below:

   Stratum 1: There were 5000 regular church attendees in the city, and their variance on the scale was estimated to be 100.
   Stratum 2: There were 10,000 occasional church attendees in the city, and their variance on the scale was estimated to be 225.
   Stratum 3: There were 5000 adults in the city who never attended church but nonetheless professed a firm belief in a Supreme Being. Their variance on the scale was estimated to be 324.
   Stratum 4: There were 5000 self-proclaimed adult atheists in the city. Their variance on the scale was estimated to be 196.

How large a sample size would the researcher have to take in order to estimate the mean religiosity of the citywide population of adults within one point, with 95 percent confidence?

4. In a sampling problem involving stratified random sampling with six strata, the researchers wanted to estimate a population total and then compute a 95 percent confidence interval on the population total. The data available for each stratum were as follows:

Population	Sample Size	Sample Mean	Sample Variance
1000	100	10	100
1000	200	20	200
1000	100	30	100
2000	200	40	200
2000	100	50	100
2000	200	60	200

Find the estimates that the researchers want.

Linear Systematic Sampling:
Basic Concepts, Estimators, and Applications

Introduction

The term *systematic sampling* has been applied to procedures that have the following two characteristics in common: (1) the elements of a population are treated as an ordered sequence, and (2) the elements are selected at a constant interval from an ordered sampling frame. In simple random sampling, the arrangement or ordering of elements in the sampling frame is irrelevant. In systematic sampling, estimates are usually affected by the choice of elements that are labeled "first," "second," etc.

Linear systematic sampling, the most commonly used systematic sampling procedure, is considered in this chapter. Chapter 6 presents theory and applications for other systematic sampling procedures which are particularly useful with special populations.

In linear systematic sampling, the elements of a population are assigned numbers from 1 to N. The number assigned to an element is called its *index*. Suppose n elements are to be sampled. The population size N is divided by the sample size n to determine the sampling interval k, defined as the integer closest to the ratio N/n. A table of random numbers is then used to select a number, r, in the range 1 to k inclusive. The element with index r is chosen as the first element in the sample. The rest of the sample consists of elements with indices that are some multiple of k units from the first sampled element. Thus the index of the second sampled element is $r + k$, the index of the third is $r + 2k$, and so on, up to the nth sampled element, which has index $r + (n - 1)k$. A diagram of linear systematic sampling is shown in Figure 6.1 (Chapter 6). An example may clarify the procedure.

Suppose a sample of two elements is selected from a population of size six. The sampling interval k is then $\frac{6}{2}$ or 3. The elements of the population have indices from 1 to 6. A random number r from the set 1, 2, 3 is chosen to determine the first sampled element. If the random number is 2, the first element in the sample will have index 2, and the second sampled element will have an index equal to $r + k = 2 + 3 = 5$. Using linear systematic sampling, three distinct samples of two elements could be selected from the population, depending on the value of r. The possible samples are listed below.

Index of first sampled element
r = 1
r = 2
r = 3

Indices of elements in the sample
1 and 4
2 and 5
3 and 6

Three facts are evident from this example. First, every element in the population appears in exactly one sample. Second, each sample has a probability of selection equal to $1/k$, the inverse of the sampling interval. Third, each element in the population has a probability of selection equal to $1/k$. The first of these facts is true for linear systematic sampling in general. The second and third are true whenever the ratio of the population size to the sample size, N/n, is an integer. The ratio N/n will be assumed to be an integer in this chapter, since it makes the mathematical theory much simpler. In practical applications, it makes no difference whether N/n is an integer.

Because linear systematic sampling is readily learned and affords operational convenience, it warrants consideration in any setting where samples must be selected from frames composed of physical records (for example, file cards) or where samples are selected "in the field" by clerical personnel or others with no formal training in sampling procedures (for example, teachers who select samples of pupils from their classes or administrative workers who select samples of employees from personnel files).

Linear systematic sampling can be used instead of simple random sampling in many applications. For some populations, linear systematic sampling is preferable to simple random sampling; for others, it is not. A major consideration in choosing between the two methods is the sample size required. Another is the number of physical locations or administrative units in which data collection must take place. When simple random sampling is used in some surveys, the number of locations or administrative units in which data are collected is a random variable, constrained only by the total number of elements sampled. With linear systematic sampling, sampled elements are usually distributed evenly over locations or administrative units. If the sampling fraction is small and the number of elements in each location or administrative unit is also small, elements might only be sampled from half or a third of the locations or units.

Linear systematic sampling is sometimes recommended for surveys that use multistage sampling designs. A multistage design might call for selection of large geographic or administrative units, called *primary sampling units,* at the first stage of sampling. As an example, counties or schools might serve as primary sampling units. Then at the second stage of sampling, elements (for example, businesses or pupils) would be chosen within the selected first-stage units. During the first stage of sampling, primary sampling units could be selected using either simple random sampling or stratified sampling. Then during the second stage of sampling, elements could be selected using linear systematic sampling within the primary sampling units that were chosen during the first stage.

Whether any sampling procedure can be applied successfully in practice depends on the cost of achieving desired estimation precision and the operational feasibility of the method. Selection among competing procedures also should be based on cost and practicability. Linear systematic sampling is compared to simple random sampling and stratified sampling in a later section of this chapter. We show that the advisability of using linear systematic sampling depends on the composition of the population being studied.

Some Illustrative Applications of Linear Systematic Sampling

Procedures for applying linear systematic sampling will vary with the population parameter to be estimated and the population under study, and no universally applicable procedures can be described. Nonetheless, some examples of linear systematic sampling in education and the social sciences are discussed below.

A technique similar to linear systematic sampling has been used by test publishers in norming standardized achievement tests. The procedure is known as *spiraled testing* and consists of arranging tests of, say, reading, arithmetic and social studies in a cyclic pattern. When test booklets are arranged in this order, every third pupil in a class takes the reading test. The pupil to the right, say, of each of these takes the arithmetic test, and the pupil to the left is given the social studies test. Which pupil takes the reading test and which takes the arithmetic test depends on the seating arrangement and the test booklet that is on top of the stack when the tests are handed out. The latter choice is random. Spiraled testing is convenient to use and frequently is statistically efficient. All pupils are tested at the same time, and instructions regarding the distribution of test booklets can be made quite simple.

Linear systematic sampling is most convenient where a population has some natural order, such as pupils already assigned to seats in classrooms. Linear systematic sampling can be used with pupils in their established seating order by instructing each teacher to select every *k*th pupil (for example, every other pupil or every third pupil), starting with pupil number *r*. Pupils could be numbered by going down rows in a classroom. The pupil with whom the sampling cycle would begin in each classroom could be determined by selecting *r* from a table of random numbers. With this procedure, all pupils in a classroom would have an equal chance of being selected; this would not be the case if the sample always began with the first pupil in the first row.

Linear systematic sampling is often efficient when it is applied to populations that are intentionally arranged in some order. For example, in a survey of homeowners to determine the percentage who possess certain appliances or to find their average yearly income, it would be useful to arrange the sampling frame in increasing order of the assessed valuation of the homeowners' properties. A linear systematic sample would then be drawn from the ordered sampling frame. This procedure would very likely result in smaller estimator variances than would linear systematic sampling from a randomly arranged list. The higher the correlation between the sampling variable and the variable used to arrange the frame, the greater the efficiency of the sampling procedure. Linear systematic sampling of purposefully arranged sampling frames is discussed in greater detail in Chapter 6, together with detailed examples.

Estimating Population Means

FORMING THE ESTIMATES

As in simple random sampling, the sample mean

$$\bar{y} = \frac{1}{n} \sum_{i=1}^{n} y_i \tag{5.1}$$

where y_i denotes the value of the sampling variable for the ith sample element, is usually used to estimate the population mean:

$$\bar{Y} = \frac{1}{N} \sum_{i=1}^{N} y_i \tag{5.2}$$

Here

y_i = value of the sampling variable for the ith element in the sampling frame

N = population size

n = sample size

ESTIMATOR BIAS AND MEAN SQUARE ERROR

With linear systematic sampling, the sample mean is an unbiased estimator of the population mean whenever N/n is an integer. Even when N/n is not an integer, the bias is negligible for population sizes likely to be encountered in practice.

Since in linear systematic sampling, each potential sample has the same probability of selection (equal to $1/k$), the mean square error of the estimated population mean equals

$$\text{MSE}(\bar{y}) = \frac{1}{k} \sum_{r=1}^{k} (\bar{y}_r - \bar{Y})^2 \tag{5.3}$$

Here

$$\bar{y}_r = \frac{1}{n} \sum_{i=0}^{n-1} y_{r+ik} \tag{5.4}$$

denotes the mean of the systematic sample with initial element r

y_{r+ik} = value of the sampling variable for the element in the sampling frame with index $r + ik$

$\bar{Y}$ = population mean

n = sample size.

In the following example, linear systematic sampling is used to estimate the mean reading achievement of all sixth-grade pupils in a medium-sized city. Although the example is fictitious, the data used are actual test scores.

Example 5.1. Estimating Mean Achievement in a School District

CONTEXT AND PROBLEM

The data used in this example are the same as those used in Example 3.1 and Example 4.2. The mean reading achievement of sixth-grade pupils in Midcity is to be estimated using linear systematic sampling. As in previous examples, the population mean is to be estimated within 0.2 grade equivalent units with 95 percent confidence. The largest sampling interval that affords desired estimation precision is to be determined.

SOLUTION AND RESULTS

In 1983, the sixth-grade enrollment in Midcity was 1180 pupils. To provide a sampling frame, an alphabetic list of sixth-graders was compiled and numbered from 1 to 1180. For purposes of this example, all possible systematic samples were selected from the sampling frame by using the SYSAMP-I computer program described in Appendix C. Sampling intervals of 2, 4, 5, 10, and 20 resulted in sampling fractions of 50, 25, 20, 10, and 5 percent respectively. Details on using the SYSAMP-I computer program can be found in Appendix C.

As in previous examples, error limits of ±0.2 grade equivalent units were assumed to correspond to ±4 raw score points.

When samples are selected using linear systematic sampling, it may be unreasonable to compute confidence intervals on the population mean. If the sampling interval is small, for example, 4, there are only four potential samples and thus four sample means. Since each value will result with probability $\frac{1}{4}$, the sample mean will not be normally distributed. Therefore, the variance of the sample mean cannot be interpreted easily as an indicator of estimation precision. Other distribution parameters, such as skewness, may be equally important. However, for the sake of comparison with previous examples, 95 percent confidence intervals with an average width of 8 raw score points will be assumed to follow from a mean square error of 4 raw score points squared. This is tantamount to assuming that the sample mean is normally distributed. With sampling intervals of 20 or more, the assumption is probably reasonable.

Mean square errors for linear systematic sampling and simple random sampling are presented for a variety of sampling fractions in Table 5.1.

TABLE 5.1
Mean Square Errors of Estimated Population Means for Simple Random Sampling and Linear Systematic Sampling[a]

Sampling Interval	Sampling Fraction, %	Mean Square Errors for Simple Random Sampling	Mean Square Errors for Linear Systematic Sampling
2	50	0.39	0.03
4	25	1.16	0.33
5	20	1.55	0.52
10	10	3.50	0.84
20	5	7.44	2.49

[a] All data are for the population of sixth-grade pupils in Midcity. Units are raw score points squared.

The efficiency of linear systematic sampling relative to simple random sampling is found by computing the ratio of mean square errors and converting to percents:

$$\text{Efficiency} = 100 \left(\frac{\text{mean square error for simple random sampling}}{\text{mean square error for linear systematic sampling}} \right)$$

Results are shown in Table 5.2.

DISCUSSION

As shown in Table 5.1, the mean square errors resulting from linear systematic sampling are one-third to one-tenth as large as those resulting from simple

TABLE 5.2
Efficiency of Linear Systematic Sampling Relative to Simple Random Sampling[a]

Sampling Interval	Sampling Fraction, %	Efficiency, %[b]
2	50	1125
4	25	354
5	20	299
10	10	418
20	5	298

[a] All data are for the population of sixth-grade pupils in Midcity.
[b] When interpreting these results and the relative efficiencies presented in subsequent tables, the reader should realize that they are subject to large fluctuations across populations. This is especially true when large relative efficiencies result from small mean square errors in the denominator.

random sampling. These data violate intuition. If the arrangement of elements in a sampling frame is random, then linear systematic sampling and simple random sampling are, on the average, equally precise. (The next section contains a theoretical discussion on this point.) Therefore, one of two conclusions must be drawn. Either these results for Midcity are fortuitous, or arranging pupils in alphabetic order results in a nonrandom ordering of achievement test scores. The latter conclusion can be argued to a point. Certainly, alphabetic arrangement would result in some twins being listed sequentially. Twins often have similar achievement test scores (this is more typical with monozygotic twins). It is also possible that alphabetic listing results in some ordering of pupils by ethnic or nationality groups. Certain surnames occur more frequently for some nationality groups than for others. Carrying a tenuous argument further, members of the same ethnic group tend to have similar socioeconomic status and, therefore, similar achievement. This is particularly true, for instance, of Japanese-American children, who tend to have higher achievement test scores. If these phenomena are present to a sufficient degree, pupils with similar achievement test scores will be grouped together, and those with less similar scores will be separated in the alphabetic list. This arrangement would tend to make linear systematic sampling more precise than simple random sampling because different samples would contain similar elements. Whether this line of reasoning explains these data or is totally irrelevant can be determined only from additional empirical studies. Investigation of linear systematic sampling in other school systems may show that the method is consistently superior to simple random sampling for populations in alphabetic order. If so, serial correlation coefficients (Cochran, 1977, pp. 219–221) could be used to test the hypotheses mentioned above.

With linear systematic sampling, estimation precision is within desired limits for all sampling fractions investigated (Table 5.1). For a sampling fraction of 5 percent, the mean square error is 2.49. An even smaller sampling fraction would probably yield a mean square error that was less than 4. The mean square error resulting from simple random sampling is less than 4 for a sampling fraction of 10 percent, but it is unacceptably large for a sampling fraction of 5 percent. Thus linear systematic sampling appears to offer substantial economies.

Efficiency Relative to Simple Random Sampling

The efficiency of linear systematic sampling relative to simple random sampling depends on the relationships among elements in the population and the order of elements in the sampling frame. Cochran (1977) and Madow and Madow (1944) provide formulas that show the relationship between efficiency and some descriptive population parameters. Cochran's formulation will be considered first.

VARIANCE WITHIN SAMPLES

Suppose elements in the sampling frame are numbered from 1 to N and that $N = nk$. Here n is the size of each sample, and k denotes both the sampling interval and the number of potential samples. The rth linear systematic sample contains the elements $y_r, y_{r+k}, y_{r+2k}, \ldots, y_{r+(n-1)k}$. The mean of the rth sample is denoted by $\bar{y}_r$. The efficiency of linear systematic sampling relative to simple random sampling depends on the variance among elements within the same sample relative to the variance among elements in the entire population.

The variance among linear systematic sample means can be written

$$V(\bar{y}) = \frac{N-1}{N} S^2 - \frac{k(n-1)}{N} S^2_{wsy} \tag{5.5}$$

where

$$S^2 = \frac{1}{N-1} \sum_{i=1}^{N} (y_i - \bar{Y})^2$$

is the variance among elements in the sampling frame and

$$S^2_{wsy} = \frac{1}{k(n-1)} \sum_{r=1}^{k} \sum_{i=0}^{n-1} (y_{r+ik} - \bar{y}_r)^2 \tag{5.6}$$

is the average variance among elements in the same linear systematic sample.

From Equation (3.6), the variance of the mean of a simple random sample equals

$$V(\bar{y}) = \frac{N-n}{Nn(N-1)} \sum_{i=1}^{N} (y_i - \bar{Y})^2 \tag{5.7}$$

Here N and n are as defined above, and $\bar{Y}$ is the population mean.

It follows that linear systematic sampling yields more precise estimates than simple random sampling only when $S^2_{wsy} > S^2$ and that the two methods are equally efficient only when $S^2_{wsy} = S^2$. That is, linear systematic sampling is more efficient than simple random sampling only when the average variance among elements in the same systematic sample is larger than the variance among elements in the entire population. This condition may be found frequently in practice. Whenever adjacent elements in the sampling frame are more similar than those some distance apart, the within-sample variance will be larger than the overall variance, and linear systematic sampling will be more efficient than simple random sampling.

In Example 5.1, linear systematic sampling was found to yield more precise results than simple random sampling. The average within-sample variance should therefore be larger than the population variance for the data in that example. These variances are computed and discussed in Example 5.2.

Example 5.2. Average Within-Sample Variance Compared to Population Variance

CONTEXT AND PROBLEM

In Example 5.1, the efficiency of linear systematic sampling relative to simple random sampling was computed for a population of sixth-graders in Midcity. The sampling variable was raw score on a test of reading achievement. Samples were drawn from an alphabetized list of all sixth-graders in the city.

TABLE 5.3
Average Within-Sample Variances S_{wsy}^2 for Linear Systematic Samples Compared with Population Variance S^{2a}

Sampling Interval	Sampling Fraction, %	Population Variance	Variance within Samples, Averaged across All Possible Samples
2	50	454	455
4	25	454	455
5	20	454	455
10	10	454	457
20	5	454	468

[a] Data are reading achievement scores for the population of sixth-grade pupils in Midcity.

In Example 5.1, the relative efficiency of linear systematic sampling and simple random sampling was determined by direct comparison of mean square errors. The problem here is to determine the more efficient sampling method by comparison of magnitudes of the average within-sample variance S_{wsy}^2 and the overall population variance S^2.

SOLUTION AND RESULTS

The SYSAMP-I computer program was used to select all linear systematic samples from the population of sixth-graders in Midcity. Several sampling intervals were used, resulting in sampling fractions of 50, 25, 20, 10, and 5 percent. The SYSAMP-I program was also used to calculate the population variance S^2 and the average within-sample variance S_{wsy}^2 for each sampling fraction. Results are shown in Table 5.3.

DISCUSSION

S_{wsy}^2 is larger than S^2 for all sampling fractions investigated. Thus, for the population of sixth-graders in Midcity, linear systematic sampling is consistently more efficient than simple random sampling. These results are in agreement with those presented in Example 5.1.

It should be noted that the population variance and the average within-sample variances in Table 5.3 are almost equal, even though linear systematic sampling is at least 300 percent as efficient as simple random sampling (Table 5.2). For linear systematic sampling to be more efficient than simple random sampling, S_{wsy}^2 must be larger than S^2, but the difference need not be great. Relative efficiency depends not only on S^2 and S_{wsy}^2 but on the population size and the size of the sampling interval as well. The larger the sampling interval, the larger must be S_{wsy}^2, to produce large gains in efficiency.

Intraclass Correlation Coefficient

Madow and Madow (1944) provide yet another expression for the variance of the means of linear systematic samples. Their formulation uses a measure of the relatedness of elements in the same linear systematic sample. The measure is called an *intraclass correlation coefficient* and is defined as

$$P_w = \frac{2}{(n-1)(N-1)S^2} \sum_{i=1}^{k} \sum_{j<u} (y_{ij} - \bar{Y})(y_{iu} - \bar{Y}) \tag{5.8}$$

where y_{ij} denotes the jth element in the ith linear systematic sample and the second summation is taken over all pairs of elements in the same systematic sample, such that j is less than u.

In terms of ρ_w, the variance among the means of linear systematic samples is equal to

$$V(\bar{y}) = \frac{S^2}{n} \frac{N-1}{N} [1 + (n-1)\rho_w] \tag{5.9}$$

Comparison of Equations (5.9) and (5.7) shows that linear systematic sampling is more efficient than simple random sampling only if $\rho_w < -1/(N-1)$. That is, the correlation coefficient for pairs of elements in the same linear systematic sample must be slightly negative if linear systematic sampling is to be more efficient than simple random sampling. (These results hold exactly only when N/n is an integer. The extension to cases where N/n is not an integer is mathematically complicated and not of practical interest.)

The intraclass correlation ρ_w can assume values between 1 and $-1/(n-1)$. A large positive value indicates small variation among elements in the same systematic sample. Since the overall variance among elements in the sampling frame is fixed, small variation within linear systematic samples corresponds to large variation between linear systematic samples. It is intuitively reasonable that large between-sample variation leads to low estimation efficiency.

Intraclass correlation coefficients for the population of sixth-graders considered in earlier examples are shown in Example 5.3.

Example 5.3. Intraclass Correlation Coefficients for Systematic Samples of Sixth-Graders

CONTEXT AND PROBLEM

For the achievement test data collected in Midcity, as discussed in Example 5.1, linear systematic sampling is more efficient than simple random sampling. Intraclass correlation coefficients for these data should therefore be negative and smaller than $-1/(N-1)$. In this example, we compute intraclass correlation coefficients for the Midcity data in order to confirm the superiority of linear systematic sampling.

SOLUTION AND RESULTS

All linear systematic samples were drawn from an alphabetic list of the sixth-graders in Midcity. The sampling intervals used were 2, 4, 5, 10, and 20. The SYSAMP-I computer program (Appendix C) was used to draw the samples and to compute intraclass correlation coefficients for each choice of sampling interval. Results are shown in Table 5.4.

TABLE 5.4

Intraclass Correlation Coefficients for Linear Systematic Samples Selected from the Population of Sixth-Grade Pupils in Midcity

Sampling Interval	Sampling Fraction, %	$\dfrac{-1}{N-1}$	Intraclass Correlation Coefficient
2	50	−0.00085	−0.0016
4	25	−0.00085	−0.0026
5	20	−0.00085	−0.0031
10	10	−0.00085	−0.0065
20	5	−0.00085	−0.0117

DISCUSSION

The intraclass correlation coefficients in Table 5.4 are small. One might therefore expect linear systematic sampling and simple random sampling to have similar efficiencies when applied to the Midcity data. However, the intraclass correlation coefficient ρ_w is multiplied by $n - 1$ in computing the variance of the sample mean [Equation (5.9)]. With a large sample size, even a small negative value for ρ_w can therefore lead to substantial reductions in estimator variance.

The data in Table 5.4 are consistent with those in Table 5.2. The intraclass correlation coefficients are smaller than $-1/(N - 1)$ for all sampling intervals investigated. Thus for these data, linear systematic sampling is uniformly more efficient than simple random sampling.

Efficiency Relative to Stratified Sampling

THEORETICAL RESULTS

Cochran (1946) provides an expression for the variance of the means of linear systematic samples that permits comparison of the efficiencies of linear systematic sampling and a particular kind of stratified sampling. Cochran's results are more readily understood if the composition of linear systematic samples is considered first. The schematic diagram shown in Figure 5.1 should be helpful.

From Figure 5.1, it is clear that linear systematic sampling is similar to stratified sampling. However, there are some important differences. If successive groups of k elements are regarded as strata, linear systematic sampling consists of selecting one element per stratum. In stratified sampling, elements are selected independently within each stratum. In contrast, all elements selected for a linear systematic sample are determined by the element sampled from the first stratum; the choice of initial element determines the entire sample.

Suppose a sample is composed of elements selected independently, with one element chosen at random from each stratum. Cochran's formula allows comparison of the variance of the means of such samples with the variance of the means of linear systematic samples. The efficiency of linear systematic sampling

Stratum Number	Sample Number			
	1	2	r	k
1	y_1	y_2	y_r	y_k
2	y_{k+1}	y_{k+2}	y_{k+r}	y_{2k}
3	y_{2k+1}	y_{2k+2}	y_{2k+r}	y_{3k}
.	.	.	.	.
.	.	.	.	.
.	.	.	.	.
n	$y_{(n-1)k+1}$	$y_{(n-1)k+2}$	$y_{(n-1)k+r}$	y_{nk}
Sample Mean	$\bar{y}_1$	$\bar{y}_2$	$\bar{y}_r$	$\bar{y}_k$

FIGURE 5.1. **Linear systematic samples selected from a population of size $N = nk$ (elements y_i contained in strata and in potential systematic samples).**

is thus compared to the efficiency of stratified random sampling where one element is selected from each stratum, all strata are of equal size, and the classification variable is the index of population elements. Since the same sample size is used for this type of stratified sampling and for linear systematic sampling, the comparison is of practical interest.

The variance of the means of linear systematic samples can be expressed as

$$V(\bar{y}) = \frac{S^2_{wst}(N - n)}{Nn}[1 + (n - 1)\rho_{wst}] \tag{5.10}$$

where

$$S^2_{wst} = \frac{1}{n(k - 1)}\sum_{j=1}^{n}\sum_{i=1}^{k}(y_{ij} - \bar{y}_{\cdot j})^2 \tag{5.11}$$

   = average variance among elements within the same stratum
y_{ij} = value of the sampling variable for the ith element in the jth stratum
$y_{\cdot j}$ = mean of the sampling variable for elements in the jth stratum,

$$P_{wst} = \frac{2}{n(n - 1)(k - 1)S^2_{wst}}\sum_{i=1}^{k}\sum_{j<u}(y_{ij} - \bar{y}_{\cdot j})(y_{iu} - \bar{y}_{\cdot u}) \tag{5.12}$$

   = correlation coefficient between deviations from stratum means of pairs of elements in the same linear systematic sample

The general formula for the variance of the means of stratified random samples can be simplified when the sizes of the populations in all strata are equal and the same sampling fraction is used in every stratum. Equation (4.5) reduces to

$$V(\bar{y}_{st}) = \frac{N - n}{Nn}S^2_{wst} \tag{5.13}$$

By comparing Equations (5.13) and (5.10), it is easy to see that linear systematic sampling is more efficient than stratified sampling with one element per stratum only when ρ_{wst} is negative and that the two methods are equally efficient only when ρ_{wst} equals zero.

Equations (5.9) and (5.10) are of similar form; the correlation coefficients ρ_w and ρ_{wst} are multiplied by the factor $n - 1$. Here too, a small negative correlation coefficient ρ_{wst} can lead to dramatic reductions in the variance of the sample mean.

Example 5.4. Linear Systematic Sampling Compared to Stratified Sampling

CONTEXT AND PROBLEM

In Example 5.1, the efficiency of linear systematic sampling was compared to that of simple random sampling. The data used were achievement test scores for an alphabetic list of sixth-graders in a medium-sized city (Midcity). In this example, the same data are used to compare the efficiency of linear systematic sampling to that of stratified sampling. In the stratified sampling used here, it is assumed that strata consist of successive groups of k elements in the alphabetically ordered population. One element is selected from each stratum, using simple random sampling. The efficiencies of linear systematic sampling and stratified sampling are to be compared by computing and examining intraclass correlation coefficients ρ_{wst}.

SOLUTION AND RESULTS

The SYSAMP-I computer program was used to compute values of ρ_{wst} for the population of sixth-graders in Midcity. Correlation coefficients ρ_{wst} were computed for sampling intervals of 2, 4, 5, 10, and 20. Results are shown in Table 5.5.

DISCUSSION

Linear systematic sampling from an alphabetic list of pupils is consistently more efficient than stratified sampling with one element per stratum for Midcity data. The correlation of deviations from stratum means, ρ_{wst}, is negative for all sampling intervals considered (Table 5.5).

TABLE 5.5
Correlation Coefficients for Deviations from Stratum Means for Linear Systematic Samples Selected from the Population of Sixth-Grade Pupils in Midcity

Sampling Interval	Sampling Fraction, %	Correlation of Deviations from Stratum Means
2	50	−0.0016
4	25	−0.0023
5	20	−0.0028
10	10	−0.0063
20	5	−0.0075

Estimation of Population Totals

Relationships between formulas used in estimating population means and population totals, described for simple random sampling in Chapter 3, also apply to linear systematic sampling. The estimated population total is equal to the product of the population size and the estimated population mean. The variance of the estimated population total is simply the square of the population size multiplied by the variance of the estimated population mean. Symbolically, these results are as follows:

If $N = nk$, an unbiased estimator of the population total Y is given by

$$y = N\bar{y} = \frac{N}{n} \sum_{i=1}^{n} y_i \tag{5.14}$$

where y_i denotes the value of the sampling variable for the ith sampled element.

If $N = nk$, estimation of the population total is unbiased, and the variance and mean square error of the estimated population total will be equal. The mean square error is given by

$$\text{MSE}(y) = \frac{N^2}{k} \sum_{r=1}^{k} (\bar{y}_r - \bar{Y})^2 = \frac{1}{k} \sum_{r=1}^{k} (y_r - Y)^2 \tag{5.15}$$

where

$$y_r = \frac{N}{n} \sum_{i=0}^{n-1} y_{r+ik} \tag{5.16}$$

denotes an estimate of the population total that is based on data from the linear systematic sample with initial element r, and all other terms are as defined earlier.

When the population size N is not an integral multiple of the sampling interval k, Equation (5.14) is a biased estimator of the population total. However, in virtually all practical applications, the population will be sufficiently large that the degree of bias will be negligible.

The following example illustrates the use of linear systematic sampling in estimating a population total. The example is contrived, and the data used are fictitious.

Example 5.5. Estimating the Need for Hospital Beds

CONTEXT AND PROBLEM

Regional health agencies are responsible for reviewing requests by hospitals to build additional facilities or purchase major equipment. Presumably, the agencies help to limit the costs of hospitalization by ensuring that hospitals in the same community do not duplicate expensive equipment or build facilities that will not be utilized. An agency in North Carolina was faced with requests from three hospitals in a city with a population of 150,000 to approve additions that would house 200 new hospital beds. The agency decided to conduct a survey of the city's population in order to determine the need for additional hospital facilities.

After much discussion and planning, staff members in the agency decided to survey households in the city in order to estimate the number of person-days likely to be spent in hospitals during the coming year. Since individuals have little basis for estimating their future needs for hospitalization, it seemed reasonable to estimate the mean number of days members of the community had spent in hospitals during the previous year and then multiply this estimated mean by the latest information on the city's population in order to project short-term future needs.

SOLUTION AND RESULTS

Since a city directory that listed every address in the city was readily available and there was no sampling frame of city residents, it was decided to use households as sampling units. The city directory listed 60,000 households and was arranged by city block and area, so that a linear systematic sample from the directory would ensure balanced coverage of all sections of the city. Since the budget available for the survey would support about 600 interviews, and there was no good basis for estimating the variance of the estimated population total, it was decided to collect as much data as the budget allowed.

SELECTION OF THE SAMPLE

The population size (number of households) was 60,000, and the desired sample size was 600; so the sampling interval to be used for linear systematic sampling was $k = 60,000/600 = 100$. The households listed in the city directory were numbered sequentially from 1 to 60,000. A table of random numbers was used to determine the index number of the first household to be sampled, by selecting a random number between 1 and 100, inclusive. That random number was 42. Households where interviews would take place were then determined by going through the city directory and recording the addresses of households numbered 42, 142, 242, 342, and so on, up to and including the household numbered 59,942.

COLLECTION AND ANALYSIS OF DATA

Interviewers were sent to all sampled households and were instructed to secure data on (1) the number of persons who had been members of the household during the entire preceding calendar year and (2) the number of days each of these persons had spent as a patient in a hospital in the city during the preceding calendar year. These data were used to estimate the mean and variance of the number of days per person spent in a city hospital during the preceding calendar year, for all residents of the city. The results were as follows: mean number of days in a city hospital per person $= \bar{y} = 0.62$, and variance of number of days in a city hospital per person $= s^2 = 0.0961$.

The most recent data on the population of the city indicated that there were 155,490 residents. The total number of days that will be spent in a city hospital during the coming year by city residents was thus estimated to be

$$y = N\bar{y} = 155,490(0.62) = 96,403.8$$

Assuming full occupancy of hospital beds every day during the coming year, the number of beds needed was estimated as the estimated total number of person-days beds would be required divided by 365 = 96,403.8/365 = 264.12. However, since the incidence of illness and accidents is not uniformly distributed over the calendar year, it was judged to be foolhardy for any community to have just enough hospital beds to meet the average demand during the year. It would be more reasonable to have enough hospital beds to support a peak demand that was, say, 30 percent higher than the average. The average demand was therefore multiplied by a factor of 1.30, and this resulted in an estimated total need of 343.36 beds, which was, of course, rounded up to 344.

ESTIMATOR VARIANCE AND CONFIDENCE INTERVALS

Unfortunately, estimation of the variance of the population total is very difficult when linear systematic sampling is used for data collection. In addition, the design used in this example complicates matters further. Instead of sampling individuals, the researchers sampled households and then collected data on individuals within selected households. This is an example of single-stage cluster sampling, which is disucssed in Chapter 7. Since the incidence of illness and accidental injury is likely to be less variable among members of the same household than among members of the general population, the cluster sampling method is likely to yield a conservative estimate of the variability of hospital use by members of the general public. Methods for determining the extent of this problem and potential solutions are discussed in Chapter 7 and will not be addressed further in this example.

If data were collected using simple random sampling, the variance of the estimated population total would likely be larger than the variance that resulted from linear systematic sampling. To simplify this example, we will assume that this increase in estimator variance would exactly compensate for the reduction in estimator variance associated with the use of single-stage cluster sampling. In other words, we will act as though data were collected using simple random sampling, instead of linear systematic sampling of clusters, and we will use Equation (3.12) to estimate the variance of the estimated population total. We cannot emphasize too strongly that this assumption is purely speculative and that alternative methods of estimation, discussed in a later section of this chapter, are preferable.

From Equation (3.12), the variance of the estimated number of person-days hospital beds would be required during the coming calendar year was estimated to be

$$v(y) = \frac{155,490^2(0.0961)}{600}\left(1 - \frac{600}{60,000}\right)$$

$$= 3,833,648.2$$

The estimated standard error is equal to the square root of the estimated variance, which in this case is 1957.97 person-days of hospital bedtime.

Based on our previous assumptions, an approximate 95 percent confidence interval on the total number of person-days of hospital demand during the coming year can be found from Equation (3.14):

$$\hat{Y}_U = 96{,}403.8 + 1.96(1957.97) = 100{,}241.4$$

and

$$\hat{Y}_L = 96{,}403.8 - 1.96(1957.97) = 92{,}566.2$$

Assuming full occupancy of hospital beds every day during the year, these figures would be divided by 365 to estimate a 95 percent confidence interval on the number of hospital beds needed. The results are an upper confidence limit of 274.6 and a lower limit of 253.6. If these limits are inflated by 30 percent to accommodate peak demand for hospital beds, the estimated confidence interval would have an upper limit of 357.0 hospital beds and a lower limit of 329.7 beds.

DISCUSSION

This example is intended to illustrate the use of linear systematic sampling in a social planning application. The context and the data were fictitious, and we made a number of simplifying assumptions that are of dubious validity in order to illustrate the construction of a confidence interval on a population total. In later sections of this chapter, we discuss methods for estimating the variance of an estimated population mean when data are collected using linear systematic sampling. This example involved linear systematic sampling of clusters rather than individuals. Single-stage cluster sampling is discussed in detail in Chapter 7.

Estimating Population Proportions

FORMING THE ESTIMATES

In many surveys, the proportion of elements in the population that have values of the sampling variable within some interval, say between C_1 and C_2, or the proportion of elements that have some attribute is of central interest. Linear systematic sampling can be used to estimate such proportions.

The sample proportion of elements that have measurements between C_1 and C_2 or that have some attribute, is given by

$$p = \frac{1}{n} \sum_{i=1}^{n} \delta(i) \tag{5.17}$$

which is an estimator of the corresponding population proportion.

$$P = \frac{1}{N} \sum_{i=1}^{N} \delta(i) \tag{5.18}$$

Here

$$\delta(i) = \begin{cases} 1 & \text{if } C_1 \le y_i \le C_2, \text{ or if the element has the attribute of interest} \\ 0 & \text{if } y_i < C_1 \text{ or } y_i > C_2, \text{ or if the element does not have the at-tribute of interest} \end{cases}$$

and defines mathematically the transformation of the value of the sampling variable for the ith sampled element y_i to 0 and 1, as described in Chapter 3.

ESTIMATOR BIAS AND MEAN SQUARE ERROR

With linear systematic sampling, the sample proportion is an unbiased estimator of the population proportion whenever N/n is an integer. When N/n is not an integer, the sample proportion is a biased estimator, but the bias is negligible for population sizes likely to be encountered in practice.

The mean square error of the sample proportion equals

$$\text{MSE}(p) = \frac{1}{k} \sum_{r=1}^{k} (p_r - P)^2 \tag{5.19}$$

where p_r denotes the sample proportion for the rth linear systematic sample and k denotes the sampling interval.

The following example illustrates the use of linear systematic sampling in estimating a population proportion. Although the example is contrived, the data used are actual test scores from a medium-sized school district.

Example 5.6. Estimates of Proportions of Low Achievers

CONTEXT AND PROBLEM

The context and problem of this example are the same as those of Example 3.3. The parameter to be estimated is the proportion of sixth-graders in Midcity who have achievement scores that are at least 1 grade equivalent unit below the national norm. Linear systematic sampling is to be used to estimate the proportion. As in Example 3.3, the population proportion is to be estimated within ± 0.05 with 95 percent confidence.

SOLUTION AND RESULTS

The 1180 sixth-graders in Midcity were listed alphabetically to form a sampling frame. The SYSAMP-I computer program was used to select all linear systematic samples for sampling intervals of 2, 4, 5, 10, and 20.

For the reading subtest of the Stanford Achievement Test, a raw score of 46 corresponds to a score of 5.0 grade equivalent units. Since the test was administered to Midcity sixth-graders during the first month of the school year, those with raw scores of 46 or lower would have scores that are at least 1 grade equivalent unit below the national norm.

For each linear systematic sample, the SYSAMP-I computer program calculates the proportion of elements with values below a specified cutoff. The largest and smallest sample proportions are shown in Table 5.6 for each sampling interval, along with the population proportion. With systematic sampling, the largest and smallest sample proportions have high probabilities of occurrence. They are thus of greater interest than the mean square error.

DISCUSSION

For simple random sampling, a sampling fraction of 25.6 percent resulted in a 95 percent confidence interval on the population proportion with a width of

TABLE 5.6
Largest and Smallest Sample Proportions of Pupils with Achievement Below a Standard, Based on Linear Systematic Sampling (Actual Population Proportion = 19.4% for Comparison)[a]

Sampling Interval	Sampling Fraction, %	Largest Sample Proportion, %	Smallest Sample Proportion, %
2	50	20.1	18.7
4	25	20.8	18.4
5	20	21.3	17.0
10	10	25.4	14.5
20	5	27.1	13.8

[a] The standard used is a reading achievement score at least one grade equivalent unit below the national norm. Data are for sixth-grade pupils in Midcity.

± 4 percent (Example 3.3). Linear systematic sampling is therefore more efficient than simple random sampling for estimation of this population proportion. Table 5.6 shows that when a sampling fraction of 20 percent is used, sample proportions for all five systematic samples are within the 95 percent confidence limits resulting from simple random sampling (23.4 and 15.4 percent). However, for sampling fractions of 10 and 5 percent, the largest and smallest sample proportions are outside the confidence interval resulting from the use of simple random sampling.

It should be realized that the effects that appear in Table 5.6 are to some degree artifactual. The probability that the range of a random variable exceeds a certain limit is directly proportional to the size of the population (Parzen, 1960, pp. 322–323). One would therefore expect the range of sample proportions for a sampling interval of 10, say, to be larger than that for a sampling interval of 5. However, for 95 percent confidence intervals and the sampling intervals considered here, it is of no consequence that the increasing range of sample proportions is partly artifactual.

Consider, for example, a sampling interval of 5. According to the probability statement used in the derivation of sample size requirements [Equation (3.24)], it is desired that the population proportion and any sample proportion differ by no more than ε with probability greater than $1 - \alpha$. Here $\alpha = 0.05$, and $1 - \alpha = 0.95$. Now if the range of sample proportions is so small that the difference between the population proportion and any sample proportion is less than ε, it is certain that Equation (3.24) is satisfied. If, however, one of the five sample proportions differs from the population proportion by more than ε, the probability statement is not satisfied. With five equally likely sample proportions, the offending proportion will appear with probability 0.20.

This argument can be applied to 95 percent confidence limits whenever the sampling interval is smaller than 20. For a sampling interval of 20, Equation (3.24) would not be violated as long as no more than one sample proportion differed from the population proportion by more than ε.

It is tempting to redefine the concept of confidence intervals when the number of potential samples is very small. One could define a probability statement analogous to Equation (3.24) that applied not to a single population parameter but to a distribution of parameters for a set of substantively similar populations. For example, one could consider all populations of sixth-graders in medium-sized school districts. Instead of considering the selection of a systematic sample from a specific population, one would consider a population to be selected at random and a systematic sample to be selected at random from the selected population. The set of populations of sixth-graders would be called a *superpopulation* (Cochran, 1946). In the expression analogous to Equation (3.24), both the population parameter and the sample statistic would be random variables. The probability statement would thus apply to the superpopulation rather than to any specific population. While this model is theoretically palatable when the number of potential samples for a single population is small, it is not of practical use when only one population is of interest.

Estimation of Sample Size Requirements

A serious problem in using systematic sampling is the lack of a predictable relationship between sample size and estimation precision. Sampling methods discussed in previous chapters allowed estimation of the sample size necessary to achieve a specified level of estimation precision from an estimate of sampling variance. Systematic sampling procedures afford no such simple relationships. Murthy (1967, p. 173) states:

> Since the sampling variance of an estimator based on a systematic sample depends much on the arrangement of the units in the population, we have seen that it is difficult to predict its behaviour with increase in sample size. . . . Hence, it is difficult to determine the sample size required to ensure a specified precision for the estimates and this could only be done on the basis of extensive empirical studies using values of some related characteristics.

Cochran (1977, p. 212) voices similar concerns:

> The performance of systematic sampling in relation to that of stratified or simple random sampling is greatly dependent on the properties of the population. . . . For some populations and some values of [sample size], the variance of the sample mean may even *increase* when a larger sample is taken—a startling departure from good behavior.

Perhaps the most helpful advice on estimation of sample size requirements is as follows. The empirical results presented here should be viewed as suggestive rather than definitive. Systematic sampling procedures should be evaluated in local situations by using data for related populations, selecting all samples, and computing mean square errors and ranges of estimates. Data from related populations (for example, in a school survey, data for last year's students) can be used to build an array of hopefully confirmatory results.

Estimation of Estimator Variance

Several methods of estimating the variance of estimators associated with linear systematic sampling have been proposed in the literature. Unfortunately, the methods proposed are biased except when populations fit specific mathematical models. The degree of bias is generally large.

Estimation of estimator variance using data from a single linear systematic sample is an attempt to determine variation between systematic samples from data within a sample. If the population from which the sample is drawn contains trends or periodicities as a function of the order of observations, variance estimates based on a single sample may be misleading. Cochran (1977, pp. 223–227) suggests variance estimation formulas appropriate to populations that have specific trend functions. Empirical studies (Johnson, 1943) have shown these methods to be highly biased in practice, overestimating estimator variances by factors as high as 6 to 1. Johnson evaluated the variance estimation methods with agricultural survey data.

Cochran (1977) first considers populations in which the order of elements in the sampling frame and the magnitude of the sampling variable are unrelated. In this case, the unbiased estimator of the variance of the mean of a simple random sample also estimates the variance of the mean of a linear systematic sample:

$$v(\bar{y}_{sy1}) = \frac{N - n}{Nn(n - 1)} \sum_{i=1}^{n} (y_i - \bar{y})^2 \tag{5.20}$$

where

y_i = value of the sampling variable for the ith element in the linear systematic sample

$\bar{y}$ = mean of the linear systematic sample

n = number of observations in the linear systematic sample

Whether Equation (5.20) will provide reasonable estimates of the variance of the mean in educational and behavioral science surveys is subject to empirical investigation. In school settings, selection of pupils in their "natural" order of arrangement in classrooms might be equivalent to simple random sampling, provided the pupils have not been grouped by ability. If pupils are arranged in alphabetic order, we have already seen that overestimates of the variance between sample means may result, since intraclass correlation coefficients are sometimes negative (Table 5.4). We would expect the same result with social surveys of households if linear systematic samples were chosen from a city directory or any other source where households are listed in order of their city addresses. It is likely that contiguous or closely spaced households would be more alike in many features and characteristics than would randomly sampled households. Thus negative intraclass correlation coefficients would be expected for households in the same linear systematic sample.

Two more variance estimators that are based on the assumption that the order of elements in the sampling frame is unrelated to the sampling variable have been suggested by Murthy (1967). The first divides a population of size $N = nk$ into $n/2$ groups of $2k$ elements each. When linear systematic sampling is used, two elements are, in effect, selected from each group. Murthy treats these observations as though they were selected from each group using simple random sampling. The resulting variance estimator is

$$v(\bar{y}_{sy2}) = \frac{N - n}{Nn^2} \sum_{i=1}^{n/2} (y_{2i} - y_{2i-1})^2 \tag{5.21}$$

where y_i denotes the ith observation in the linear systematic sample. Equation (5.21) applies only when N/n is an integer, and the sample size n is an even number. Since the estimator is biased, it seems reasonable to test it empirically whatever the relationship between the population size and the sample size.

Murthy's second variance estimator is based on differences between successive elements in the linear systematic sample. It is equal to

$$v(\bar{y}_{sy3}) = \frac{N-n}{2Nn(n-1)} \sum_{i=1}^{n-1} (y_{i+1} - y_i)^2 \tag{5.22}$$

Empirical results for these two estimators show them to be seriously biased and to have relative mean square errors between 6 and 38 percent when applied to an agricultural survey. Relative mean square errors between 4 and 32 percent resulted from their use in a survey of population (Murthy, 1967, p. 160). Extensive empirical studies are necessary to determine the behavior of these variance estimators when used with data from surveys in education or the social sciences.

Other approaches to estimation of the variance of the means of linear systematic samples have been suggested by Cochran (1946) and Yates (1948). These methods utilize balanced differences of the form

$$d_{3i} = \tfrac{1}{2}y_i - y_{i+1} + \tfrac{1}{2}y_{i+2} \qquad \text{for } i = 1, 2, \ldots, n-2$$

$$d_{5i} = \tfrac{1}{2}y_i - y_{i+1} + y_{i+2} - y_{i+3} + \tfrac{1}{2}y_{i+4} \qquad \text{for } i = 1, 2, \ldots, n-4$$

and

$$d_{9i} = \tfrac{1}{2}y_i - y_{i+1} + y_{i+2} - y_{i+3} + y_{i+4} - y_{i+5} + y_{i+6} - y_{i+7} + \tfrac{1}{2}y_{i+8}$$

$$\text{for } i = 1, 2, \ldots, n-8$$

The corresponding variance estimators are

$$v(\bar{y}_{sy4}) = \frac{N-n}{1.5Nn(n-2)} \sum_{i=1}^{n-2} d_{3i}^2 \tag{5.23}$$

$$v(\bar{y}_{sy5}) = \frac{N-n}{3.5Nn(n-4)} \sum_{i=1}^{n-4} d_{5i}^2 \tag{5.24}$$

and

$$v(\bar{y}_{sy6}) = \frac{N-n}{7.5Nn(n-8)} \sum_{i=1}^{n-8} d_{9i}^2 \tag{5.25}$$

Empirical results provided by Yates (1948) show the estimator $v(\bar{y}_{sy6})$ to have smaller bias than $v(\bar{y}_{sy2})$ for populations of soil temperatures and air temperatures. Intuitively, estimators based on balanced differences should prove to be more accurate than $v(\bar{y}_{sy1})$, $v(\bar{y}_{sy2})$, and $v(\bar{y}_{sy3})$ when used with populations that have gradual linear trends. Empirical studies are needed to determine which of the estimators given by Equations (5.20) through (5.25) are most useful when applied to data from surveys in education and the social sciences.

Yet another estimation method proposed by Murthy (1967) sacrifices some of the convenience and potential economy of linear systematic sampling. In this method, m linear systematic samples, each of size n/m, are selected instead of one linear systematic sample of size n. The method requires that n/m be an integer and that m be smaller than n. Each sample is drawn with a random starting element selected without replacement from the integers 1 through k', where $k' = mk$ and $k = N/n$. This sampling procedure is equivalent to simple random sampling of m elements from a population of size k'. The variance estimation formula for simple random sampling can be used, with the mean of each linear systematic sample treated as one observation:

$$v(\bar{y}_{sy7}) = \frac{k' - m}{k'm(m - 1)} \sum_{i=1}^{m} (\bar{y}_i - \bar{y})^2 \tag{5.26}$$

where $\bar{y}_i$ denotes the mean of the ith linear systematic sample and

$$\bar{y} = \frac{1}{m} \sum_{i=1}^{m} \bar{y}_i$$

is used as an estimator of the population mean.

Equation (5.26) should yield more precise estimates of the variance among the means of linear systematic samples than any of the other methods discussed above. However, there are added costs in gaining this precision. Equation (5.26) probably sacrifices precision in estimating the population mean to gain precision in estimating the variance of the estimator. As with other methods for estimating sampling error, an extensive empirical study with data from surveys in education and the social sciences is required to determine the precision and bias of this estimator.

Summary

Despite difficulties in determining required sample size and estimating estimator variance, linear systematic sampling should be quite useful in a number of surveys in education and the social sciences. When the population is large and samples must be selected from physical records without the benefit of a computerized sampling frame, linear systematic sampling (or one of the alternative systematic sampling procedures discussed in Chapter 6) may be the only practical alternative. When the natural order of a sampling frame arranges elements so that those close together are more alike on the sampling variable than are those far apart, linear systematic sampling could prove to be far more efficient than simple random sampling or even stratified sampling.

In several examples that used achievement test data from a medium-sized city school system, we found linear systematic sampling to be superior to simple random sampling and stratified random sampling with one observation per stratum. These results should pique your curiosity, but they should be generalized with great caution. On the average, when the elements of a population are randomly arranged, linear systematic sampling and simple random sampling should be equally efficient. The superiority of linear systematic sampling when selecting pupils from a list arranged in alphabetic order is thus unexpected. We would be surprised to see these results confirmed with data from other school systems. However, the SYSAMP-I computer program contained in Appendix C should facilitate research that could answer this question.

Exercises

Two psychologists who were conducting research on cognitive functioning in young children devised a maze task for use in their work. The task required each child to begin at an indicated starting position and then draw a continuous path through the maze, avoiding dead ends, until he or she arrived at an indicated ending position. The psychologists recorded the time (in seconds) needed by each child to complete the task. Since the maze was small and simple, the average child finished in a little over 60 seconds.

Times (in seconds) needed to complete the maze are shown in the following table for the first 100 children tested. The 100 children will be treated as a population, and we will use their data to examine the behavior of linear systematic sampling.

Child No.	Time	Child No.	Time	Child No.	Time
1	30	35	21	69	98
2	86	36	71	70	44
3	43	37	95	71	75
4	65	38	21	72	38
5	104	39	40	73	71
6	16	40	37	74	81
7	54	41	41	75	55
8	38	42	96	76	59
9	39	43	71	77	83
10	55	44	45	78	93
11	32	45	41	79	60
12	80	46	76	80	51
13	47	47	67	81	41
14	57	48	67	82	30
15	91	49	91	83	40
16	53	50	55	84	73
17	76	51	32	85	41
18	42	52	80	86	44
19	32	53	57	87	66
20	56	54	80	88	83
21	13	55	87	89	47
22	34	56	66	90	76
23	54	57	59	91	51
24	78	58	57	92	84
25	55	59	46	93	49
26	49	60	93	94	68
27	42	61	96	95	43
28	80	62	67	96	54
29	84	63	67	97	96
30	37	64	40	98	84
31	51	65	69	99	80
32	41	66	67	100	107
33	40	67	84		
34	52	68	62		

The problems to be solved using the data in this table are described below. Use simple random sampling or linear systematic sampling with a sampling interval $k = 5$, as required by each problem.

1. The size of the population is 100, and the sampling interval is 5. What is the size n of each linear systematic sample?

2. Select *all* linear systematic samples of the size determined in Exercise 1 from the population of size 100. For *each* linear systematic sample, estimate the population mean time children require to solve the maze.

3. Use the results of Exercise 2 to calculate the mean square error (MSE) of the sample mean time children require to solve the maze.

4. Suppose that you wanted to estimate the population mean time to solve the maze within ±2 seconds, with 95 percent confidence. Would the sample size you used be sufficient? Justify your answer by citing appropriate data.

5. Use data for the entire population to calculate the variance of the individual times the children took to solve the maze, and then use this value to find the variance of the mean of simple random samples of the size used in Exercise 1.

6. Using your results from Exercises 3 and 5, what is the efficiency of linear systematic sampling relative to simple random sampling for samples of size n?

7. Using data from the linear systematic sample that contains the time needed to solve the maze by the *first* child in the sampling frame, *estimate* the variance of the mean of linear systematic samples of size n, using the following three estimators:

    A. $v(\bar{y}_{sy1})$
    B. $v(\bar{y}_{sy3})$
    C. $v(\bar{y}_{sy4})$

Compare the estimates provided by these three estimators with the actual value of the mean square error of the mean of linear systematic samples of size n (the answer you found in Exercise 3) to decide which of the three estimation methods appears to be most accurate. (Realize that you cannot find a definitive answer to this question using data from just one sample.)

6

Further Topics in Systematic Sampling

Introduction

Several modifications of linear systematic sampling and alternative systematic sampling procedures have been proposed in the literature. Some of these procedures are particularly efficient when applied to populations in which the sampling variable increases as a function of the order of the elements in the sampling frame. Others provide unbiased estimation in cases where linear systematic sampling is biased.

The theory of linear systematic sampling, as described in Chapter 5, was first investigated by Madow and Madow (1944). However, the method was used in survey research long before its statistical theory was developed (Madow, 1970). Cochran (1946) presented theoretical results on the efficiency of linear systematic sampling as a function of the ordering of populations. He also suggested a modification of linear systematic sampling, which has desirable statistical properties. Yates (1948) proposed a modification that is particularly efficient when applied to sampling frames in which the sampling variable increases as the element index increases. Other systematic sampling procedures, also efficient for such populations, were developed by Lahiri (1954) and Sethi (1965). These procedures are discussed in this chapter, and their efficiency is evaluated in several illustrative applications.

Several empirical investigations in this chapter are largely without precedent. Apparently, sophisticated systematic sampling methods have not been applied in educational research. The SYSAMP-I computer program (Appendix C) provides statistics for comparing modifications of linear systematic sampling and alternative systematic sampling procedures.

Alternative Systematic Sampling Procedures

PROCEDURES FOR GENERAL POPULATIONS

Linear systematic sampling has several properties that are theoretically troublesome. Although these properties rarely cause serious trouble in practice, they require cumbersome statistical treatment. These problems can often be avoided by using other systematic sampling procedures.

Linear systematic sampling leads to unbiased estimation of means, totals, and proportions only when the ratio of population size to sample size, N/n, is an integer. When N/n is not an integer, the sample mean, the conventional estimator of the population total, and the sample proportion are biased estimators. However, the magnitude of the bias is negligible for population sizes likely to be encountered in practice. Two systematic sampling procedures discussed in this section lead to unbiased estimation even when N/n is not an integer.

A second undesirable property of linear systematic sampling arises when N/n is not an integer. The sampling interval k equals the integer closest to N/n, so the population size N does not equal nk. Some systematic samples therefore have one element more than others, and sample size is a random variable that depends on the first element in the sample. While this complicates some formulas for estimator variance, it is not of concern in practice. However, circular systematic sampling avoids this problem altogether.

MODIFIED LINEAR SYSTEMATIC SAMPLING

Cochran (1963) proposed a modification of linear systematic sampling that results in unbiased estimation of the population mean even when N/n is not an integer. With Cochran's method, the sample mean is always an unbiased estimator, and the product of the population size N and the sample mean is an unbiased estimator of the population total. The initial sample element is chosen at random, with an index between 1 and N. If the index of the first sampled element is r, elements with indices $r + k, r + 2k, \ldots$ and $r - k, r - 2k, \ldots$ make up the rest of the sample. That is, the elements that compose the sample have indices that are multiples of k units from the initial element, counting both "forward" and "backward" in the ordered sampling frame.

Suppose a sample of three elements is to be selected from a population of seven. Since $\frac{7}{3} = 2.33$, $k = 2$ is used as the sampling interval. A random number between 1 and 7 (inclusive) is chosen to identify the initial sample element. If the index of the initial sample element is 4, for example, indices of the remaining sample elements are $4 + 2 = 6$ and $4 - 2 = 2$.

One unfortunate consequence of both linear systematic sampling and modified linear systematic sampling is that some samples have one more element than the rest, unless N/n is an integer. In the example above, if the index of the first sampled element was 1, the remaining elements would have indices 3, 5, and 7; the sample size would thus equal 4. If the index of the first element were 2, the sample size would be 3.

CIRCULAR SYSTEMATIC SAMPLING

This method (Lahiri, 1954) leads to unbiased estimation of population means, totals, and proportions and equal sample sizes for all choices of initial sample element. Another advantage of the method is that each element in the population has the same probability of selection whether or not N/n is an integer. With linear systematic sampling and modified linear systematic sampling, population elements have equal selection probabilities only when N/n is an integer.

In circular systematic sampling, the elements in the sampling frame are treated as though arranged around a circle, with the element having index 1 following the element having index N. An initial sample element with index between 1 and N is chosen randomly. The remaining $n - 1$ sample elements are selected as if going around a circle, taking every element whose index is some multiple of k units from the initial element. If the initial element has index r, elements in the sample will have indices

$$r + ik \quad \text{if } r + ik \leq N$$

and

$$r + ik - N \quad \text{if } r + ik > N$$

where $i = 0, 1, 2, \ldots, n - 1$. Circular systematic sampling is illustrated schematically in Figure 6.1.

PROCEDURES FOR SPECIALLY ORDERED POPULATIONS

As noted in Chapter 5, the efficiency of systematic sampling is particularly sensitive to the ordering of elements in the sampling frame. If the sampling variable follows trends or cycles as a function of the order of the elements in the sampling frame, systematic sampling will be efficient sometimes and inefficient sometimes. Efficiency will depend on the nature of the trend, the choice of systematic sampling procedure, and the size of the sampling interval.

Several systematic sampling procedures provide efficient estimation when the index of elements in the sampling frame has a monotonic relationship with the sampling variable. They are most efficient when the relationship is approximately linear and provide exact estimation for the theoretically interesting case where the relationship is perfectly linear. That is, if

$$y_i = a + bi \tag{6.1}$$

where y_i is the value of the sampling variable for the element in the sampling frame with index i, then the mean of any sample would equal the population mean. Similarly, if Equation (6.1) were satisfied, the product of the sample mean and the population size would equal the population total, and any sample proportion would equal its corresponding population proportion. Equation (6.1) almost never holds in practice. The methods described in this section are still useful, however, since they are efficient when there is an approximately linear relationship between the sampling variable and the indices of elements in the sampling frame.

Often, an approximately linear relationship between the sampling variable and element index can be created. Suppose, for each element in the sampling frame, we knew the value of an auxiliary variable that was highly correlated with the sampling variable. We could then construct a new sampling frame with

Linear Systematic Sampling

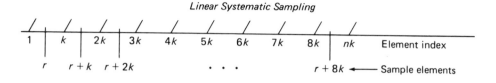

Circular Systematic Sampling

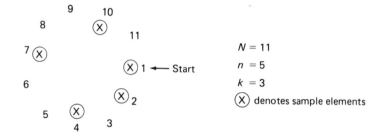

Balanced Systematic Sampling

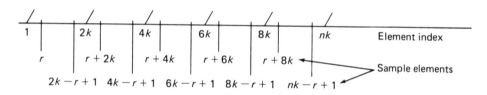

Centrally Located Samples

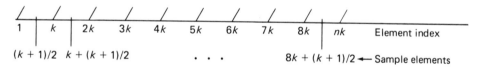

FIGURE 6.1. **Schematic diagrams of various systematic sampling procedures.**

elements arranged in increasing order of the auxiliary variable. If the auxiliary variable increased as a function of the element index, the sampling variable would increase in the same way. The relationship might be approximately linear. The systematic sampling procedures described below depend on such linear relationships in order to provide efficient estimation.

BALANCED SYSTEMATIC SAMPLING

Sethi (1965) proposed a systematic sampling procedure that is particularly useful with populations arranged such that the sampling variable increases or decreases linearly with the indices of the elements in the sampling frame. The method is called *balanced systematic sampling* and is as follows.

Suppose N/n equals an integer, k, and the desired sample size n is an even number. The ordered sampling frame is divided into $n/2$ groups; two elements are sampled from each group, with the first element the same distance from the lower boundary of its group as the second element is from the upper boundary of its group. A random number, r, is chosen between 1 and k; the sampled elements are those with indices $r - 1$ from the upper and the lower boundaries of each group. The elements selected from the first group of $2k$ are those with indices r and $2k - r + 1$. From the second group, the elements selected have indices $2k + r$ and $4k - r + 1$, and so on. This process is illustrated in Figure 6.1, and a numerical example is given below.

Assume that four elements are to be selected from 20; $N/n = k$ and, in this case, is $\frac{20}{4} = 5$. Two groups of 10 elements each would be formed, since $n/2 = 2$ and $2k = 10$. A random number r between 1 and 5 would be chosen to determine the sampled elements. Suppose $r = 3$. The first group would consist of elements with indices 1 to 10, and the elements sampled from the first group would have indices $r = 3$ and $2k - r + 1 = 2(5) - 3 + 1 = 8$. Note that both sampled elements would have indices that were $r - 1 = 2$ elements from the ends of the group, that is, from the elements with indices 1 and 10. The second group would contain elements with indices 11 to 20. The first element sampled from the second group would have index $r + 2k = 3 + 2(5) = 13$. The second element sampled from this group would have index $4k - r + 1 = 4(5) - 3 + 1 = 18$. Note that these elements, too, would have indices that were $r - 1 = 2$ elements from the ends of their group, that is, from the elements with indices 11 and 20.

CENTRALLY LOCATED SAMPLES

When the sampling variable and the indices of elements in the sampling frame have an approximately linear relationship, centrally located sampling (Madow, 1953) may provide efficient estimation of the population mean, total, or proportion. The method is as follows.

The sampling interval k is defined as the integer that is closest to the ratio of the population size to the desired sample size, N/n, just as in linear systematic sampling. If k is an odd number, the first sampled element is the one with index equal to $(k + 1)/2$. If k is an even number, the first sampled element is either the one with index equal to $k/2$ or the one with index equal to $(k + 2)/2$. The choice is made randomly, with each option having a probability $\frac{1}{2}$. As in linear systematic sampling, every kth element following the initial element is selected for the sample until the sampling frame has been exhausted.

A slight modification of this method, appropriate only when k is an even number, would be to choose as the first sampled element either the one with index $k/2$ or the one with index $(k + 2)/2$. Each would have a 50 percent chance of being selected. After the first element had been selected, the rest of the sample would be constructed by alternating the choice for successive groups of k elements in the sampling frame. That is, if the first sampled element had an index of $k/2$, the next one would have an index of $k + (k + 2)/2$, and then the next would have an index of $2k + k/2$, and so on, until the sampling frame had been exhausted. If the first sampled element was the one with an index of $(k + 2)/2$, the next sampled element would have an index of $k + k/2$, and then the next would have an index of $2k + (k + 2)/2$, continuing in this alternating pattern until the sampling frame had been exhausted.

Although we will evaluate the centrally located sampling procedure in its original form, as proposed by Madow (1953), the modification suggested here appears sensible in situations where the indices of the elements in the sampling frame and the sampling variable have a linear or strictly monotonic relationship. Reference to Figure 6.1 will show that in this situation, sampling of elements with indices $k/2$, $k + k/2$, etc., would underestimate the population mean, total, or proportion, whereas sampling of elements with indices $(k + 2)/2, k + (k + 2)/2$, etc., would overestimate the population mean, total, or proportion.

LINEAR SYSTEMATIC SAMPLING WITH END CORRECTIONS

Yates (1948) proposed a modification of linear systematic sampling termed *end corrections*. It ensures that the sample mean will be equal to the population mean and that the product of the sample mean and the population size will be equal to the population total whenever Equation (6.1) holds precisely. The procedure can also increase estimation efficiency when the sampling variable and the indices of elements in the sampling frame have an approximately linear relationship. Yates's modification consists of weighting the first and last elements in the sample before estimators of the population mean or total are computed. Yates derived proper weights for the case when the population size N is an integral multiple of the sample size, n. Weights were derived by Jaeger (1970) for the case where N is not an integral multiple of n.

Formulas for end corrections are given below. Yates's end corrections, for the case where N is an integral multiple of n, are given by Equation (6.2). When N is not an integral multiple of n, appropriate end corrections depend on the index of the first sampled element. There are two cases. The population size N always equals the product of the sample size n and the sampling interval k plus a remainder, R; that is, $N = nk + R$. When the first sampled element has an index between 1 and R inclusive (the first case), the systematic sample will have $n' = n + 1$ elements, and the end corrections given by Equation (6.3) should be used. When the first sampled element has an index between $R + 1$ and k inclusive (the second case), the systematic sample will have n elements, and the end corrections given by Equation (6.4) should be used.

Use the following weights for end corrections:

1. If N is an integral multiple of n, multiply the value of the sampling variable for the first sampled element by

$$1 + \frac{n(2r - k - 1)}{2(n - 1)k} \tag{6.2a}$$

and the value of the sampling variable for the last sampled element by

$$1 - \frac{n(2r - k - 1)}{2(n - 1)k} \tag{6.2b}$$

where r denotes the index of the first sampled element.

2. If N is not an integral multiple of n and the first sampled element has an index r between 1 and R inclusive, multiply the value of the sampling variable for

the first sampled element by

$$1 + \frac{n'(2r - R - 1)}{2(n' - 1)k} \tag{6.3a}$$

and the value of the sampling variable for the last sampled element by

$$1 - \frac{n'(2r - R - 1)}{2(n' - 1)k} \tag{6.3b}$$

where $n' = n + 1$.

3. If N is not an integral multiple of n and the first sampled element has an index r between $R + 1$ and k inclusive, multiply the value of the sampling variable for the first sampled element by

$$1 + \frac{n(2r - k - R - 1)}{2(n - 1)k} \tag{6.4a}$$

and the value of the sampling variable for the last sampled element by

$$1 - \frac{n(2r - k - R - 1)}{2(n - 1)k} \tag{6.4b}$$

Equations (6.2a) and (6.2b) are a special case of Equations (6.4a) and (6.4b). When N is an integral multiple of n, $N = nk$ and $R = 0$; in that case, Equations (6.4a) and (6.2b) reduce to Equations (6.2a) and (6.2b).

As an example of the use of end corrections, assume a population of size $N = 11$ and a desired sample size of $n = 3$. N can be written as $11 = 3(3) + 2 = nk + R$. For samples beginning with elements having indices of 1 or 2, Equations (6.3a) and (6.3b) provide proper end corrections; for the sample beginning with the element indexed 3, Equations (6.4a) and (6.4b) should be used. Appropriate end corrections for this example are shown in Table 6.1.

A SCHEME FOR REORDERING SAMPLING FRAMES

Cochran (1963) suggested a scheme for reordering sampling frames prior to linear systematic sampling. In this scheme, elements in the sampling frame are

TABLE 6.1
End Corrections for a Population of Size 11 and a Desired Sample Size of 3

Index of Initial Sample Element	Correction for Initial Sample Element	Correction for Last Sample Element
$r = 1$	$1 - \frac{4}{18} = 0.7778$	$1 + \frac{4}{18} = 1.2222$
$r = 2$	$1 + \frac{4}{18} = 1.2222$	$1 - \frac{4}{18} = 0.7778$
$r = 3$	$1 + 0 = 1.0000$	$1 - 0 = 1.0000$

first arranged in increasing order of a variable correlated with the sampling variable. After dividing the sampling frame into successive groups of k elements, the order of the elements within every second group is reversed. Linear systematic sampling is then applied to the reordered sampling frame. This procedure results in small estimator variances for some populations. The procedure is particularly advantageous when the sampling variable is highly correlated with the variable used for reordering elements in the sampling frame.

Estimating Population Means

FORMING THE ESTIMATES

When data are collected through any of the systematic sampling procedures described in the previous section, the sample mean

$$\bar{y} = \frac{1}{n} \sum_{i=1}^{n} y_i$$

can be used as an estimator of the population mean

$$\bar{Y} = \frac{1}{N} \sum_{i=1}^{N} y_i$$

Depending on the relationship between the size of the population N and the sample size n, the estimator might or might not be unbiased. Circumstances leading to unbiased estimation are discussed above for each sampling procedure.

MEAN SQUARE ERROR

For each of the sampling procedures described in the previous section, the mean square error (MSE) of the estimated population mean is given by

$$\text{MSE}(\bar{y}) = \frac{1}{n_s} \sum_{i=1}^{n_s} (\bar{y}_i - \bar{Y})^2 \tag{6.5}$$

where n_s denotes the number of systematic samples that can be selected with the sampling procedure for a specified set of values of the design parameters (for example, sample size n or sampling interval k) and $\bar{y}_i$ denotes the mean of the sampling variable for the ith systematic sample.

The following example illustrates the application of each of the systematic sampling procedures to the problem treated in many earlier examples—estimating the mean achievement of sixth-graders in a medium-sized school system. In addition, the procedures are applied to the problem of estimating the mean achievement of sixth-graders in individual schools. Working with very small populations presents particular problems and severely tests the feasibility of any probability sampling procedure.

The relative efficiency of each systematic sampling procedure is evaluated in each application, using simple random sampling as a benchmark.

Example 6.1. Efficiencies of Systematic Sampling Procedures for Estimation of Population Means

CONTEXT AND PROBLEM

The data used in this example are reading achievement scores for sixth-graders in Midcity. Achievement scores are to be analyzed for pupils throughout the city and for pupils in each of four elementary schools. The object of this example is to determine, for these populations, the efficiencies of the systematic sampling procedures described in the second section of this chapter. Also to be determined are the sampling fractions necessary to estimate mean achievements at desired levels of precision.

Population means are to be estimated within 0.2 grade equivalent units with 95 percent confidence. As in Example 5.1, we assume that a mean square error equal to 4 raw score points squared defines an acceptable level of precision.

SOLUTION AND RESULTS

Six combinations of systematic sampling procedure and ordering of elements in the sampling frame were used with each of five populations of pupils. The populations consisted of sixth-graders in each of four elementary schools and all six-graders in Midcity.

The mean square error of estimate for linear systematic sampling, applied to the citywide population of sixth-graders when listed in alphabetic order, was computed in Example 5.1. That mean square error is repeated here to allow comparison with those of other systematic sampling procedures. Mean square errors and efficiencies were computed for several systematic sampling procedures applied to populations of sixth-graders arranged in increasing order of their scores on the Lorge-Thorndike ability test. The sampling methods used with this arrangement of pupils were linear systematic sampling with and without the use of end corrections, balanced systematic sampling, and centrally located sampling. Linear systematic sampling was also applied to populations of sixth-graders, first arranged in increasing order of their Lorge-Thorndike scores, then with their order reversed in alternate strata (successive groups of k elements in the sampling frame).

Simple random sampling and stratified sampling with one element per stratum (described in Chapter 5) were evaluated for each population in order to compute variances of estimated population means. These variances were compared to mean square errors resulting from each systematic sampling procedure to determine relative efficiencies. Sampling intervals of 2, 4, 5, 10, and 20 were used for the population of all sixth-graders in Midcity and sampling intervals of 2, 3, 4, and 5 were used for the populations of sixth-graders in individual schools. Thus sampling fractions varied between 5 and 50 percent for the citywide population and between 20 and 50 percent for the school populations. Results of these analyses are presented in Tables 6.2 through 6.11.

DISCUSSION

PRECISION OF ESTIMATES OF THE DISTRICT MEAN

Some sampling methods were consistently more precise than others when applied to the population of sixth-grade pupils in Midcity (Table 6.2). Stratified

TABLE 6.2
Mean Square Errors of Estimated Population Means for Systematic Sampling Procedures[a]

Sampling Procedure	Mean Square Errors Arising When Sampling Interval and Sampling Fraction Are				
	2 50%	4 25%	5 20%	10 10%	20 5%
1. Procedures applied when population is in alphabetic order					
1.1 Simple random sampling	0.39	1.16	1.55	3.50	7.44
1.2 Stratified sampling, one element per stratum	0.36	1.15	1.49	3.39	11.58
1.3 Linear systematic sampling	0.03	0.33	0.52	0.84	2.49
2. Procedures applied to population arranged in increasing order of ability test scores					
2.1 Stratified sampling, one element per stratum	0.07	0.22	0.28	0.68	5.35
2.2 Linear systematic sampling	0.29	0.23	0.10	0.72	2.18
2.3 Linear systematic sampling with end corrections	0.25	0.22	0.11	0.59	1.68
2.4 Balanced systematic sampling	0.003	0.006	0.11	0.89	5.80
2.5 Centrally located systematic samples	0.33	0.26	0.16	1.41	0.56
3. Linear systematic sampling applied to the population in increasing order of ability test scores with order reversed in alternate strata	0.00	0.01	0.09	0.69	2.51

[a] All data are for the population of sixth-grade pupils in Midcity. Units are raw score points squared on the Stanford Achievement Test.

sampling of pupils arranged in increasing order of their Lorge-Thorndike scores was more efficient than stratified sampling of pupils arranged in alphabetic order. With either arrangement of the sampling frame, stratified sampling was more efficient than simple random sampling, but the increase in efficiency was negligible when pupils were arranged in alphabetic order.

With a sampling fraction of 10 percent and the sampling frame arranged in alphabetic order, simple random sampling and stratified sampling provided marginally acceptable precision. Using the same sampling fraction of 10 percent, stratified sampling from the population arranged in increasing order of pupils' Lorge-Thorndike scores provided a mean square error that was less than 1 raw score point squared. However, using a sampling fraction of 5 percent, the method did not meet the specified precision requirement of a mean square error smaller than 4.

Results for systematic sampling methods were not consistent across sampling fractions; that is, no systematic sampling procedure was "best" for all sampling fractions. However, linear systematic sampling applied to the population arranged in increasing order of pupils' Lorge-Thorndike scores, with their order reversed in alternate strata, was consistently superior to simple random sampling and stratified sampling from a sampling frame in increasing order or with alphabetic arrangement of pupils. Linear systematic sampling of the frame arranged in increasing order of Lorge-Thorndike scores, with order reversed in alternate strata, and balanced systematic sampling of the frame arranged in increasing order provided excellent precision for sampling fractions of 20 percent or more. Balanced systematic sampling of the frame arranged in increasing order afforded the greatest sampling precision (among all sampling methods considered) for a sampling fraction of 25 percent. However, it was inferior to at least one other sampling method for all other sampling fractions and was particularly inefficient for a sampling fraction of 5 percent. Although centrally located sampling of the frame arranged in increasing order of Lorge-Thorndike scores provided acceptable precision for all sampling fractions investigated, its mean square errors varied erratically. The method was much more precise for a sampling fraction of 5 percent than for a sampling fraction of 10 percent. It also provided more precise estimates for a sampling fraction of 20 percent than for sampling fractions of 50 or 25 percent.

A well-behaved sampling procedure should always afford greater precision when the sample size is increased. For sixth-grade pupils in Midcity, this was the case only for linear systematic sampling of the frame arranged in increasing order of pupil's ability test scores, with their order reversed in alternate strata and for balanced systematic sampling of pupils arranged in increasing order of their ability test scores without reversing their order in alternate strata.

Five systematic sampling procedures provided acceptable precision for a sampling fraction of 5 percent (all four linear systematic sampling procedures and centrally located sampling). These methods provided real economy compared to simple random sampling and stratified sampling with one element per stratum, since they required testing only half as many pupils.

The efficiencies of systematic sampling relative to simple random sampling and stratified sampling with one element per stratum, when applied to the citywide population of sixth-graders, are shown in Table 6.3. Relative to simple random sampling, the efficiencies of systematic sampling procedures varied from 117 to more than 9999 percent. Large values of relative efficiency in Table 6.3 and in subsequent tables should be interpreted with caution, since they usually result from a small mean square error in the denominator of the efficiency ratio. For any sampling procedure, and for systematic sampling procedures especially, mean square errors will fluctuate from population to population. Thus a sampling procedure with a mean square error of 0.003, say, for the sixth-grade population in Midcity might have a mean square error of 0.005 for another similar population. This small increase in mean square error might correspond to a reduction in relative efficiency from 2000 to 1200 percent (assuming the mean square error in the numerator was equal to 0.06). If one sampling procedure has an efficiency of 9999 percent relative to another for one population, we can safely conclude that the first procedure will be more efficient than the second when applied to some other population. However, we would not expect the magnitude of the relative efficiency to be the same for other populations.

TABLE 6.3

Efficiencies of Systematic Sampling Procedures Relative to Simple Random Sampling and Stratified Sampling with One Element per Stratum[a]

Sampling Procedure	Relative Efficiencies, % for Indicated Methods When Sampling Interval and Sampling Fraction Are				
	2 50%	4 25%	5 20%	10 10%	20 5%
1. Linear systematic sampling applied to the population arranged in alphabetic order					
Efficiencies relative to					
Simple random sampling	1125	354	299	418	298
Stratified sampling	1036	349	289	405	464
2. Procedures applied to the population arranged in increasing order of ability test scores					
2.2 Linear systematic sampling					
Efficiencies relative to					
Simple random sampling	133	500	1500	485	341
Stratified sampling	26	94	275	95	245
2.3 Linear systematic sampling with end corrections					
Efficiencies relative to					
Simple random sampling	156	539	1367	593	442
Stratified sampling	30	101	251	116	318
2.4 Balanced systematic sampling					
Efficiencies relative to					
Simple random sampling	> 9999	> 9999	1468	392	128
Stratified sampling	2782	3825	269	77	92
2.5 Centrally located systematic samples					
Efficiencies relative to					
Simple random sampling	117	450	958	248	1322
Stratified sampling	23	84	176	49	950
3. Linear systematic sampling applied to the population arranged in increasing order of ability test scores, with order reversed in alternate strata					
Efficiencies relative to					
Simple random sampling	> 9999	> 9999	1786	505	296
Stratified sampling	> 9999	3143	327	99	215

[a] All data are for the population of sixth-grade pupils in Midcity.

It is worth noting that all systematic sampling methods were more efficient than simple random sampling for all sampling fractions investigated, regardless of the order of pupils in the sampling frame. The efficiencies of systematic sampling relative to stratified sampling with one element per stratum varied from 26 to more than 9999 percent. For 15 of 25 combinations of sampling method and sampling fraction, systematic sampling was more efficient than stratified sampling with one element per stratum. In contrast, for a sampling fraction of 50 percent, three of the systematic sampling methods had mean square errors that were more than three times as large as corresponding mean square errors for stratified sampling with one element per stratum. Centrally located sampling was inferior to stratified sampling for three of the five sampling fractions investigated. For sampling fractions of practical interest, the efficiencies of other systematic sampling procedures relative to stratified sampling with one element per stratum ranged from just under 100 to more than 200 percent. Ordering sixth-graders by their Lorge-Thorndike scores prior to sampling pro-

TABLE 6.4
Mean Square Errors of Estimated Population Means for Systematic Sampling Procedures[a]

	Mean Square Errors Arising When Sampling Interval and Sampling Fraction Are			
	2	3	4	5
Sampling Procedure	50%	33%	25%	20%
1. Procedures applied when population is in alphabetic order				
1.1 Simple random sampling	9.58	18.98	29.84	36.94
1.2 Stratified sampling, one element per stratum	9.04	18.95	24.88	34.69
1.3 Linear systematic sampling	0.24	6.00	32.14	52.57
2. Procedures applied to population arranged in increasing order of ability test scores				
2.1 Stratified sampling, one element per stratum	2.76	5.13	7.21	8.29
2.2 Linear systematic sampling	0.50	3.38	8.59	9.17
2.3 Linear systematic sampling with end corrections	0.03	4.41	14.47	10.85
2.4 Balanced systematic sampling	0.81	8.34	2.03	22.94
2.5 Centrally located systematic samples	2.15	5.54	10.76	19.04
3. Linear systematic sampling applied to the population in increasing order of ability test scores with order reversed in alternate strata	1.49	12.67	1.04	16.36

[a] All data are for the population of sixth-grade pupils in School 2 in Midcity. Data are raw score points squared on the Stanford Achievement Test.

TABLE 6.5

Efficiencies of Systematic Sampling Procedures Relative to Simple Random Sampling and Stratified Sampling with One Element per Stratum [a]

Sampling Procedure	Relative Efficiencies, %, for Indicated Methods When Sampling Interval and Sampling Fraction Are			
	2 50%	3 33%	4 25%	5 20%
1. Linear systematic sampling applied to the population arranged in alphabetic order				
Efficiencies relative to				
Simple random sampling	3997	316	93	70
Stratified sampling	3772	316	77	66
2. Procedures applied to the population arranged in increasing order of ability test scores				
2.2 Linear systematic sampling				
Efficiencies relative to				
Simple random sampling	1912	562	347	403
Stratified sampling	550	152	84	90
2.3 Linear systematic sampling with end corrections				
Efficiencies relative to				
Simple random sampling	> 9999	430	206	340
Stratified sampling	8580	116	50	76
2.4 Balanced systematic sampling				
Efficiencies relative to				
Simple random sampling	1189	228	1469	161
Stratified sampling	342	62	355	36
2.5 Centrally located systematic samples				
Efficiencies relative to				
Simple random sampling	445	343	277	194
Stratified sampling	128	93	67	44
3. Linear systematic sampling applied to the population arranged in increasing order of ability test scores with order reversed in alternate strata				
Efficiencies relative to				
Simple random sampling	643	150	2853	226
Stratified sampling	185	41	690	51

[a] All data are for the population of sixth-grade pupils in School 2 in Midcity.

duced large increases in estimation precision. If we were to risk generalization from these findings, we would suggest that savings in the number of pupils tested would more than offset the added costs of sampling from an ordered list.

PRECISION OF ESTIMATES OF SCHOOL MEAN ACHIEVEMENT

Achievement test data for pupils in four different schools were used to evaluate the precision of systematic sampling for small populations. Efficiencies relative to simple random sampling and stratified sampling with one element per stratum were also evaluated. Since the school populations were small, unstable results were expected. However, it was thought that analysis of results for several populations might reveal some consistent relationships.

For three of the four schools investigated (see Tables 6.4 through 6.11), neither simple random sampling nor stratified sampling from sampling frames of pupils arranged in alphabetic order provided acceptable precision (defined as a mean square error of 4 or less). For the fourth school (School 21), stratified

TABLE 6.6
Mean Square Errors of Estimated Population Means for Systematic Sampling Procedures[a]

Sampling Procedure	Mean Square Errors Arising When Sampling Interval and Sampling Fraction Are			
	2 50%	3 33%	4 25%	5 20%
1. Procedures applied when population is in alphabetic order				
1.1 Simple random sampling	7.33	14.90	21.77	28.44
1.2 Stratified sampling, one element per stratum	6.92	15.95	19.33	30.49
1.3 Linear systematic sampling	1.97	5.92	26.67	11.14
2. Procedures applied to population arranged in increasing order of ability test scores				
2.1 Stratified sampling, one element per stratum	1.18	2.47	3.29	3.84
2.2 Linear systematic sampling	0.67	3.43	1.02	0.82
2.3 Linear systematic sampling with end corrections	1.75	2.29	1.59	1.52
2.4 Balanced systematic sampling	2.12	2.94	1.89	7.40
2.5 Centrally located systematic samples	0.42	6.42	1.71	0.46
3. Linear systematic sampling applied to the population in increasing order of ability test scores with order reversed in alternate strata	0.52	4.51	0.61	0.77

[a] All data are for the population of sixth-grade pupils in school 5 in Midcity. Units are raw score points squared on the Stanford Achievement Test.

TABLE 6.7

Efficiencies of Systematic Sampling Procedures Relative to Simple Random Sampling and Stratified Sampling[a]

Sampling Procedure	Relative Efficiencies, %, for Indicated Methods When Sampling Interval and Sampling Fraction Are			
	2 50%	3 33%	4 25%	5 20%
1. Linear systematic sampling applied to the population arranged in alphabetic order				
Efficiencies relative to				
Simple random sampling	372	252	82	255
Stratified sampling	351	269	72	274
2. Procedures applied to the population arranged in increasing order of ability test scores				
2.2 Linear systematic sampling				
Efficiencies relative to				
Simple random sampling	1095	434	2134	3479
Stratified sampling	176	72	322	369
2.3 Linear systematic sampling with end corrections				
Efficiencies relative to				
Simple random sampling	418	650	1369	1866
Stratified sampling	67	108	207	252
2.4 Balanced systematic sampling				
Efficiencies relative to				
Simple random sampling	346	507	1150	384
Stratified sampling	56	84	174	52
2.5 Centrally located systematic samples				
Efficiencies relative to				
Simple random sampling	1786	232	1276	6206
Stratified sampling	287	38	193	837
3. Linear systematic sampling applied to the population arranged in increasing order of ability test scores with order reversed in alternate strata				
Efficiencies relative to				
Simple random sampling	1411	331	3567	3738
Stratified sampling	227	55	539	504

[a] All data are for the population of sixth-grade pupils in School 5 in Midcity.

sampling from an alphabetic list of sixth-graders provided a mean square error that was just under 4, using a sampling fraction of 50 percent. Stratified sampling from frames of sixth-graders arranged in increasing order of their Lorge-Thorndike ability test scores provided acceptable precision for all four schools when the sampling fraction was 50 percent. For two schools (School 5 and School 21), this method was acceptably precise for sampling fractions as small as 20 percent.

For the populations of sixth-graders in each school, mean square errors for simple random sampling and stratified sampling with one element per stratum showed an orderly progression with changes in sampling fraction. In contrast, the progressions of mean square errors for most of the systematic sampling procedures were erratic and inconsistent. The most extreme example of this inconsistency was found in the results for linear systematic sampling in School 21. When the sampling frame was arranged in alphabetic order, mean square errors were 0.31, 36.94, 0.85, and 12.66 for sampling fractions of 50, 33, 25, and

TABLE 6.8
Mean Square Errors of Estimated Population Means for Systematic Sampling Procedures[a]

Sampling Procedure	Mean Square Errors Arising When Sampling Interval and Sampling Fraction Are			
	2 50%	3 33%	4 25%	5 20%
1. Procedures applied when population is in alphabetic order				
1.1 Simple random sampling	7.71	15.27	24.01	29.72
1.2 Stratified sampling, one element per stratum	7.16	15.55	25.04	26.47
1.3 Linear systematic sampling	4.01	35.90	30.22	4.39
2. Procedures applied to population arranged in increasing order of ability test scores				
2.1 Stratified sampling, one element per stratum	1.45	2.48	4.87	5.93
2.2 Linear systematic sampling	0.70	1.56	1.87	8.98
2.3 Linear systematic sampling with end corrections	2.11	5.31	6.26	3.34
2.4 Balanced systematic sampling	0.84	1.92	6.12	12.14
2.5 Centrally located systematic samples	0.38	1.49	1.75	1.23
3. Linear systematic sampling applied to the population in increasing order of ability test scores with order reversed in alternate strata	1.36	1.82	3.89	5.03

[a] All data are for the population of sixth-grade pupils in School 19 in Midcity. Units are raw score points squared on the Stanford Achievement Test.

TABLE 6.9

Efficiencies of Systematic Sampling Procedures Relative to Simple Random Sampling and Stratified Sampling with One Element per Stratum [a]

	Relative Efficiencies, %, for Indicated Methods When Sampling Interval and Sampling Fraction Are			
Sampling Procedure	2 50%	3 33%	4 25%	5 20%
1. Linear systematic sampling applied to the population arranged in alphabetic order				
Efficiencies relative to				
Simple random sampling	192	43	79	648
Stratified sampling	179	43	83	577
2. Procedures applied to the population arranged in increasing order of ability test scores				
2.2 Linear systematic sampling				
Efficiencies relative to				
Simple random sampling	1095	977	1287	331
Stratified sampling	205	159	261	66
2.3 Linear systematic sampling with end corrections				
Efficiencies relative to				
Simple random sampling	336	288	383	889
Stratified sampling	69	47	78	177
2.4 Balanced systematic sampling				
Efficiencies relative to				
Simple random sampling	922	795	392	245
Stratified sampling	173	129	80	49
2.5 Centrally located systematic samples				
Efficiencies relative to				
Simple random sampling	2050	1026	1371	2416
Stratified sampling	385	167	278	482
3. Linear systematic sampling applied to the population arranged in increasing order of ability test scores with order reversed in alternate strata				
Efficiencies relative to				
Simple random sampling	567	838	617	591
Stratified sampling	106	136	125	118

[a] All data are for the population of sixth-grade pupils in School 19 in Midcity.

20 percent, respectively. The large value of the mean square error for a sampling fraction of 33 percent reflects sample means of 78.9, 68.2, and 64.6. The population mean was 70.6 raw score points. Apparently, alphabetizing these pupils created a pattern in their test scores that was matched by a sampling interval of 3. When a sampling interval of 4 (sampling fraction of 25 percent) was used, linear systematic sampling resulted in sample means of 70.3, 69.9, 72.1, and 70.1. Since these sample means were similar and all were close to the population mean of 70.6, the mean square error was desirably small (0.85).

In view of the inconsistencies described above, the data in Tables 6.4 through 6.11 must be interpreted with great caution. Considering all four schools, linear systematic sampling of sixth-graders arranged in increasing order of their Lorge-Thorndike scores, with their order reversed in alternate strata, was the most efficient of the sampling methods investigated. It provided acceptable precision with a sampling fraction of 25 percent for Schools 2 and 19 and with a sampling fraction of 20 percent for Schools 5 and 21. However, for three

TABLE 6.10

Mean Square Errors of Estimated Population Means for Systematic Sampling Procedures[a]

Sampling Procedure	Mean Square Errors Arising When Sampling Interval and Sampling Fraction Are			
	2 50%	3 33%	4 25%	5 20%
1. Procedures applied when population is in alphabetic order				
1.1 Simple random sampling	3.88	7.73	11.88	15.80
1.2 Stratified sampling, one element per stratum	3.46	7.41	11.04	13.99
1.3 Linear systematic sampling	0.31	36.94	0.85	12.66
2. Procedures applied to population arranged in increasing order of ability test scores				
2.1 Stratified sampling, one element per stratum	0.78	1.85	2.83	3.71
2.2 Linear systematic sampling	0.40	0.25	0.52	2.98
2.3 Linear systematic sampling with end corrections	0.95	0.68	1.95	4.68
2.4 Balanced systematic sampling	0.10	1.47	5.92	4.27
2.5 Centrally located systematic samples	0.21	0.49	0.70	9.30
3. Linear systematic sampling applied to the population in increasing order of ability test scores with order reversed in alternate strata	0.003	1.85	5.04	3.01

[a] All data are for the population of sixth-grade pupils in School 21 in Midcity. Units are raw score points squared on the Stanford Achievement Test.

TABLE 6.11

Efficiencies of Systematic Sampling Procedures Relative to Simple Random Sampling and Stratified Sampling with One Element per Stratum[a]

Sampling Procedure	Relative Efficiencies, %, for Indicated Methods When Sampling Interval and Sampling Fraction Are			
	2 50%	3 33%	4 25%	5 20%
1. Linear systematic sampling applied to the population arranged in alphabetic order				
Efficiencies relative to				
Simple random sampling	1250	21	1390	125
Stratified sampling	1114	20	1292	110
2. Procedures applied to the population arranged in increasing order of ability test scores				
2.2 Linear systematic sampling				
Efficiencies relative to				
Simple random sampling	963	3121	2276	530
Stratified sampling	194	750	542	124
2.3 Linear systematic sampling with end corrections				
Efficiencies relative to				
Simple random sampling	409	1230	610	337
Stratified sampling	82	271	145	79
2.4 Balanced systematic sampling				
Efficiencies relative to				
Simple random sampling	3884	526	201	370
Stratified sampling	783	126	48	87
2.5 Centrally located systematic samples				
Efficiencies relative to				
Simple random sampling	1775	1567	1686	170
Stratified sampling	358	376	401	40
3. Linear systematic sampling applied to the population arranged in increasing order of ability test scores with order reversed in alternate strata				
Efficiencies relative to				
Simple random sampling	> 9999	418	236	524
Stratified sampling	> 9999	100	56	123

[a] All data are for the population of sixth-grade pupils in School 21 in Midcity.

of the four schools, it produced larger mean square errors when sample size was increased. This inconsistency makes the generality of these results suspect.

Linear systematic sampling of sixth-graders arranged in increasing order of their ability test scores was second in efficiency. This method provided acceptable precision with a sampling fraction of 33 percent in two schools and with a sampling fraction of 20 percent in the other two. Unlike sampling of pupils arranged in increasing order of their ability test scores, with order reversed in alternate strata, the results for this method were generally consistent for different sampling fractions. Shifts in efficiency from school to school were similar for this method and for stratified sampling of pupils arranged in increasing order of their ability test scores.

When sixth-graders were arranged in increasing order of their Lorge-Thorndike scores, centrally located sampling and linear systematic sampling with use of end corrections provided precise estimates for some school populations but were unacceptable for others. The inconsistency of these methods precludes their use with these small populations.

The within-school efficiencies of systematic sampling procedures relative to simple random sampling and stratified sampling with one element per stratum are shown in Tables 6.5, 6.7, 6.9, and 6.11. These data are consistent with the direct comparisons of mean square errors discussed above. In almost all cases, systematic sampling procedures were more efficient than simple random sampling. The efficiencies of systematic sampling relative to stratified sampling with one element per stratum varied from 20 to more than 9999 percent. The details of this variation are consistent with direct comparisons of mean square errors.

The analyses of Midcity data showed systematic sampling to be relatively efficient. However, it should not be assumed that these methods will show similar efficiencies in all school districts. The sampling methods discussed in preceding chapters are sensitive to population variances. Systematic sampling procedures are sensitive to many population parameters in addition to variances, and these parameters will surely show marked differences across school districts.

In using systematic sampling, it is essential to gain some experience with the population of interest. Computing estimates of population means and variances of estimators for data from related populations or for related variables that have been measured for the entire population of interest would be prudent and informative. The SYSAMP-I computer program will facilitate such investigations.

Estimator Bias

Although the theory of circular systematic sampling was discussed in an earlier section of this chapter, no empirical results have been shown for this method. The variances of sample means and proportions resulting from linear systematic sampling and circular systematic sampling will be approximately equal when the methods are applied to the same population. The mean square errors of these estimators might differ, since the estimators will provide unbiased estimation when used with circular systematic sampling, but estimation will often be biased when they are used with linear systematic sampling. However, estimator bias will be negligible in most cases. As an illustration, results of analy-

TABLE 6.12
Estimator Bias Resulting from Linear Systematic Sampling[a]

Sampling Interval	Sampling Fraction,%	School			
		2	5	19	21
2	50	0.01	−0.02	−0.03	0.00
3	33	−0.06	0.05	−0.09	−0.08
4	25	−0.10	0.07	0.07	0.00
5	20	0.00	0.00	0.00	−0.01

[a] Data are for the populations of sixth-grade pupils in alphabetic order in Schools 2, 5, 19, and 21 of Midcity. Units are raw score points on the Stanford Achievement Test.

ses of reading achievement scores for populations of sixth-graders in the individual schools examined in the last example are shown in Table 6.12.

Estimator bias for these populations in no case exceeded one-tenth of a raw score point, a negligible value compared to the mean square errors of estimation shown in Tables 6.4, 6.6, 6.8, and 6.10. Values of estimator bias were not computed for the entire population of sixth-graders in Midcity. The larger population size and correspondingly larger sample sizes would have resulted in biases that were even smaller than those shown in Table 6.12.

We asserted earlier that linear systematic sampling results in unbiased estimation when the population size is an integral multiple of the sample size, that is, when N/n is an integer. This claim is supported by empirical results shown in Table 6.12. Schools 2, 5, and 19 have sixth-grade enrollments of 55, 65, and 55, respectively. The estimator of mean sixth-grade achievement for these schools was unbiased for a sampling interval of 5.

Estimation of Population Totals

The procedures for estimating population totals that are discussed in Chapter 5 (for linear systematic sampling) can also be applied to the alternative systematic sampling procedures described in this chapter. An estimator of the population total is given by the product of the population size and the estimated population mean. The mean square error of the estimated population total is given by the square of the population size multiplied by the mean square error of the estimated population mean. Symbolically, these relationships are as follows:

The estimated population total is equal to

$$y = N\bar{y} = \frac{N}{n} \sum_{i=1}^{n} y_i \tag{6.6}$$

where

N = population size
n = sample size
y_i = value of the sampling variable for the ith sampled element

For each of the systematic sampling procedures described earlier in this chapter, the mean square error of the estimated population total is given by

$$\text{MSE}(y) = \frac{N^2}{n_s} \sum_{i=1}^{n_s} (y_i - \bar{Y})^2 \qquad (6.7)$$

where n_s denotes the number of systematic samples that can be selected with the sampling procedure, given a specified set of values of the design parameters (for example, sample size n or sampling interval k). All other terms are as defined previously.

The following example illustrates the use of each of the systematic sampling methods discussed in this chapter in estimating a population total. The example is contrived, and the data used are fictitious. In order to provide a detailed description of the mechanics of each sampling method, we have made the population size unrealistically small.

Example 6.2. Estimation of the Total Time Needed to Complete a Maze Task

CONTEXT AND PROBLEM

In the Exercises section of Chapter 5, we described a maze task that was used by two psychologists who were conducting research on cognitive functioning in children. The task required each child to begin at an indicated starting position and then draw a continuous path through the maze, avoiding dead ends, until he or she arrived at an indicated ending position. The psychologists recorded the time (in seconds) needed by each child to complete the task.

In this example, we assume that the psychologists want to estimate the total length of time it will take a population of 49 children to complete the maze task. Their interest in this parameter arises from their need to reserve time in a shared laboratory facility so that they can collect data. The problem is to estimate the total time parameter using a number of different systematic sampling procedures.

The psychologists have access to scores on the Performance Scale of the Wechsler Intelligence Scales for Children (WISC) for all children in the population. They could therefore arrange the sampling frame in increasing order of the children's WISC Performance scores. If WISC Performance scores are highly correlated with times to complete the maze task, systematic sampling from the ordered frame should improve the efficiency of estimation.

Although the psychologists would know maze completion times only for the children they sampled and measured, we will tabulate maze completion times for the entire population of 49 children. Knowing the value of the population parameter will facilitate an evaluation of the alternative systematic sampling procedures used in this example. This departure from reality is useful for pedagogic purposes, but you should remember that the situation is fictitious and would never be found in practice.

The following tables list maze completion times (in seconds) and IQ scores on the Performance Scale of the WISC for the population of 49 children. Table 6.13 lists the sampling frame in its "natural," presumably random, order, and

TABLE 6.13

Times (in seconds) Needed to Complete a Maze Task and Scores on the Performance Scale of the Wechsler Intelligence Scale for Children, for a Population of 49 Children

Child No.	Time	WISC Perf. IQ	Child No.	Time	WISC Perf. IQ
1	54	115	26	88	123
2	56	90	27	42	92
3	53	93	28	102	146
4	57	97	29	48	102
5	76	108	30	78	120
6	85	129	31	87	124
7	44	88	32	57	107
8	93	116	33	26	82
9	104	139	34	50	102
10	41	102	35	26	90
11	95	126	36	57	95
12	69	105	37	61	109
13	64	105	38	67	107
14	85	121	39	24	65
15	89	105	40	56	103
16	79	109	41	94	125
17	62	109	42	80	119
18	48	102	43	69	105
19	31	84	44	43	100
20	82	118	45	92	123
21	93	137	46	30	89
22	99	133	47	72	119
23	67	107	48	106	150
24	63	117	49	78	107
25	76	106			

Table 6.14 lists the frame in increasing order of children's WISC Performance scores.

SOLUTION AND RESULTS

To compare the results of applying the alternative systematic sampling procedures described in this chapter, we will assume that the psychologists used a sampling interval of $k = 5$ or a desired sample size of $n = 10$ for all procedures where one or the other of these design parameters must be specified. Furthermore, we will assume that the index of the first sampled element was 5 for those systematic sampling procedures that require random selection of an index between 1 and k and that the index of the first sampled element was 35 for those sampling procedures that require random selection of an index between 1 and the population size N.

SAMPLING FROM THE POPULATION IN NATURAL ORDER

We will first consider the application of alternative systematic sampling procedures to the population of children arranged in their natural order, that is, to the data shown in Table 6.13.

TABLE 6.14

Times (in seconds) Needed to Complete a Maze Task and Scores on the Performance Scale of the Wechsler Intelligence Scale for Children, for a Population of 49 Children Arranged in Increasing Order of Their Wechsler Performance Scores

Child No.	Time	WISC Perf. IQ	Child No.	Time	WISC Perf. IQ
39	24	65	49	78	107
33	26	82	5	76	108
19	31	84	16	79	109
7	44	88	17	62	109
46	30	89	37	61	109
2	56	90	1	54	115
35	26	90	8	93	116
27	42	92	24	63	117
3	53	93	20	82	118
36	57	95	47	72	119
4	57	97	42	80	119
44	43	100	30	78	120
34	50	102	14	85	121
18	48	102	45	92	123
10	41	102	26	88	123
29	48	102	31	87	124
40	56	103	41	94	125
43	69	105	11	95	126
12	69	105	6	85	129
13	64	105	22	99	133
15	89	105	21	93	137
25	76	106	9	104	139
32	57	107	28	102	146
38	67	107	48	106	150
23	67	107			

Linear Systematic Sampling. If the first sampled element had an index of 5 and the sampling interval equaled 5, the indices and maze completion times of children selected through linear systematic sampling from the frame shown in Table 6.13 would be as follows:

Child No. (Index)	Time
5	76
10	41
15	89
20	82
25	76
30	78
35	26
40	56
45	92
Total	616

The average time the children in this sample required to complete the maze task was $\frac{616}{9} = 68.44$ seconds. From Equation (6.6), the estimated total time the population of 49 children would require to complete the maze task would be $49(68.44) = 3353.78$ seconds.

Modified Linear Systematic Sampling. With this method, the first sampled child would have an index of 35, and additional children would be sampled by going forward and backward in the sampling frame, at an interval of 5. Going forward in the sampling frame, we would select children with indices of 40 and 45. Going backward, we would select children with indices of 30, 25, 20, 15, 10, and 5. This procedure would result in a sample that was identical to the one selected through linear systematic sampling. Of course, the estimated total time the 49 children would require to complete the maze task would be the same as that resulting from linear systematic sampling, 3353.78 seconds.

Circular Systematic Sampling. With this method, we would treat the sampling frame as though it were arranged around a circle, with the children listed from 1 to 49 and then the first child following the forty-ninth. We would again assume that the randomly chosen index of the first sampled child was 35. We would select children by using a sampling interval of 5 until the desired sample size of 10 had been realized. With this process, the sample of 10 children would be selected in the following order:

Child No. (Index)	Time
35	26
40	56
45	92
1	54
6	85
11	95
16	79
21	93
26	88
31	87
Total	755

The average time the children in this sample required to complete the maze task was $\frac{755}{10} = 75.50$ seconds. From Equation (6.6), the estimated total time the population of 49 children would require to complete the maze task would be $49(75.50) = 3699.50$ seconds.

SAMPLING FROM THE POPULATION IN INCREASING ORDER

We will now apply alternative systematic sampling procedures to the sampling frame of children listed in increasing order of their WISC Performance scores, as shown in Table 6.14.

Linear Systematic Sampling. If we applied linear systematic sampling to the data shown in Table 6.14 in accordance with the rules specified above, we would first sample the child listed fifth. We would then sample every fifth child thereafter, until we had exhausted the sampling frame. With this procedure, the sampled children would have the following index numbers and times for completion of the maze task:

Child No. (Index)	Time
46	30
36	57
10	41
13	64
23	67
37	61
47	72
26	88
22	99
Total	579

The average time the children in this sample required to complete the maze task was $\frac{579}{9} = 64.33$ seconds. From Equation (6.6), the estimated total time the population of 49 children would require to complete the maze task would be 49(64.33) = 3152.33 seconds.

Balanced Systematic Sampling. This sampling method cannot be applied to the sampling frame of 49 children because the ratio of the population size to the desired sample size, $N/n = \frac{49}{10}$, is not an integer. However, for purposes of illustration, we will "bend the rules" by modifying the index of the last sampled child.

The balanced systematic sampling procedure first requires that we divide the sampling frame into $n/2$ groups of elements. In this case, $n/2 = \frac{10}{2} = 5$. Each group will contain $2k = 2(5) = 10$ elements. The first group of elements would be those listed first through tenth in the frame. The second group would be those listed eleventh through twentieth; the third group, those listed twenty-first through thirtieth; the fourth group, those listed thirty-first through fortieth; and the fifth group, in our case, would consist of the nine elements listed forty-first through forty-ninth.

Consistent with our earlier specifications, we will assume that the random number we chose to determine the sample to be selected, r, was equal to 5. We are to sample elements that are $r - 1 = 5 - 1 = 4$ places from the boundaries of each group. Applying this rule, we would sample children listed fifth (1 + 4 = 5) and sixth (10 − 4 = 6) from the first group, fifteenth (11 + 4 = 15) and sixteenth (20 − 4 = 16) from the second group, twenty-fifth (21 + 4 = 25) and twenty-sixth (30 − 4 = 26) from the third group, thirty-fifth (31 + 4 = 35) and thirty-sixth (40 − 4 = 36) from the fourth group, and finally, forty-fifth (41 + 4 = 45) and, with our modification, forty-sixth (49 − 3 = 46) from the fifth group. In selecting a unit that is only three places from the upper boundary of the fifth group, we have reasoned that if N/n were an integer, there would have been 50 elements in the population. In that case, we would have selected the forty-sixth element in the sampling frame.

Applying these rules to the data listed in Table 6.14, the indices and maze completion times of the sampled children would be as follows:

Child No. (Index)	Time
46	30
2	56
10	41
29	48
23	67
49	78
47	72
42	80
22	99
21	93
Total	664

The average time the children in this sample required to complete the maze task was $\frac{664}{10} = 66.40$ seconds. From Equation (6.6), the estimated total time the population of 49 children would require to complete the maze task would be $49(66.40) = 3253.60$ seconds.

Centrally Located Samples. Since the sampling interval k is equal to an odd number in this example, there is only one centrally located sample. The first element in the sample should have an ordered index of $(k + 1)/2 = (5 + 1)/2 = 3$, so the third child in the ordered sampling frame (Table 6.14) will be the first member of our sample. Every fifth child will be selected thereafter. Applying these rules, the indices and maze completion times of the children who compose the centrally located sample would be as follows:

Child No. (Index)	Time
19	31
27	42
34	50
43	69
32	57
16	79
24	63
14	85
11	95
28	102
Total	673

The average time the children in this sample required to complete the maze task was $\frac{673}{10} = 67.30$ seconds. From Equation (6.6), the estimated total time the population of 49 children would require to complete the maze task would be $49(67.30) = 3297.70$ seconds.

Linear Systematic Sampling with End Corrections. When end corrections are used with linear systematic sampling, the sample selection procedure is not affected at all. Precisely the same sample is chosen whether end corrections are used or not.

End corrections are multiplicative weights that are applied to the values of the sampling variable for the first and last elements sampled.

In this case, we would use the sample listed above for linear systematic sampling from the ordered frame contained in Table 6.14. We would then apply appropriate weights to the maze completion times of the children sampled first and last, prior to computing the sample average maze completion time.

In this example, the population size, 49, is not an integral multiple of the desired sample size, 10. We can define the population size as a multiple of the sampling interval plus a remainder: $N = nk + R$. For our data, $49 = 9(5) + 4$; the remainder R equals 4. Since we have specified that the random number used to determine the initially sampled element, r, equals 5, and since r is greater than R, we must use Equations (6.4a) and (6.4b) to define appropriate end corrections.

Given our design parameters, the end corrections are

$$1 + \frac{10[2(5) - 5 - 4 - 1]}{2(9 - 1)(5)} = 1 + 0 = 1.0$$

and

$$1 - \frac{10[2(5) - 5 - 4 - 1]}{2(9 - 1)(5)} = 1 - 0 = 1.0$$

Since both of these weights equal 1.0, the application of end corrections would not change the values of the sampling variable for the children sampled first and last; results for linear systematic sampling are exactly the same, with and without the use of end corrections: the estimated total time needed by the 49 children to complete the maze task is 3152.33 seconds.

DISCUSSION AND SUMMARY

In this example, a number of systematic sampling procedures have been applied to the problem of estimating a population total. In contrast to Example 6.1, we have stressed the mechanics of using each procedure. All the procedures have been applied to a sampling frame in natural order and to a frame that has been reordered in accordance with the values of a variable that should improve the efficiency of estimation.

Although several of the systematic sampling procedures resulted in exactly the same sample (largely because we specified various design parameters, including values that we would normally select randomly), other methods produced results that were noticeably different. Estimates of the total time needed by 49 children to complete a maze task are summarized below for various systematic sampling procedures. These procedures were applied to the sampling frame in its natural order and to the reordered frame.

Sampling Procedure	Estimated Total Time, Seconds
Sampling frame in natural order	
Linear systematic sampling	3353.78
Modified linear systematic sampling	3353.78
Circular systematic sampling	3699.50
Sampling frame in increasing order	
Linear systematic sampling	3152.33
Balanced systematic sampling	3253.60
Centrally located samples	3297.70
Linear with end corrections	3152.33

Since this is an unrealistic situation and we know the value the sampling variable for every child in the population, we can compute the actual value of the population parameter. The total time needed by the 49 children to complete the maze task would be 3298 seconds. Comparing the actual value of the population total and the range of estimates listed above, we can note that centrally located systematic sampling from the list of children arranged in increasing order of their WISC Performance scores resulted in near-perfect estimation. Circular systematic sampling of the naturally ordered list produced results with the largest estimation error.

Of course, these findings cannot be generalized. Nor is it reasonable to draw any conclusions on the relative accuracy of the various sampling methods in this specific application on the basis of our limited analyses. To determine which sampling method is best in this application, we would have to select all possible systematic samples of each type and then compute the estimator bias and a mean square error for each method. Since such an undertaking would require a great deal of computation and would be of limited pedagogical value, we shall leave it for the *very* curious reader.

Estimating Population Proportions

FORMING THE ESTIMATES

It was shown in Chapter 3 that formulas for estimation of population means and the variances of their estimators could easily be extended to estimation of population proportions and the variances of their estimators. The extension is accomplished by coding the sampling variable so that elements that possess a specified attribute or have values of the sampling variable within some interval, say between C_1 and C_2, are given codes of one, and all other elements are given codes of zero. The sample mean of these coded values equals the sample proportion, and the mean of coded values for the entire population equals the population proportion.

The same method can be applied when samples are drawn using systematic sampling procedures. The sample proportion (the sample mean of coded values) is used to estimate the population proportion (the population mean of coded values).

Example 6.3. Efficiencies of Systematic Sampling Procedures for Estimation of Population Proportions

CONTEXT AND PROBLEM

The efficiencies of systematic sampling procedures in estimating proportions of low-achieving pupils are investigated in this example. The populations and sampling procedures are the same as those used in Example 6.1.

The five populations are composed of sixth-graders in each of four Midcity elementary schools and all sixth-graders in Midcity. Proportions of pupils having reading achievement scores at least 1 grade equivalent unit below the national norm are to be estimated. As in Example 5.6, population proportions are to be estimated within ± 0.05 with 95 percent confidence. The largest sampling intervals that provide desired estimation precision are to be determined.

SOLUTION AND RESULTS

Six combinations of systematic sampling procedure and population ordering were used with each of the five populations of pupils. Proportions of low achievers were estimated for populations of pupils arranged in increasing order of their scores on the Lorge-Thorndike ability test. The sampling procedures used with this arrangement of pupils were linear systematic sampling with and without use of end corrections, balanced systematic sampling, and centrally located sampling. Linear systematic sampling was also applied to sampling frames of sixth-graders, first arranged in increasing order of their Lorge-Thorndike scores, and then the order was reversed for every other group of k students. Results for linear systematic sampling applied to the citywide population of sixth-graders arranged in alphabetic order, presented in Example 5.6, are also shown here.

The SYSAMP-I computer program (Appendix C) was used to calculate proportions of low achievers for each systematic sample and for populations. The largest and smallest sample proportions are shown in Tables 6.15 through 6.19 each of six systematic sampling procedures. With systematic sampling, the largest and smallest sample proportions have high probabilities of occurrence. They are therefore better indicators of estimation precision than is the mean square error.

DISCUSSION

RESULTS FOR THE MIDCITY POPULATION

In Example 3.2, simple random sampling (SRS) was used to estimate the percentage of sixth-graders in Midcity with achievement scores at least 1 grade equivalent unit below the national norm. Using a sampling fraction of 26 percent, upper and lower 95 percent confidence limits on the population percentage were found to be 23.4 and 15.4 percent, respectively. All systematic sampling procedures considered here were thus more efficient than simple random sampling for estimation of this population percentage (Table 6.15). For a sampling fraction of 20 percent, every sample percentage produced by every systematic sampling procedure fell within the 95 percent confidence interval resulting from the SRS procedure. With a sampling fraction of 10 percent, all sample

TABLE 6.15
Largest and Smallest Estimates of Percentages of Sixth-Grade Pupils in Midcity with Reading Achievement Scores at Least 1 Grade Equivalent Unit Below National Norms

	Population Percentage = 19.4% Sampling Interval and Sampling Fraction				
Sampling Procedure	2 50%	4 25%	5 20%	10 10%	20 5%
1. Linear systematic sampling applied to the population arranged in alphabetic order					
Largest sample percentage	20.1	20.8	21.3	25.4	27.1
Smallest sample percentage	18.7	18.4	17.0	14.5	13.8
2. Procedures applied to the population arranged in increasing order of ability test scores					
2.2 Linear systematic sampling					
Largest sample percentage	20.2	20.8	22.1	22.2	23.7
Smallest sample percentage	18.6	17.8	17.0	15.4	13.8
2.3 Linear systematic sampling with end corrections					
Largest sample percentage	20.2	20.8	22.1	22.2	23.7
Smallest sample percentage	18.6	17.8	17.0	15.4	13.8
2.4 Balanced systematic sampling					
Largest sample percentage	19.6	20.1	22.0	23.7	31.0
Smallest sample percentage	19.2	18.4	17.8	17.8	15.5
2.5 Centrally located samples					
Largest sample percentage	20.3	19.7	22.1	22.2	20.3
Smallest sample percentage	18.6	19.4	22.1	15.4	15.3
3. Linear systematic sampling applied to population arranged in increasing order of ability test scores with order reversed in alternate strata					
Largest sample percentage	19.6	20.1	22.1	23.7	23.7
Smallest sample percentage	19.2	18.4	17.9	17.8	11.9

percentages produced by three of the systematic sampling procedures were within the SRS confidence limits. Those three procedures were linear systematic sampling of the population arranged in increasing order of Lorge-Thorndike scores, centrally located sampling, and linear systematic sampling of the population arranged in increasing order, with use of end corrections. Using a sampling fraction of 5 percent, almost all the sample percentages resulting from centrally located sampling were within the SRS confidence interval. The smallest sample percentage was slightly smaller than the lower confidence limit resulting from use of SRS.

For some systematic sampling procedures, the mean square errors of estimated population means varied erratically as sampling fractions were increased (Table 6.2). In contrast, the largest and smallest systematic sample percentages tended to show orderly variations with increases in sampling fractions. Also in contrast with the estimation of population means, similar efficiencies were afforded by most of the systematic sampling methods when population percentages were estimated.

RESULTS FOR INDIVIDUAL SCHOOLS

The percentages of pupils with achievement scores of at least 1 grade equivalent unit below the national norm were 18.2 and 23.1 percent, respectively, for Schools 2 and 5 of Midcity. If this 5-point difference in percentages is to be detected with certainty, the ranges of sample proportions for these schools must be less than ± 2.5 percent. Clearly, larger ranges would produce some sample results for which the percentage of low achievers was larger in School 2 than in School 5. A range of sample percentages equal to ± 2.5 percent might therefore be judged acceptable for these schools. However, this range is narrower than the ± 5 percent suggested for estimates of the districtwide percentage of low achievers. It would also require that larger sampling fractions be used. To be consistent with the discussion of districtwide results, we will assume that percentages of low achievers in schools must be estimated within ± 5 percent, that is, within ± 0.05. This choice is both arbitrary and subjective, although it is probably reasonable.

Results for School 2. For School 2, a sampling fraction of 50 percent resulted in sample percentages with an approximate range of ± 4 percent for all systematic sampling procedures investigated (Table 6.16). With a sampling fraction of 33 percent, linear systematic sampling of the frame of sixth-graders arranged in alphabetic order resulted in a range of percentages from 16.7 to 21.1 percent. These values are 1.5 percent lower and 3 percent higher than the population value. These results must be attributed to chance and should not be considered replicable. With a sampling fraction of 33 percent, all other systematic sampling procedures produced intolerable estimation errors. Several methods provided equally good results for sampling fractions of 50 and 25 percent. However, these results did not hold for all four schools.

Results for School 5. With a sampling fraction of 50 percent, some systematic sampling procedures provided acceptably narrow ranges of sample percentages when applied to the sixth-grade population in School 5. Linear systematic sampling of the frame arranged in increasing order of Lorge-Thorndike scores, centrally located sampling, and linear systematic sampling of the ordered frame with the use of end corrections all resulted in sample percentages that were within ± 2 percent of the population value. The other methods produced ranges of percentages that were far wider. For smaller sampling fractions, all sampling methods provided very wide ranges of percentages when applied to the population of sixth-graders in School 5 (Table 6.17).

TABLE 6.16

Largest and Smallest Estimates of Percentages of Sixth-Grade Pupils in School 2 in Midcity with Reading Achievement Scores at Least 1 Grade Equivalent Unit Below National Norms

	Population Proportion = 18.2% Sampling Interval and Sampling Fraction			
	2	3	4	5
Sampling Procedure	50%	33%	25%	20%
1. Linear systematic sampling applied to the population arranged in alphabetic order				
Largest sample percentage	21.4	21.1	28.6	27.3
Smallest sample percentage	14.8	16.7	14.2	0.0
2. Procedures applied to the population arranged in increasing order of ability test scores				
2.2 Linear systematic sampling				
Largest sample percentage	22.2	22.2	28.6	36.4
Smallest sample percentage	14.3	11.1	14.3	9.1
2.3 Linear systematic sampling with end corrections				
Largest sample percentage	22.2	26.3	28.6	36.4
Smallest sample percentage	14.3	11.1	14.3	9.1
2.4 Balanced systematic sampling				
Largest sample percentage	21.4	22.2	21.4	33.3
Smallest sample percentage	14.3	11.1	14.3	8.3
2.5 Centrally located samples				
Largest sample percentage	22.2	11.1	28.6	36.4
Smallest sample percentage	14.8	11.1	14.3	36.4
3. Linear systematic sampling applied to population arranged in increasing order of ability test scores with order reversed in alternate strata				
Largest sample percentage	22.2	22.2	21.4	36.4
Smallest sample percentage	14.3	11.1	14.3	9.1

Results for School 19. Using a sampling fraction of 50 percent, all sampling methods produced unusually narrow ranges of sample percentages when applied to data for sixth-graders in School 19 (Table 6.18). For this sampling fraction, all sample percentages were within 0.2 percent of the population value. Linear systematic sampling of the sampling frame of pupils arranged in increasing order of their Lorge-Thorndike scores, balanced systematic sampling, and

TABLE 6.17

Largest and Smallest Estimates of Percentages of Sixth-Grade Pupils In School 5 In Midcity with Reading Achievement Scores at Least 1 Grade Equivalent Unit Below National Norms

| | Population Proportion = 23.1% Sampling Interval and Sampling Fraction | | | |
| | 2 | 3 | 4 | 5 |
Sampling Procedure	50%	33%	25%	20%
1. Linear systematic sampling applied to the population arranged in alphabetic order				
Largest sample percentage	28.1	27.3	37.5	30.8
Smallest sample percentage	18.2	19.1	18.8	15.4
2. Procedures applied to the population arranged in increasing order of ability test scores				
2.2 Linear systematic sampling				
Largest sample percentage	24.2	27.3	35.3	30.8
Smallest sample percentage	21.9	18.2	12.5	15.4
2.3 Linear systematic sampling with end corrections				
Largest sample percentage	24.2	27.3	35.3	30.8
Smallest sample percentage	21.9	18.2	12.5	7.7
2.4 Balanced systematic sampling				
Largest sample percentage	31.3	27.3	31.3	28.6
Smallest sample percentage	15.6	18.2	12.5	7.1
2.5 Centrally located samples				
Largest sample percentage	25.0	18.2	18.8	30.8
Smallest sample percentage	21.9	18.2	12.5	30.8
3. Linear systematic sampling applied to population arranged in increasing order of ability test scores with order reversed in alternate strata				
Largest sample percentage	30.3	28.6	31.3	30.8
Smallest sample percentage	15.6	18.2	12.8	7.7

centrally located samples provided percentages that were within 0.4 percent of the population percentage, using a sampling fraction of 33 percent. For a sampling fraction of 25 percent, these methods resulted in sample percentages that were within ± 4 percent of the population value.

Results for School 21. With a sampling fraction of 50 percent, three systematic sampling procedures provided ranges of sample percentages that were within ± 2 percent of the population percentage of low-achieving sixth-graders in School 21 (Table 6.19). Linear systematic sampling of the frame of pupils arranged in increasing order of their ability test scores, with and without use of end corrections, and centrally located sampling resulted in acceptably narrow ranges of sample percentages. These methods and two others (balanced systematic sampling and linear systematic sampling of the frame arranged in increasing order of ability test scores with order reversed in alternate strata) provided even

TABLE 6.18

Largest and Smallest Estimates of Percentages of Sixth-Grade Pupils in School 19 in Midcity with Reading Achievement Scores at Least 1 Grade Equivalent Unit Below National Norms

	Population Proportion = 10.9% Sampling Interval and Sampling Fraction			
Sampling Procedure	2 50%	3 33%	4 25%	5 20%
1. Linear systematic sampling applied to the population arranged in alphabetic order				
Largest sample percentage	11.1	22.2	21.4	18.2
Smallest sample percentage	10.7	5.6	0.0	0.0
2. Procedures applied to the population arranged in increasing order of ability test scores				
2.2 Linear systematic sampling				
Largest sample percentage	11.1	11.1	14.3	18.2
Smallest sample percentage	10.7	10.5	7.1	9.1
2.3 Linear systematic sampling with end corrections				
Largest sample percentage	11.1	11.1	14.3	18.2
Smallest sample percentage	10.7	5.6	0.0	0.0
2.4 Balanced systematic sampling				
Largest sample percentage	10.7	11.1	14.3	16.7
Smallest sample percentage	10.7	11.1	7.1	8.3
2.5 Centrally located samples				
Largest sample percentage	11.1	11.1	14.3	9.1
Smallest sample percentage	11.1	11.1	7.1	9.1
3. Linear systematic sampling applied to population arranged in increasing order of ability test scores with order reversed in alternate strata				
Largest sample percentage	11.1	11.1	15.4	18.2
Smallest sample percentage	10.7	5.6	7.1	9.1

TABLE 6.19

Largest and Smallest Estimates of Percentages of Sixth-Grade Pupils in School 21 in Midcity with Reading Achievement Scores at Least 1 Grade Equivalent Unit Below National Norms

| | Population Proportion = 13.6% Sampling Interval and Sampling Fraction | | | |
Sampling Procedure	2 50%	3 33%	4 25%	5 20%
1. Linear systematic sampling applied to the population arranged in alphabetic order				
Largest sample percentage	19.6	17.7	24.0	20.0
Smallest sample percentage	7.7	5.7	7.7	9.5
2. Procedures applied to the population arranged in increasing order of ability test scores				
2.2 Linear systematic sampling				
Largest sample percentage	15.7	14.7	16.0	19.1
Smallest sample percentage	11.5	11.4	11.5	5.0
2.3 Linear systematic sampling with end corrections				
Largest sample percentage	13.7	14.7	16.0	19.1
Smallest sample percentage	11.5	11.4	7.7	5.0
2.4 Balanced systematic sampling				
Largest sample percentage	17.3	14.7	15.4	20.0
Smallest sample percentage	9.6	11.8	11.5	10.0
2.5 Centrally located samples				
Largest sample percentage	15.7	14.7	11.5	14.3
Smallest sample percentage	11.8	14.7	11.5	14.3
3. Linear systematic sampling applied to the population arranged in increasing order of ability test scores with order reversed in alternate strata				
Largest sample percentage	17.7	14.7	16.0	20.0
Smallest sample percentage	9.6	11.8	11.5	9.5

narrower ranges of sample percentages using a sampling fraction of 33 percent. Centrally located sampling resulted in sample percentages that were 2 percent too low and about 1 percent too high for sampling fractions of 25 and 20 percent, respectively.

Summary of Results for Schools. Some systematic sampling procedures provided acceptably narrow ranges of sample percentages for Schools 19 and 21, using sampling fractions smaller than 33 percent. However, these results were not consistent for all four schools. Using a sampling fraction of 50 percent, some procedures resulted in marginally acceptable ranges of sample percentages for all schools. These results are consistent with findings on estimation of mean sixth-grade achievements for these schools. When estimating achievement means, sampling fractions of 50 percent resulted in tolerable mean square errors, and smaller sampling fractions produced erratic results.

Summary

Because the behavior of systematic sampling procedures depends so much on the arrangement of elements in the frame being sampled, it is virtually impossible to develop accurate theory-based predictions of their performance. Consequently, empirical results for systematic sampling are of far greater interest than are those for other sampling methods. Our summary of this chapter therefore contains more discussion of empirical findings than of theoretical issues.

EMPIRICAL RESULTS

For the populations of pupils investigated in Examples 6.1 and 6.3, systematic sampling procedures were frequently more efficient than simple random sampling or stratified sampling with one element per stratum. When samples were drawn from populations arranged in alphabetic order, systematic sampling offered convenience and substantial gains in efficiency (relative efficiencies above 300 percent compared to simple random sampling).

When pupils were arranged in increasing order of their scores on the Lorge-Thorndike ability test, stratified sampling with one element per stratum was more efficient than linear systematic sampling for some sampling fractions. It was also more efficient than many other systematic sampling procedures. Using a sampling fraction of 10 percent to estimate the mean achievement of sixth-grade pupils in a medium-sized city, both linear systematic sampling and stratified sampling with one element per stratum provided excellent precision.

When pupils were arranged in alphabetic order, neither simple random sampling nor stratified sampling with one element per stratum provided acceptably precise estimates of the mean achievement of pupils in each of three small elementary schools. However, these methods did provide acceptable precision for a large elementary school. The sampling fractions investigated varied between 20 and 50 percent. Several systematic sampling methods provided precise estimates of pupils' mean achievement, provided the sampling frame was arranged in increasing order of the pupils' ability test scores. Some systematic sampling procedures provided acceptable precision using sampling fractions as small as 20 percent, but the results were inconsistent and may be unreliable.

Results for estimation of percentages of low achievers paralleled those for estimation of mean achievement. With a sampling fraction of 10 percent, several systematic sampling methods produced acceptably precise estimates of the percentage of pupils in a school district with reading achievement scores at least 1 grade equivalent unit below the national norm. When the percentage of low achievers was estimated for an individual school, a sampling fraction of 50 percent was needed to obtain acceptable levels of precision. Results were mixed when smaller sampling fractions were used, with some systematic sampling procedures providing acceptably precise estimates. These findings for smaller sampling fractions cannot be considered reliable.

THEORETICAL CONSIDERATIONS

Acceptable methods of estimating sample size requirements are unavailable for systematic sampling procedures. The relationship between sample size and estimation precision is generally unpredictable and frequently unstable. Several methods for estimating the precision of estimators have been proposed in the literature and were reviewed in Chapter 5. Most of these methods have shown a large degree of bias and poor precision when applied to a variety of natural populations. Their utility in surveys in education and the social sciences is subject to empirical investigation.

A computer program for systematic sampling (SYSAMP-I) was designed to facilitate evaluation of the efficiencies of systematic sampling procedures relative to each other and relative to simple random sampling and stratified sampling. The program can be used to compute estimates of population means, estimates of the proportion of elements in a population with values below a specified cutoff, mean square errors of estimators, and relative efficiencies. These results should aid in the selection of sampling procedures for studies in education and the social sciences. The SYSAMP-I computer program is described in detail in Appendix C.

Exercises

In Chapter 5, we presented a set of exercises that were based on data collected by two psychologists who were conducting research on cognitive functioning in young children. Their research was described as follows:

The psychologists devised a maze task for use in their work. The task required each child to begin at an indicated starting position and then draw a continuous path through the maze, avoiding dead ends, until he or she arrived at an indicated ending position. The psychologists recorded the time (in seconds) needed by each child to complete the task. Since the maze was small and simple, the average child finished in a little over 60 seconds.

In Example 6.3, we used some of the data on children's times to complete the maze task to estimate a population total. We employed a variety of systematic sampling procedures. In that example, we assumed that the psychologists had access to scores on the Performance Scale of the Wechsler Intelligence Scale for Children (WISC) for all children in the population and that they had arranged the sampling frame of children in increasing order of their WISC Performance scores. We will make the same assumption here, for the entire population of 100 children, and use the resulting sampling frame to apply and evaluate a number of systematic sampling procedures.

Times (in seconds) needed to complete the maze are shown in the following table for the population of 100 children, arranged in increasing order of their WISC Performance scores:

Child No.	Time	Child No.	Time	Child No.	Time
1	13	35	56	69	91
2	21	36	41	70	69
3	16	37	46	71	67
4	21	38	37	72	76
5	41	39	66	73	80
6	32	40	84	74	84
7	40	41	73	75	71
8	40	42	49	76	66
9	32	43	55	77	80
10	40	44	52	78	67
11	41	45	59	79	51
12	41	46	75	80	57
13	30	47	55	81	91
14	30	48	44	82	80
15	34	49	55	83	83
16	32	50	57	84	96
17	40	51	60	85	81
18	43	52	42	86	84
19	41	53	57	87	84
20	42	54	78	88	86
21	38	55	67	89	83
22	54	56	55	90	80
23	39	57	65	91	93
24	44	58	71	92	87
25	47	59	67	93	93
26	54	60	80	94	76
27	37	61	38	95	95
28	68	62	49	96	96
29	43	63	59	97	98
30	47	64	67	98	96
31	54	65	71	99	104
32	51	66	53	100	107
33	51	67	76		
34	45	68	62		

The problems to be solved using the data in this table are described below.

1. Using a sampling interval of 5, select *all* linear systematic samples from the population of size 100. For *each* linear systematic sample, estimate the mean time the population of 100 children requires to solve the maze.

2. Use the results of Exercise 1 to calculate the mean square error (MSE) of the sample mean time children require to solve the maze.

3. Using your results from Exercise 2 and the variance of the means of simple random samples that you calculated for this population in Exercise 5 of Chapter 5, calculate the efficiency of linear systematic sampling from the ordered sampling frame relative to simple random sampling. (Recall that the variance of the means of simple random samples does not depend on the order of elements in a sampling frame. Thus the value you calculated for the population of children in their natural order is the same as the variance for this sampling frame.)

4. Since the sampling interval k is an odd number, there is only one centrally located sample. Select that sample and use the resulting data to estimate the population mean. Compare this value with the actual population mean of 60.15 seconds. Does centrally located sampling provide a good estimate of the population mean?

5. Again assuming a sampling interval of 5 and a resulting sample size of 20, select all balanced systematic samples. Compute an estimate of the population mean from the data provided by each balanced systematic sample.

6. Using the results of Exercise 5, calculate the mean square error of the balanced systematic sample means.

7. Using the results of Exercise 6 and the variance of the means of simple random samples that you calculated in Exercise 5 of Chapter 5, compute the efficiency of balanced systematic sampling relative to simple random sampling. Do you think balanced systematic sampling is efficient in this application?

Single-Stage Cluster Sampling: Procedures, Estimation Formulas, and Applications

Basic Concepts and Applications of Cluster Sampling

DEFINITION OF CLUSTER SAMPLING

The sampling methods described in preceding chapters used sampling units that were basic elements of a population, for example, individual children. In cluster sampling, sampling units are not basic elements but are groups or collections of elements. These groups or collections of elements are termed *clusters*.

In most applications of cluster sampling, the clusters used are naturally occurring groups. In surveys of consumer behavior, for example, homes are frequently used as sampling units. In educational research studies where pupils are the basic elements of interest, classrooms and schools form convenient clusters of pupils. Other possibilities for clusters are artificially contrived groups such as voters living in specified areas of a city or groups of pupils who have surnames beginning with the same letter. In most surveys, naturally occurring clusters afford far greater administrative convenience than would these contrived groups. Pupils can be identified readily by classroom and school, and individual consumers can easily be assembled for interviewing on a household-by-household basis.

SOME APPLICATIONS OF CLUSTER SAMPLING IN EDUCATION

Because public schooling in the United States is so highly organized, cluster sampling can be applied readily in many educational research studies. School districts form convenient clusters for studies in which the elements of interest are school administrators, such as principals or district-level personnel. Schools form readily accessible and easily identifiable clusters when estimates of various parameters are needed for populations of teachers or pupils. When estimates of population parameters are desired for pupils in the primary grades, it will often be convenient to sample classrooms, since most such pupils are taught in single classrooms having only one teacher throughout the school day.

Research studies in higher education might also make use of cluster sampling. Institutions, such as colleges and universities, or individual departments and professional schools within such institutions, could be used as clusters in studies where parameters are to be estimated for populations of students, faculty, or administrators.

The following are some examples of the ways cluster sampling might be used in educational research studies.

ESTIMATING THE MEAN YEARS OF EXPERIENCE OF TEACHERS THROUGHOUT A STATE

Suppose that the average number of years of teaching experience was to be estimated for elementary school teachers in public schools throughout a state. Elementary schools could be defined operationally as any school that (1) had pupils enrolled in at least one of the grades kindergarten through grade 6, or (2) had pupils who would, by virtue of their age, be enrolled in one of the grades kindergarten through grade 6 if the school separated pupils by grade level.

In many states, the department of education would not have an up-to-date, statewide list of all elementary school teachers. Without such a list, it would not be possible to select a sample of teachers through any of the sampling procedures described in earlier chapters. Since all state departments of education are likely to have a statewide list of elementary schools, the use of single-stage cluster sampling, with schools as clusters, would be feasible.

In this application, a survey researcher could select a simple random sample of schools from the statewide sampling frame of elementary schools. Once a sample of elementary schools had been drawn, every teacher in those schools would be asked to report the number of years he or she had been teaching. Of course it would be necessary to provide a clear, operational definition of "number of years of teaching experience." Although the term seems unambiguous, without a precise definition, the survey would result in inconsistent data. For example, some teachers would include only full-time experience, while others might include part-time work. Some teachers might include the student teaching that was a part of their training, while others would not.

Collection of data from all the elements in each sampled cluster (in this case, all the teachers in each sampled elementary school) is an essential feature of single-stage cluster sampling. In more complex cluster sampling procedures, such as two-stage cluster sampling, elements are sampled from each sampled cluster.

ESTIMATING THE MEAN ACHIEVEMENT OF PUPILS IN A SCHOOL DISTRICT

We have perhaps described this use of sampling to the point of tedium, but it lends itself very well to single-stage cluster sampling. We will therefore consider it in some detail later in this chapter. Suppose the mean achievement of all fourth-grade pupils in a school district is to be estimated. One approach to this problem would be to use classrooms as clusters. The sampling frame would consist of a list of all fourth-grade class sections in the school district. If the district contains nongraded schools, the sampling frame could be defined to include classrooms containing pupils who were of the same age as most fourth-graders. A probability sample of classrooms would be selected from the frame, perhaps using simple random sampling or stratified sampling, with schools as strata. Then the pupils in all sampled classrooms would be tested. An estimator that was appropriate to the sampling design would be applied to the resulting test scores in order to estimate the mean achievement of all fourth-graders in the school district.

Using classrooms as clusters would certainly be convenient. First, it is likely that a sampling frame of classrooms containing fourth-grade pupils would be available in virtually all school districts. Construction of a special sampling frame would therefore be avoided. Second, classrooms are naturally occurring clusters, so all pupils in a given classroom would be accessible for data collection in the same place and at the same time. Third, since group achievement testing disrupts the normal instructional program in a classroom, the least disruptive approach would be to test every pupil in some classrooms and avoid testing in others.

ESTIMATING THE PROPORTION OF SCHOOL PRINCIPALS WITH DOCTORAL DEGREES

Suppose that researchers in a state department of education wanted to estimate the proportion of school principals in the state who held earned doctoral degrees. A cluster-sampling approach, using school districts as clusters, would be particularly effective in a state such as California or New York, where there are several thousand school districts.

A list of all school districts in the state would constitute the sampling frame. To use single-stage cluster sampling, a survey researcher would draw a probability sample of school districts from the frame and then survey all school principals in each sampled district. The statewide proportion of school principals with earned doctoral degrees would be estimated from the sample data by applying a formula that was consistent with the method used to sample school districts. Although simple random sampling could be used to select districts, other approaches, such as linear systematic sampling or stratified sampling of districts using enrollment categories as strata, would be possible.

SOME APPLICATIONS OF CLUSTER SAMPLING IN SOCIAL SCIENCE RESEARCH

Survey research in the social sciences, like survey research in education, often involves working with well-defined institutions. Examples include sampling of clinics, care centers, social welfare agencies, laboratories, and political units such as cities or counties. These kinds of institutions could be used as clusters in research studies where parameters are to be estimated for populations of clients, patients, or citizens. In the case of laboratories, the populations might consist of drug samples, animals, or tissue samples. In single-stage cluster sampling designs, a probability sample of institutions would be selected, and data would then be collected from (or for) all elements within the sampled institutions.

AUDITING OF WELFARE PROGRAMS

A persistent problem in public welfare programs is ensuring that all recipients of benefits are, under the relevant laws, eligible to receive benefits. In one analysis of this problem, federal government agencies have studied the eligibility of Medicare recipients in nursing homes throughout the nation. The study design involved the selection of a probability sample of nursing homes and an assessment of the eligibility of all Medicare recipients within the sampled nursing homes. The sampling design for this study was thus single-stage cluster sampling with nursing homes used as clusters. Data were collected from all elements, that is, Medicare recipients, within sampled clusters. Since state governments license nursing homes, a sampling frame of nursing homes was readily available for each state. It would have been impossible to sample individual Medicare recipients in nursing homes because no government agency would have an appropriate list of these residents.

SURVEY ON EMPLOYMENT

The U.S. Bureau of the Census conducts a monthly survey to secure data on employment and unemployment in the United States. The final stage in the Bureau's sampling plan involves selection of a probability sample of households. Interviewers collect data from all members of the labor force within sampled households. Thus households are used in clusters in this survey, and data are collected from all elements (members of the labor force) within sampled clusters. This is a single-stage cluster sampling design. Census records, city directories, and other readily available sources are used to develop a sampling frame of households. Time and cost constraints would make it totally infeasible to develop an up-to-date sampling frame of individual members of the labor force. Cluster sampling of households is thus the only practical alternative in this study.

AGE DISTRIBUTION OF MARRIED COUPLES

In a study to determine whether the age at which couples are getting married in the United States is changing, it would not be feasible to sample couples directly. A nationwide list of newlyweds is simply not available. A logical approach would be to select a probability sample of marriage license bureaus from an entire state or from the entire nation (depending on the purpose of the study) and then to record, for each sampled marriage bureau, the ages of all couples who applied for marriage licenses within a specified time period. This is an example of single-stage cluster sampling in which marriage license bureaus serve as clusters, and couples are the elements of interest.

SOME ADVANTAGES AND DISADVANTAGES OF CLUSTER SAMPLING

We have alluded to some of the advantages of cluster sampling in the examples discussed above. In many cases, cluster sampling is the only feasible approach because sampling frames of individual elements are not available. In other cases, although it might be possible to develop a sampling frame of elements, the resulting benefits would not warrant the time required or the cost.

In interview surveys, the use of cluster sampling sometimes affords significant economies. Very often, the major data-collection expense is for travel to scattered interview sites. When cluster sampling is used, groups of potential respondents are usually in close proximity, and travel costs are reduced.

The administrative convenience afforded by cluster sampling has already been described in our discussion of applications in educational surveys. Because data are collected in some clusters and not in others, disruption of normal institutional routines is minimized. In addition, administration of the survey (particularly for mail surveys) is simplified when distribution of survey materials is restricted to a sample of sites or institutional units.

The relative simplicity and convenience of cluster sampling carry a price. Often, the elements in a cluster are similar in the characteristic that is measured by the sampling variable. For example, schools generally draw pupils from economically homogeneous neighborhoods, and many schools group students on the basis of their past achievement and ability test scores. Therefore, the current achievement test scores of pupils in the same classroom or school will not vary as greatly as will scores in an entire school district or state. This similarity of elements within a cluster means that more elements must be sampled to achieve a given level of estimation precision. Thus in many applications, cluster sampling will be statistically less efficient than simple random sampling. The more similar the elements within a cluster, the lower will be the efficiency of cluster sampling. In the extreme case, where elements within a cluster are identical on the characteristic measured by the sampling variable, sampling a cluster is no more informative than sampling one element; the additional elements in that cluster provide no additional information.

The statistical efficiency of cluster sampling will vary widely across applications in education and the social sciences. The examples discussed later in this chapter provide information on the behavior of single-stage cluster sampling in school testing applications. However, the results shown in these examples cannot be assumed to generalize to other settings.

Another unfortunate characteristic of cluster sampling is the relative complexity of the associated estimation formulas. Although estimation of means, totals, and proportions is not difficult, the formulas for calculating estimator variances and the procedures for estimating these variances are complex. Computer programs for computation of single-stage cluster sampling estimates and variances, contained in Appendixes D-1 and D-2, should reduce the tedium of computation.

Cluster Sampling Methods and Estimators

In most applications of cluster sampling in education and the behavioral sciences, the clusters used will be of unequal sizes. For example, enrollments vary from classroom to classroom and from school to school, the numbers of labor-force members vary among households, and the numbers of Medicare recipients vary among nursing homes. Using any of these institutions as clusters will therefore result in variable cluster sizes.

When clusters vary in size, the formulas for estimating population means, totals, proportions, and estimator variances are considerably more complicated than are those used with clusters that are equal in size. With unequal cluster sizes, not only the sample statistic of interest but the number of elements in the sample varies randomly; for example, the proportion of ineligible welfare recipients and the total number of welfare recipients in the sample vary randomly.

Several single-stage cluster sampling and estimation methods have been proposed for situations in which the sizes of clusters vary. A review of relevant literature is provided by Murthy (1967, pp. 293–316), and the most important formulas are given by Som (1973, pp. 59–80). This chapter contains descriptions of four methods of estimation and three methods of sample selection. Procedures for selection of clusters and formulas for computing estimates are presented in this section. The next section contains empirical results derived by applying the various methods of sampling and estimation to a school testing program. These results will facilitate a comparison of single-stage cluster sampling and estimation methods.

The sampling and estimation methods discussed in this section are random sampling of clusters with unbiased estimation (RSC unbiased), random sampling of clusters with ratio estimation (RSC ratio), sampling of clusters with probabilities proportional to cluster sizes (PPS sampling) with unbiased PPS estimation, and sampling of clusters with probabilities proportional to the values of an auxiliary variable (PPES sampling) with unbiased PPES estimation.

Table 7.1 summarizes some important properties of these sampling procedures and estimators. These properties are discussed at length in the balance of this section.

RANDOM SAMPLING OF CLUSTERS WITH UNBIASED ESTIMATION (RSC UNBIASED)

ESTIMATION OF A POPULATION MEAN

Suppose that a population contains N clusters (such as households) and that the ith cluster contains M_i elements (such as wage earners). If n clusters are selected from a sampling frame through simple random sampling, an unbiased and consistent estimator of the average per element (such as the average income per wage earner) is given by

$$\hat{\bar{Y}} = \frac{N}{nM_0} \sum_{i=1}^{n} M_i \bar{y}_i \qquad (7.1)$$

where M_0 denotes the total number of elements in the population and is equal to

$$M_0 = \sum_{i=1}^{N} M_i$$

$\bar{y}_i$ denotes the average of the sampling variable for elements in the ith cluster and is equal to

$$\bar{y}_i = \frac{1}{M_i} \sum_{j=1}^{M_i} y_{ij} \qquad (7.2)$$

with y_{ij} denoting the value of the sampling variable for the jth element in the ith cluster. A hypothetical example should help to make the notation clear.

Suppose that a city contains five nursing homes and that researchers want to estimate the average age of nursing home residents in the city. Nursing homes are to be used as clusters.

Assume that the number of residents and the average age of residents in each nursing home are as given in Table 7.2. Using the notation of Equations

TABLE 7.1
Properties of Single-Stage Cluster Sampling Methods and Estimation Procedures

Sampling Procedure	Estimation Method	Bias	Estimator Variance	Consistency	Estimation of Estimator Variance	Sample Size Required
Simple random sampling of clusters (RSC)	Unbiased estimation (7.1)[a]	Unbiased	Frequently large (7.5)	Consistent	(7.7)	(7.8)
Simple random sampling of clusters (RSC)	Ratio estimation (7.9)	Biased, but bias is usually insignificant (7.12)	Often small (7.10)	Consistent	(7.11)	(7.13)
Sample clusters with probabilities proportional to their sizes (PPS)	Unbiased PPS (7.15)	Unbiased	Moderate to small (7.16)	Not consistent, but a problem only for large sampling fractions	(7.17)	(7.18)
Sampling clusters with probabilities proportional to values of an auxiliary variable (PPES)	Unbiased PPES (7.19)	Unbiased	Often small (7.20)	Not consistent, but a problem only for large sampling fractions	(7.21)	(7.22)

[a] Numbers in parentheses refer to equation numbers in the text.

TABLE 7.2

Hypothetical Numbers of Residents and Average Ages of Residents in Five Nursing Homes

Nursing Home	Number of Residents	Average Age of Residents
1	51	76.2
2	34	81.4
3	104	79.5
4	76	72.0
5	61	78.9
Total	326	

(7.1) and (7.2), the number of residents in the first nursing home is $M_1 = 51$, the number in the second nursing home is $M_2 = 34$, etc. The average age of residents in the first nursing home is $\bar{y}_1 = 76.2$, the average age of residents in the second home is $\bar{y}_2 = 81.4$, etc.

Suppose that the average age of nursing home residents in the city is to be estimated by collecting data in only two homes ($n = 2$). Suppose further that nursing homes numbered 1 and 4 are selected through simple random sampling. The average age of nursing home residents for the city can then be estimated by applying Equation (7.1). M_0, the total number of elements (that is, nursing home residents) in the population, is found by summing the numbers of residents in each of the five homes. The total is $M_0 = 326$. Using the tabulated data for Nursing Homes 1 and 4 in Equation (7.1), an estimate of the average age of nursing home residents for the entire city is given by

$$\hat{\bar{Y}} = \frac{N}{nM_0}(M_1\bar{y}_1 + M_4\bar{y}_4)$$

$$= \frac{5}{2(326)}[51(76.2) + 76(72.0)]$$

$$= 71.8 \text{ years}$$

The actual average age of nursing home residents in the city can also be calculated from the data given in Table 7.2. The total of the ages of residents in the ith nursing home is equal to the product of their average age and the number of residents in the home. That is,

$$y_i = M_i\bar{y}_i \tag{7.3}$$

where y_i denotes the total of the ages of residents in the ith nursing home. The total of the ages of all nursing home residents in the city is given by

$$Y = \sum_{i=1}^{N} y_i$$

and the actual average age of nursing home residents in the city is given by

$$\bar{\bar{Y}} = \frac{1}{M_0} \sum_{i=1}^{N} y_i = \frac{1}{M_0} \sum_{i=1}^{N} M_i \bar{y}_i \tag{7.4}$$

which, for the data in Table 7.2, is 77.3 years. Using a sample of two nursing homes, the citywide average age of nursing home residents was underestimated by 5.5 years, which is a substantial error.

ESTIMATOR VARIANCE

The variance of the unbiased estimator [Equation (7.1)] is equal to

$$V(\hat{\bar{Y}}) = \frac{N^2(1-f)}{(N-1)nM_0^2} \sum_{i=1}^{N} (y_i - \bar{Y})^2 \tag{7.5}$$

where

f = sampling fraction for clusters, n/N

y_i = total of the sampling variable over all elements in the ith cluster

$$\bar{Y} = \frac{1}{N} \sum_{i=1}^{N} \sum_{j=1}^{M_j} y_{ij} \tag{7.6}$$

= population mean of the sampling variable, *per cluster*

Equation (7.5) is similar to the expressions for the variance of a sample mean formed through simple random sampling [Equations (3.7) and (3.8)]. The multiplier N^2/M_0^2 is necessary here since the variance sought is that of the estimated mean per element rather than that of the estimated mean per cluster.

In many applications, and particularly when cluster sizes are highly variable, the variance of the RSC-unbiased estimator will be unacceptably large. Detailed consideration of Equation (7.5) will reveal the cause of these large estimator variances. If cluster sizes vary greatly but the average of the sampling variable is about the same for each cluster, the totals for clusters will vary greatly. Thus, the sum

$$\sum_{i=1}^{N} (y_i - \bar{Y})^2$$

will be large.

Empirical results presented in the next section show that unbiased estimation of achievement test averages, using schools as clusters, leads to impractically large estimator variances. These large variances are due to large variations in school enrollments and are consistent with the analysis of Equation (7.5) given above.

An unbiased estimator of the variance given in Equation (7.5) is

$$v(\hat{\bar{Y}}) = \frac{N^2(1-f)}{(n-1)nM_0^2} \sum_{i=1}^{n} (y_i - \bar{y})^2 \tag{7.7}$$

where $\bar{y}$ is the average of the totals in sampled clusters and equals

$$\frac{1}{n} \sum_{i=1}^{n} y_i$$

Like the unbiased estimator of the population mean, this variance estimator will often exhibit large fluctuations from sample to sample.

DETERMINATION OF SAMPLE SIZE

A formula for determining the required sample size can be derived from Equation (7.5). To estimate a population mean per element within an error limit ε with $100(1 - \alpha)$ percent confidence, the approximate number of clusters that would have to be sampled is given by

$$n = \frac{(NtS/\varepsilon M_0)^2}{1 + (1/N)(NtS/\varepsilon M_0)^2} \tag{7.8}$$

where
    ε = error limit, which may be exceeded only with probability α
    t = value on the abscissa of a normal distribution with mean 0 and variance of 1; $100\,\alpha/2$ percent of the distribution lies to the right of t

$$S^2 = \frac{1}{(N-1)} \sum_{i=1}^{N} (y_i - \bar{Y})^2$$

       = variance among *cluster totals*
  $\bar{Y}$ = population mean *per cluster*

Because simple random sampling of clusters with unbiased estimation is often statistically inefficient, alternative methods of sampling and estimation have been developed. Three alternatives are presented below.

RANDOM SAMPLING OF CLUSTERS WITH RATIO ESTIMATION (RSC RATIO)

ESTIMATION OF A POPULATION MEAN

In the RSC-ratio method, clusters are chosen through simple random sampling, but a ratio estimator is used. Ratio estimators are quotients of two variables, each of which varies randomly from element to element or, in the case of cluster sampling, from cluster to cluster. Ratio estimation is frequently an efficient technique. Although it can be used with a variety of sampling procedures, it is considered here only with simple random sampling of clusters.

When n clusters are selected at random from a population of N, the RSC-ratio estimator of the population mean per element is given by

$$\hat{\bar{Y}}_R = \frac{\sum\limits_{i=1}^{n} y_i}{\sum} \tag{7.9}$$

All terms have been defined previously.

The RSC-ratio estimator equals the sum of the cluster totals divided by the sum of the cluster sizes, where the sums range over all clusters in the sample. The sum in the numerator and the sum in the denominator of Equation (7.9) both vary from sample to sample.

Ratio estimators are frequently biased, and their variances can only be approximated. However, the degree of bias usually is negligible for sample sizes likely to be encountered in practice. The ratio estimator, like the unbiased estimator, is consistent. Consistency is of practical interest when large sampling fractions are required to secure adequate precision.

ESTIMATOR VARIANCE

The variance of the RSC-ratio estimator of the population mean is approximated by

$$V(\hat{\bar{Y}}_R) \cong \frac{N^2(1 - f)}{(N - 1)M_0^2 n} \sum_{i=1}^{N} M_i^2 (\bar{y}_i - \bar{\bar{Y}})^2 \qquad (7.10)$$

where

$\bar{\bar{Y}}$ = population mean *per element*
N = number of clusters in the population
n = number of clusters in the sample
M_i = number of elements in the ith cluster
f = sampling fraction for *clusters, n/N*
M_0 = number of elements in the population

The variance of the RSC-ratio estimator is frequently much smaller than the variance of the unbiased estimator. If the mean per element within clusters, $\bar{y}_i$, does not vary greatly, the squared differences $(\bar{y}_i - \bar{\bar{Y}})^2$ will be small, causing the term

$$\sum_{i=1}^{N} M_i^2 (\bar{y}_i - \bar{\bar{Y}})^2$$

in Equation (7.10) to be small. The empirical results presented in the next section show that the variance of the RSC-ratio estimator was consistently smaller than the variance of the RSC-unbiased estimator when both were applied to the same population.

An estimator of the variance of the RSC-ratio estimator is given by

$$v(\hat{\bar{Y}}_R) = \frac{N^2(1 - f)}{(n - 1)nM_0^2} \sum_{i=1}^{n} M_i^2 (\bar{y}_i - \hat{\bar{Y}}_R)^2 \qquad (7.11)$$

As is the RSC estimator of the population mean, this estimator is biased. The bias of the estimated variance is inversely proportional to the sample size n and is a serious problem only for small sample sizes.

ESTIMATOR BIAS

The bias of the RSC-ratio estimator is inversely proportional to the number of clusters sampled. An approximation to the bias is given by Lord (1959, p. 252) as

$$\text{Bias} \cong \left(\frac{S_M}{\bar{M}}\right)^3 \sqrt{V(\bar{y}_i)} \frac{\rho}{n} \tag{7.12}$$

where

$$S_M = \frac{\sqrt{\sum_{i=1}^{N} (M_i - \bar{M})^2}}{N - 1}$$

= standard deviation of the number of elements per cluster
$V(\bar{y}_i)$ = population variance, among clusters, of the average per element

$$= \frac{1}{N} \sum_{i=1}^{N} (\bar{y}_i - \bar{\bar{Y}})^2$$

$\bar{M} = \dfrac{1}{N} \sum\limits_{i=1}^{N} M_i$ = average number of elements per cluster
(average cluster size)

ρ = correlation coefficient between the mean per element within clusters, $\bar{y}_i$, and the number of elements within clusters, M_i

The RSC-ratio estimator is unbiased if the mean per element within clusters is uncorrelated with cluster size [Equation (7.12)].

When applied to the problem of estimating the mean achievement of pupils in a school district, with schools used as clusters, the RSC-ratio estimator will therefore be unbiased if the mean achievement of pupils within schools is uncorrelated with school enrollment. Mollenkopf (1956) found correlations between mean achievement and school size equal to 0.00 for ninth-grade pupils in 99 schools and 0.01 for twelfth-grade pupils in 106 schools. Empirical results presented in the next section are consistent with these small correlations.

DETERMINATION OF SAMPLE SIZE

Using simple random sampling and ratio estimation, the approximate number of clusters that must be sampled in order to estimate the population mean within an error limit ε, with $100(1 - \alpha)$ percent confidence, is given by

$$n = \frac{S_{My}^2 (Nt/\varepsilon M_0)^2}{1 + (1/N)S_{My}^2 (Nt/\varepsilon M_0)^2} \tag{7.13}$$

where

$$S_{My}^2 = \frac{1}{(N - 1)} \sum_{i=1}^{N} M_i^2 (\bar{y}_i - \bar{\bar{Y}})^2 \tag{7.14}$$

Other terms are as defined above.

SAMPLING WITH PROBABILITIES PROPORTIONAL TO CLUSTER SIZES (PPS)

The sampling procedures considered to this point assume that samples have been selected "without replacement." When *sampling without replacement,* a sampling unit is removed from the population once it has been selected. The population from which the remaining sampling units are drawn consists of the original population less those units already selected. Thus, if a sampling unit appears in any sample, it appears only once.

The alternative sampling procedure described in this section returns a sampling unit to the population after its value on the sampling variable has been recorded. Thus, the same sampling unit could be selected more than once. Procedures of this type are termed *sampling with replacement.* For example, if sampling with replacement were used to select two nursing homes from five, the sample could consist of two different homes (say, Nursing Homes 1 and 4) or a single home selected twice (say, two selections of Nursing Home 3).

The sampling method considered in this section, in addition to being "with replacement," makes the probability of selecting a cluster proportional to its size, that is, makes the probability proportional to the number of elements it contains. A larger cluster thus has a greater probability of selection than has a smaller cluster. An example may clarify this concept.

Table 7.2 lists the number of residents in each of five nursing homes. If nursing homes are used as clusters to estimate the average age of residents, the numbers of residents are the cluster sizes. On any given draw, the probability of selecting a nursing home can be made proportional to the number of residents it contains, by sampling with replacement and by defining the selection probabilities as follows. Make the probability of selecting the ith nursing home on any given draw

$$\text{Prob \{selecting } i\text{th home\}} = \frac{\text{number of residents in } i\text{th nursing home}}{\text{number of residents in all nursing homes}}$$

Thus for the data in Table 7.2, the probability of selecting Nursing Home 1 is $51/326 = 0.16$; the probability of selecting Nursing Home 2 is $34/326 = 0.10$; and so on. As expected, the probabilities of selection for all five nursing homes sum to 1.

More generally, the probability of selecting a cluster on any given draw can be made proportional to its size by defining the probability as

$$\text{Prob \{selecting a cluster\}} = \frac{\text{number of elements in cluster}}{\text{number of elements in population}}$$

and sampling clusters with replacement.

ESTIMATION OF A POPULATION MEAN

If $\bar{y}_i$ denotes the mean of the sampling variable in the ith cluster, an unbiased estimate of the population mean is given by the sample average of cluster means:

$$\hat{\bar{Y}}_{pps} = \frac{1}{n} \sum_{i=1}^{n} \bar{y}_i \qquad (7.15)$$

The subscript pps denotes that clusters have been selected with probabilities proportional to their sizes.

ESTIMATOR VARIANCE

The PPS estimator, although unbiased, is not consistent. If the number of clusters, n, in the sample is set equal to the number of clusters, N, in the population, the variance of the estimator will not equal zero. This can be seen from the following formula for estimator variance:

$$V(\hat{\bar{Y}}_{pps}) = \frac{1}{M_0 n} \sum_{i=1}^{N} M_i(\bar{y}_i - \bar{\bar{Y}})^2 \qquad (7.16)$$

As long as the clusters differ in their mean per element, $\bar{y}_i$, most of the terms $(\bar{y}_i - \bar{\bar{Y}})^2$ will be greater than zero, and estimator variance will be positive. Since the variance becomes smaller as n increases, lack of consistency does not impose a serious problem unless precision requirements are so stringent that a large sampling fraction must be used.

As will be seen from the empirical results in the next section, the inconsistency of the PPS estimator (and a fourth estimator to be described below) makes the choice of estimation method depend upon the required sampling fraction n/N. For large sampling fractions, the RSC-unbiased estimator and the RSC-ratio estimator may afford greater precision than the PPS estimator, whereas the opposite may be true for small sampling fractions.

An unbiased estimator of the PPS estimator variance is given by

$$v(\hat{\bar{Y}}_{pps}) = \frac{1}{(n-1)n} \sum_{i=1}^{n} (\bar{y}_i - \hat{\bar{Y}}_{pps})^2 \qquad (7.17)$$

The empirical results presented in the next section show this variance estimator to be considerably more stable than the variance estimator for the RSC-unbiased method when applied to achievement test data from a medium-sized city.

SELECTION OF CLUSTERS

The selection of clusters with probabilities proportional to their sizes can be quite simple. One method is to list, in no particular order, all clusters in the population, along with the "cumulative size" of each cluster. The cumulative size of a cluster equals its own size plus the sizes of all clusters that precede it in the list. Once the cumulative sizes of clusters have been tabulated, a random number, selected from a table, is multiplied by a factor proportional to the number of elements in the population. The ith cluster is selected if the random number times the factor is between the cumulative sizes of the ith cluster and the $(i-1)$st cluster. A different random number is chosen for selection of the next cluster, and the process is repeated until the desired number of clusters has been sampled. An example should clarify the procedure.

In Table 7.2, hypothetical numbers of residents were listed for five nursing homes. The cumulative numbers of residents for these nursing homes are shown in Table 7.3.

Two of the five nursing homes may be selected with probabilities proportional to the numbers of residents they contain, as follows. A three-digit number is read from any three adjacent columns of a random number table (see Appendix A). The number will lie between 000 and 999. The selected random number is multiplied by 0.326 (the number of elements in the population, 326, divided by 1000). This gives a random number between 000 and 326. Suppose that the number 400 is read from a random number table; when multiplied by 0.326, the resulting value is 130.4. If 130.4 is compared to the cumulative numbers of residents listed in Table 7.3, it is seen to be greater than the cumulative number of residents of Nursing Home 2 and less than the cumulative number of residents of Nursing Home 3. Nursing Home 3 is thus selected. To choose the next home, the process is repeated. Since nursing homes are being sampled with replacement, Nursing Home 3 might be selected a second time. Assume that the next random number is 104. This number is multiplied by 0.326, and the resulting value is 33.9. This value is less than the cumulative number of residents of Nursing Home 1. Hence, Nursing Home 1 is chosen. If more than two nursing homes were to be sampled, the process would be repeated.

DETERMINATION OF SAMPLE SIZES

Compared to formulas we have given earlier, the formula for determining the required sample size is somewhat simpler when clusters are sampled with probabilities proportional to their sizes. Since sampling of clusters is done with replacement, a finite-population correction does not enter the formula for estimator variance. This in turn simplifies the sample size formula.

To estimate a population mean with an error limit ε at $100(1 - \alpha)$ percent confidence, the number of clusters that must be sampled is

$$n = \frac{1}{M_0} \left(\frac{t}{\varepsilon}\right)^2 \sum_{i=1}^{N} M_i(\bar{y}_i - \bar{\bar{Y}})^2 \qquad (7.18)$$

All terms are as defined above.

TABLE 7.3
Hypothetical Cumulative Numbers of Residents in Five Nursing Homes

Nursing Home	Number of Residents	Cumulative Number of Residents
1	51	51
2	34	85
3	104	189
4	76	265
5	61	326

SAMPLING WITH PROBABILITIES PROPORTIONAL TO AN AUXILIARY VARIABLE (PPES)

Sampling of clusters with probabilities proportional to their sizes can be generalized. In fact, probabilities of selection can be made proportional to any cluster characteristic. Such a characteristic is termed an *auxiliary variable*. Population means, totals, and proportions can often be estimated efficiently by an appropriate choice of cluster sampling probabilities. What constitutes an appropriate choice is perhaps best discussed after considering the formula for estimator variance.

Sampling clusters with probabilities proportional to the values of an auxiliary variable is called *PPES sampling*. We will retain this label to be consistent with the general sampling theory literature (for example, Cochran, 1963, pp. 252 ff), although the name is misleading. The initials PPES represent "sampling with *probabilities proportional* to *expected cluster sizes*." The method can be used with any auxiliary variable, which need not be related to cluster size.

ESTIMATION OF A POPULATION MEAN

If clusters are sampled with replacement and z_i denotes the probability that the ith cluster is sampled on any given draw, the unbiased PPES estimator of the population mean is given by

$$\hat{\bar{Y}}_{\text{ppes}} = \frac{1}{nM_0} \sum_{i=1}^{N} \frac{y_i}{z_i} \qquad (7.19)$$

where y_i denotes the total of the sampling variable for all elements in the ith cluster, and other terms are as defined above. Although unbiased, the PPES estimator is not consistent. However, inconsistency will be a problem only when large sampling fractions must be used to meet precision requirements.

ESTIMATOR VARIANCE

The variance of the PPES estimator is given by

$$V(\hat{\bar{Y}}_{\text{ppes}}) = \frac{1}{nM_0^2} \sum_{i=1}^{N} z_i \left(\frac{y_i}{z_i} - Y \right)^2 \qquad (7.20)$$

where Y denotes the total of the sampling variable over all elements in the population. Other terms are as defined above.

We can use Equation (7.20) to determine the most desirable cluster selection probabilities. The selection probabilities z_i should be chosen to make the estimator variance as small as possible. From Equation (7.20), it is clear that this is accomplished if the $z_i = y_i/Y$, in which case $Y = y_i/z_i$. In words, the most desirable probability of selection for the ith cluster equals the total of the sampling variable for all elements in the cluster divided by the total of the sampling variable over all elements in the population. If such probabilities could be used, the sample mean would equal the population mean for all PPES samples, and estimator variance would equal zero. In practical situations, the selection probabilities cannot be made to equal y_i/Y, since the y_i are not known. However, probabilities close to the optimal values can often be determined.

An unbiased estimator of the variance of the PPES estimator is given by

$$v(\hat{\bar{Y}}_{\text{ppes}}) = \frac{1}{(n-1)nM_0^2} \sum_{i=1}^{n} \left(\frac{y_i}{z_i} - M_0\hat{\bar{Y}}_{\text{ppes}}\right)^2 \tag{7.21}$$

When applied to achievement test scores for sixth-graders in a medium-sized city, this variance estimator was found to be more stable than those of other cluster sampling methods. Empirical results are shown in the next section.

DETERMINATION OF SAMPLE SIZE

For PPES sampling, the number of clusters that should be sampled to estimate a population mean within an error limit ε with $100(1-\alpha)$ percent confidence is

$$n = \left(\frac{t}{\varepsilon M_0}\right)^2 \sum_{i=1}^{N} z_i \left(\frac{y_i}{z_i} - Y\right)^2 \tag{7.22}$$

where

Y_i = total of the sampling variable for all elements in ith cluster
Y = total of the sampling variable over all elements in the population
z_i = probability of selection for the ith cluster

All other terms are as defined above.

It should be noted that all the formulas for determining required sample sizes [Equations (7.8), (7.13), (7.18), and (7.22)] include the values of population parameters. In the formula for sample size using the RSC-unbiased method [Equation (7.8)], the population value of the variance among cluster totals is needed. In the other formulas for sample size, either totals or means are required for every cluster in the population. In practice, of course, these cluster means and totals will not be available. If they were, sampling would not be necessary to estimate population means.

Two alternatives are available for estimating required sample sizes. First, means and totals for an auxiliary variable, rather than for the sampling variable, can be used in the sample size formulas. For example, the clusters used in estimating school testing parameters are likely to be schools or classrooms, and achievement test means and the sizes of these clusters will be quite similar from one school year to the next. The previous year's test data could thus be used as an auxiliary variable in the sample size formulas. To be sure, Ms. Jones's class of fourth-graders will not be the same size or have the same mean achievement as her class last year. However, the distribution of average achievements and sizes over classes and schools probably will be similar from one school year to the next.

A second approach to estimation of sample size requires that current data be collected. In Equations (7.8), (7.13), and (7.18), the population variance S^2, or the cross-product term involving cluster sizes and means, could be replaced by estimates based on data from a few clusters. The sum of squares in Equation (7.22) could be estimated similarly. While this solution is feasible, it is not altogether satisfactory. In many applications, a sizable fraction of the population would have to be sampled to form stable estimates of variances. In most instances, the sample sizes necessary to estimate variances would be larger than those required to estimate population means. Empirical results on the stability of variance estimates are discussed in the next section.

Empirical Comparisons of Cluster Sampling Procedures

In this section, we apply the four cluster sampling and estimation procedures described above to the problem considered in many of the examples in preceding chapters: estimation of the mean reading achievement of sixth-grade pupils in Midcity. The data used in this section are the reading achievement scores and ability test scores for sixth-graders that were used in these earlier examples. A cluster is defined as all sixth-graders enrolled in a single school.

We compare cluster sampling methods in several ways. First, estimator variances are computed for a sample size equal to 10 of the 21 Midcity elementary schools. Computation of the estimator variance is shown in detail for simple random sampling of clusters with unbiased estimation. Estimator variances are shown for the other three methods but without detailed calculations. Next, we use the Monte Carlo method to examine empirical distributions of estimates of mean achievement for the school district. Again, 10 schools are sampled from the 21 in the district. Third, we present graphs showing estimator variances as a function of the number of schools sampled. We show that no cluster sampling procedure is most efficient for all sampling fractions. Fourth, we investigate the distributions of variance estimators using the Monte Carlo method. We again use independent samples containing data from 10 schools. Finally, our results on the distributions of means and estimated variances are combined to explore the accuracy of confidence statements for the four sampling and estimation methods.

The analyses discussed in this section are intended to be illustrative. We suspect that the results presented here are without precedent in the educational research literature. However, they are based on a single data set, and their generalizability is therefore limited.

Selection of the best cluster sampling procedure in any application depends on several factors, including the kinds of data that are available for all clusters in the population, the types of clusters that could be formed, and the variability of the sampling variable, both within and between clusters. Although we recognize that it will not be feasible in many applications, replication of the kinds of analyses discussed in this section, using data from the population of interest, would give information on the best choice of a cluster sampling procedure.

The computer programs described in Appendixes D-1 and D-2 should facilitate replication of the analyses discussed in this section and might increase the feasibility of conducting the kind of detailed investigation illustrated here.

VARIANCES OF ESTIMATORS OF MEAN ACHIEVEMENT

In Example 3.1, data from Midcity were used to estimate the mean reading achievement of all sixth-graders in the school district. The estimate was formed using simple random sampling and unbiased estimation. Here, we use the same data to calculate variances of estimates of mean reading achievement. We use single-stage cluster sampling, with all sixth-graders in an elementary school forming a cluster. We assume sample sizes equal to 10 of 21 Midcity elementary schools.

Variances are shown for each of the four sampling and estimation methods discussed above: simple random sampling of clusters with unbiased estimation, simple random sampling of clusters with ratio estimation, PPS sampling and estimation, and PPES sampling and estimation. For the PPES method, we select schools with probabilities proportional to the totals of the raw scores earned by fifth-grade pupils on the Cooperative School and College Ability Tests (SCAT). The SCAT data are for the same school year as the reading achievement data for sixth-graders. The average SCAT scores for fifth-graders and the average reading achievements for sixth-graders are highly correlated across schools. SCAT score totals for schools should therefore be useful in PPES sampling.

Sixth-grade reading achievement means, sixth-grade enrollments, totals of sixth-grade reading achievement scores, and mean SCAT scores for fifth-graders in each of the Midcity schools are shown in Table 7.4. These data are used in all formulas for estimator variances.

Equation (7.5) is used to calculate the variance of mean achievement for the RSC-unbiased method. Evaluation of Equation (7.5) requires the sum of sixth-grade achievement scores in each school and the sum of achievement scores for the entire district. The total of achievement scores in the ith school is equal to the product of the average achievement and the sixth-grade enrollment in that school ($M_i \bar{y}_i$). The total for the district equals the sum of $M_i \bar{y}_i$ over schools. Totals

TABLE 7.4

Sixth-Grade Mean Reading Achievements, Totals of Sixth-Grade Achievement Scores, Mean Fifth-Grade SCAT Scores, and Sixth-Grade Enrollments for Elementary Schools in Midcity

School Number	Mean Reading Achievement, Grade 6	Total Reading Achievements, Grade 6	Mean SCAT Score, Grade 5	Enrollment, Grade 6
1	66.1071	3702	33.5357	56
2	66.8303	4344	32.9600	65
3	71.2676	5060	38.0600	71
4	56.0862	3253	33.8113	58
5	64.5745	3035	34.2895	47
6	71.0909	4692	37.8356	66
7	74.8909	4119	36.7000	55
8	70.6731	6997	37.6944	99
9	74.5088	4247	39.0577	57
10	68.1250	2725	37.1875	40
11	70.0169	4131	36.1020	59
12	72.5694	5225	39.9012	72
13	58.8605	2531	35.3600	43
14	66.3492	4180	36.2000	63
15	70.7105	2687	36.9237	38
16	65.8235	3357	34.4200	51
17	70.9804	3620	35.1489	51
18	67.5610	2770	33.5135	41
19	82.2069	2384	40.7576	29
20	65.6081	4855	35.0167	74
21	51.1429	2506	30.1765	49

for each school are listed in Table 7.4. The sum of these totals equals 80,420, and the mean of sixth-grade reading achievement, *per school*, is 3829.5.

Next, the variance of the total achievement per school must be calculated. This variance equals the sum of squares of the differences between total achievement in each school and the districtwide mean achievement per school, divided by one less than the number of schools in the district:

$$V(y_i) = \frac{1}{N-1} \sum_{i=1}^{N} (y_i - \bar{Y})^2 \tag{7.23}$$

The data necessary to calculate this variance are given in Table 7.4. Using these data, the variance equals 26,448,270/20 = 1,322,436.

For a sample of 10 schools, the sampling fraction is $\frac{10}{21}$. The total number of elements in the population, 1184 pupils, can be found by summing the sixth-grade enrollments for each school. All the terms required to calculate estimator variance are now available. If we use Equation (7.5), the variance is

$$V(\hat{\bar{Y}}) = \frac{21^2(1 - \frac{10}{21})}{10(1184^2)(20)} (26,448,720)$$

$$= 21.79 \text{ (raw score points)}^2$$

Taking the square root of the estimator variance, we find a standard deviation of 4.67 raw score points. If the estimator of the mean is normally distributed, the width of an approximate 95 percent confidence interval equals 4 times the standard deviation. In this case, the width of the interval is 18.68 raw score points.

Equation (7.10) is used to calculate the variance of mean achievement when using RSC-ratio estimation, Equation (7.16) is used when sampling schools with probabilities proportional to sixth-grade enrollments (PPS), and Equation (7.20) is used when sampling schools with probabilities proportional to fifth-grade SCAT score totals (PPES). The data necessary to evaluate these equations are provided in Table 7.4. Since the computations are similar to those for random sampling of clusters with unbiased estimation, only the final results will be given. Variances of estimated mean achievements for sixth-grade pupils in Midcity are shown in Table 7.5.

The results in Table 7.5 show RSC-unbiased estimation to be the least efficient method for these data. The PPES method, sampling schools with probabilities proportional to their fifth-grade SCAT score totals, yields the most precise estimator, followed by RSC-ratio estimation and then by the PPS method, sampling schools with probabilities proportional to their sixth-grade enrollments.

The RSC-ratio estimator is biased; all other estimators are unbiased. Comparison of the mean square error of the RSC-ratio estimator and the variances of the other estimators is therefore more appropriate than direct comparison of variances. For these data, however, the bias of the RSC-ratio estimator is very close to zero; for samples of 10 schools, it equals 0.000142 [Equation (7.12)], and the mean square error is essentially equal to the variance. For samples of 10

TABLE 7.5

Variances of Estimators of Mean Achievement for Sixth-Grade Pupils in Midcity[a]

Method of Sampling and Estimation	Estimator Variance
Simple random sampling, unbiased estimation	21.791
Simple random sampling, ratio estimation	1.802
Sampling proportional to sixth-grade enrollments (PPS)	3.622
Sampling proportional to total fifth-grade SCAT scores (PPES)	1.353

[a] Sample sizes of 10 schools from a population of 21.

schools, the order of precision of the four cluster sampling and estimation methods evaluated here is as given in Table 7.5.

DISTRIBUTIONS OF CLUSTER SAMPLING ESTIMATORS: A MONTE CARLO STUDY

It is usually assumed in finite-population sampling theory that estimators are approximately normally distributed. This assumption permits comparisons of estimation precision by comparing variances or mean square errors.

Earlier, we compared the precision of four cluster sampling procedures by computing their estimator variances (Table 7.5). To verify the results of this comparison, we computed empirical distributions of the four estimators by using a Monte Carlo procedure. One hundred independent samples of 10 schools each were selected from the 21 schools of Midcity. When we used simple random sampling of schools, we computed a ratio estimate and an unbiased estimate of the population mean from the data of each sample. We drew separate samples of schools for the PPS method and the PPES method, so that 300 independent samples were drawn in all.

A unique computer program, described in Appendix D-2, was used to select samples and to compute estimates. The computer program accepts data for an entire population of clusters and selects independent samples of clusters using a pseudo-random number generator. The program then computes estimates from each sample drawn.

The mean reading achievement of all sixth-grade pupils in Midcity is 67.92 raw score points. Mean values and the smallest and largest values of the estimates resulting from use of the four cluster sampling estimators are presented in Table 7.6.

Distributions of estimates for the four procedures are shown in Figure 7.1. Again, these results were computed from 100 independent samples of 10 schools each.

The results shown in Table 7.6 and Figure 7.1 are consistent with the variances shown in Table 7.5. These data clearly illustrate that estimators with

TABLE 7.6
Four Cluster Sampling and Estimation Procedures Applied to Midcity Data[a]

	Results from 100 Applications of Procedure		
Procedure	Largest of 100 Estimates	Mean of 100 Estimates	Smallest of 100 Estimates
Simple random sampling with unbiased estimation	80.88	68.39	60.48
Simple random sampling with ratio estimation	71.40	68.04	64.28
PPS sampling with probabilities proportional to sixth-grade enrollments	72.12	67.99	63.28
PPES sampling with probabilities proportional to fifth-grade SCAT score totals	70.78	68.09	65.51

[a] Range and mean of 100 estimates of mean sixth-grade reading achievement, samples of 10 schools. Units are raw score points.

large variances sometimes produce inaccurate estimates. Note, for example, the largest observed estimate for simple random sampling of clusters with unbiased estimation (80.88). The actual mean achievement, 67.92 raw score points, corresponds to a score of 6.6 grade equivalent units. The grade equivalent score corresponding to 80.88 is 7.4. Had the sample that resulted in the largest observed value been the only one drawn, Midcity's sixth-graders would have been

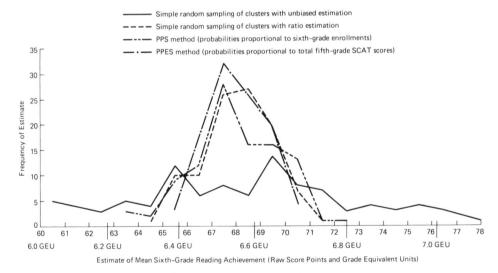

FIGURE 7.1. **Distributions of estimates of mean reading achievement for sixth-grade pupils in Midcity (data from 100 independent samples of 10 schools each).**

credited with a reading achievement mean that was much higher than they actually exhibited.

The mean values of estimates resulting from the four procedures are nearly equal. These results are consistent with sampling theory, since three of the procedures have unbiased estimators and the fourth has negligible estimator bias. The spread of the distributions shown in Figure 7.1 and the largest and smallest estimates listed in Table 7.6 leave little doubt that RSC-unbiased estimation is least precise for these data. The PPES method is seen to be most precise, although RSC-ratio estimation also affords small estimator variance.

VARIANCE AS A FUNCTION OF SAMPLE SIZE

The RSC-unbiased estimator and the RSC-ratio estimator are consistent, whereas estimators associated with the PPS method and the PPES method are not. Therefore, if the sample size equals the population size, the RSC-unbiased estimator and the RSC-ratio estimator have variances of zero. For a sample size of 21 schools, the RSC-unbiased estimator and the RSC-ratio estimator are more precise than the PPS and PPES estimators.

The data in Table 7.5 show the PPS and PPES estimators to be more precise than the RSC-unbiased estimator and the PPES estimator to be more precise than the RSC-ratio estimator for samples of 10 schools selected from a population of 21. Since the variances of all four estimators decrease as the sample size is increased, at some sample sizes the RSC-ratio estimator and the RSC-unbiased estimator become more precise than the PPS and PPES estimators. Data showing these sample sizes are presented in Figures 7.2 and 7.3. Graphs of estimator variances are shown as functions of the number of schools sampled.

The computer program described in Appendix D-1 was used to compute the data shown in Figures 7.2 and 7.3. The program provides variances and standard errors of estimators for each sample size. Other statistics provided by the CLUSAMP-I computer program are described in Appendix D-1, along with a program listing, specimen output, and specifications for program input.

Figures 7.2 and 7.3 present similar data but differ in scale. Figure 7.2 clearly illustrates the relative sizes of the variance of the RSC-unbiased estimator and those of the other three estimators over the entire range of potential sample sizes. Except for sample sizes of 20 or 21 schools, random sampling of schools with unbiased estimation is less precise than the other three sampling and estimation methods.

Estimator variances are also shown in Figure 7.3, but in this figure, the scale of the variance functions has been enlarged. Figure 7.3 permits more detailed comparisons of the precisions of the RSC-ratio estimator, the PPS estimator, and the PPES estimator. It is clear that both random sampling of clusters with ratio estimation and the PPES method are uniformly superior to the PPS method. For sample sizes of 12 or fewer schools, the PPES method is superior to all three alternatives. For sample sizes of 13 or more schools, simple random sampling with ratio estimation is best. For these data then, the most efficient sampling and estimation method depends on the required precision of estimation and the consequent sample size.

We doubt the generalizability of these results to other cluster sampling applications in education or the behavioral sciences. However, results compara-

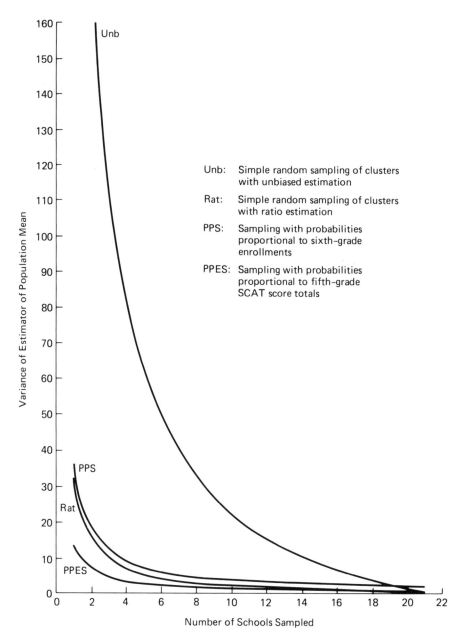

FIGURE 7.2. **Variances of estimators of mean achievement versus number of schools sampled (schools in Midcity used as clusters).**

ble to these can be developed for other applications, by using the CLUSAMP-I computer program in Appendix D-1. In school testing applications, test data for any recent school year can be used as program input, and any of several alternative auxiliary variables could be used for the PPES procedure. School records might contain data for an auxiliary variable that is more closely related to school achievement means than are ability test means for pupils in a different grade. If

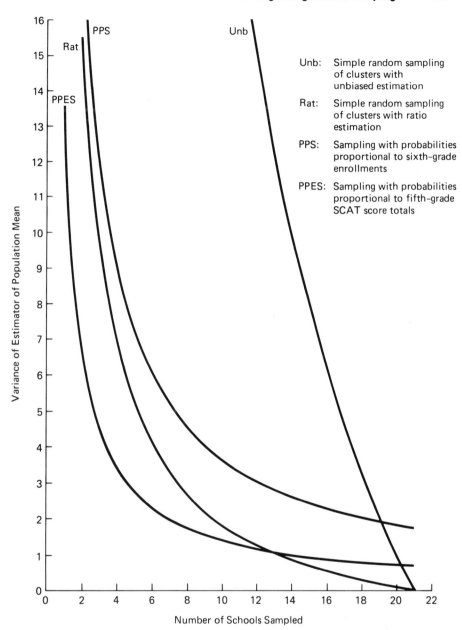

FIGURE 7.3. **Variances of estimators of mean achievement versus number of schools sampled (schools in Midcity used as clusters).**

such data are available, the PPES procedure should be even more efficient than these results show it to be.

ESTIMATION OF ESTIMATOR VARIANCE

In most applications of sampling methods, sample data are used not only to estimate a population parameter but also to estimate the variance of the estima-

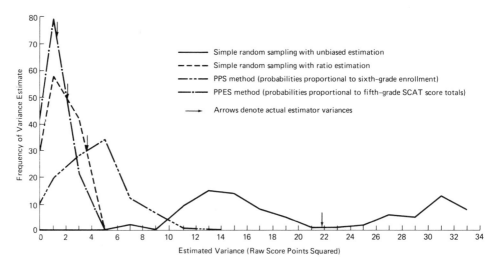

FIGURE 7.4. **Distributions of estimated variances of estimated mean achievements for single-stage cluster samples of 10 schools (data from Midcity).**

tor. Because information for every school in the population was available, we were able to compute actual estimator variances in the preceding section. This unrealistic situation would not arise in an actual application.

Estimates of estimator variances, like estimates of population means, totals, and proportions, fluctuate from sample to sample. We will therefore discuss the stability of variance estimators. Because estimates of variance are used for several purposes, the stability of these estimators is important. Variance estimates are sometimes used to determine required sample sizes. More frequently, they are used in computing confidence limits on population parameters.

Mathematical theory regarding the stability of variance estimators is not highly developed. In the case of simple random sampling without replacement, limited theory exists. An expression for the coefficient of variation of the sample variance of a population total is given by Hansen et al. (1953, Vol. II, p. 101). Results and approximations for sampling *with* replacement are also given for simple random sampling and for some extensions of simple random sampling (single-stage cluster sampling and some stratified sampling procedures) (Hansen et al., 1953, Vol. I, pp. 134 and 427–436; Vol. II, pp. 99–105 and 236–238). Rao and Bayless (1969) and Bayless and Rao (1970) have conducted some limited empirical studies of the precision of estimators of variances of population totals for some single-stage cluster sampling methods.

In this section, we report results of a Monte Carlo study of the stability of variance estimators. The CLUSAMP-II computer program (Appendix D-2) was used to compute variance estimates for estimators of the population mean. Estimates were computed for the RSC-unbiased estimator, the RSC-ratio estimator, the PPS estimator, and the PPES estimator. A single variance estimate was computed for each estimator of the mean, for each of 100 independent samples. The resulting distributions of variance estimates indicate the stability of variance estimators.

Reading achievement scores for sixth-grade pupils in Midcity are used to compute distributions of variance estimates. Samples of 10 schools were selected from the 21 elementary schools in the city. Equations (7.7) and (7.11) were used to compute variance estimates for the RSC-unbiased estimator and the RSC-ratio estimator. Estimated variances were computed for the PPS method and the PPES method by using Equations (7.17) and (7.21). Variance estimates for each estimation method were computed from each sample of 10 schools. One hundred samples were drawn independently for RSC-unbiased estimation and RSC-ratio estimation. Additional sets of 100 samples each were drawn for PPS estimation and PPES estimation.

Distributions of estimates of estimator variances are shown in Figure 7.4. The graphs in Figure 7.4, like those in Figure 7.1, are frequency polygons. Each point in the figure represents the number of samples for which a variance estimate fell within an interval two units wide, immediately surrounding the point. On the graph for the RSC-unbiased estimator, the point that shows 15 samples with a variance estimate of 13 indicates that there were 15 samples with variance estimates between 12 and 14. Other plotted points should be interpreted similarly.

Sample statistics computed from distributions of estimated variances are shown in Table 7.7. Actual variances, presented in Table 7.5, are reproduced here for ease of comparison.

For the Midcity data, estimators of the variance of estimated population means were unstable (Figure 7.4 and Table 7.7). While the means of the estimates shown in Table 7.7 were close to actual variances, the ranges of variance estimates were wide. For the RSC-ratio estimator, estimated variances ranged from one-fourth to almost twice the actual variance. Estimated variances of other estimators showed similarly wide ranges. However, variance estimators for some cluster sampling procedures were far more stable than those for others (Figure 7.4).

TABLE 7.7
Estimates of Estimator Variances: Four Cluster Sampling and Estimation Procedures[a]

Procedure	Results from 100 Applications of Procedure			
	Largest of 100 Estimates	Mean of 100 Estimates	Smallest of 100 Estimates	Actual Variance of Estimator
Simple random sampling with unbiased estimation	37.73	22.03	6.48	21.79
Simple random sampling with ratio estimation	3.27	1.80	0.44	1.80
PPS sampling with probabilities proportional to sixth-grade enrollments	10.30	3.71	0.55	3.62
PPES sampling with probabilities proportional to fifth-grade SCAT score totals	3.24	1.32	0.20	1.36

[a] Ranges and means of 100 estimates of variance of estimated mean sixth-grade reading achievement, samples of 10 schools in Midcity.

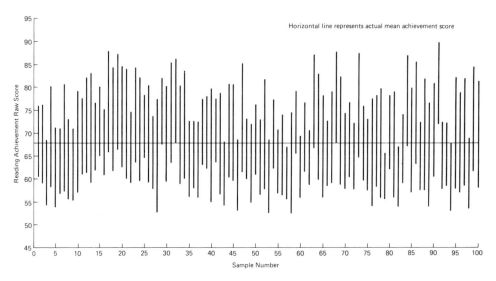

FIGURE 7.5. **Confidence intervals on the mean reading achievement of sixth-grade pupils in Midcity, computed using RSC single-stage cluster sampling of 10 schools from a population of 21, with unbiased estimation of the mean (raw scores on the Stanford Achievement Test.)**

Methods that provided the greatest precision in estimating a population mean also showed the greatest precision in estimating the variance of estimators. For samples of 10 schools, the PPES method provided the most stable variance estimator. Variance estimators for RSC-ratio estimation were almost as stable. The PPS method and the RSC-unbiased method provided very unstable variance estimates.

ACCURACY OF CONFIDENCE STATEMENTS

For each cluster sampling and estimation method discussed in this chapter, approximate confidence limits on the population mean are given by equations of the form

$$\hat{\bar{Y}}_U = \hat{\bar{Y}} + t\sqrt{v(\hat{\bar{Y}})}$$

$$\hat{\bar{Y}}_L = \hat{\bar{Y}} - t\sqrt{v(\hat{\bar{Y}})}$$

(7.24)

where

$\hat{\bar{Y}}$ = estimate of the population mean

$v(\hat{\bar{Y}})$ = sample estimate of the variance of $\hat{\bar{Y}}$, computed using one of Equations (7.7), (7.11), (7.17), or (7.21)

t = abscissa on Student's t-distribution, with $100(\alpha/2)$ percent of the distribution to its right, and degrees of freedom equal to one less than the number of clusters in the sample

An estimate of the population mean, an estimate of estimator variance, and an abscissa on Student's t-distribution are used in computing confidence intervals [Equation (7.24)]. The use of an abscissa from the t-distribution assumes that the estimator of the mean is normally distributed and the variance estimator has a chi-square distribution. If either of these assumptions is violated, confidence statements might not be accurate. In any case, a particular confidence interval might or might not include the population mean. Ninety-five percent of a very large number of 95 percent confidence intervals should include the population mean; if the percentage is larger or smaller, confidence statements are said to be inaccurate.

The Monte Carlo method can be used to compute empirical distributions of confidence intervals. When confidence intervals are computed from independently selected samples and the population mean is known, we can observe the proportion of intervals that include the mean. An assessment of the accuracy of confidence statements can be made by comparing the proportion of confidence intervals expected to include the mean and the proportion that actually includes the mean. We followed this procedure using achievement test data from Midcity; results are presented below.

Distributions of confidence intervals on the mean reading achievement of sixth-grade pupils in Midcity were computed from independent samples of 10 schools. Confidence intervals were computed from 100 samples of schools for each of the cluster sampling procedures discussed earlier. The CLUSAMP-II computer program (Appendix D-2) was used to calculate estimates for use in Equation (7.24). The t value for a 95 percent confidence level and nine degrees of freedom is 2.26.

Distributions of 100 confidence intervals are shown in Figures 7.5 through 7.8 for RSC-unbiased estimation, RSC-ratio estimation, PPS estimation, and PPES estimation, respectively. In each figure, confidence intervals were plotted vertically around the actual mean reading achievement of sixth-graders in Midcity.

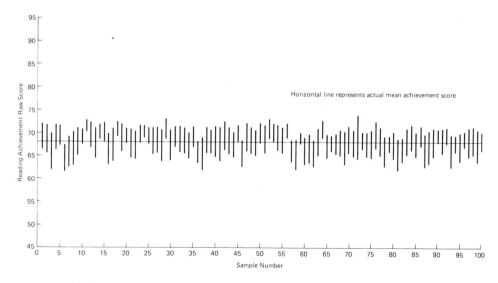

FIGURE 7.6. **Confidence intervals on the mean reading achievement of sixth-grade pupils in Midcity, computed using RSC single-stage cluster sampling of 10 schools from a population of 21, with ratio estimation of the mean (raw scores on the Stanford Achievement Test)**

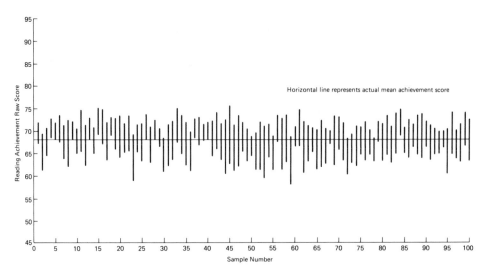

FIGURE 7.7. **Confidence intervals on the mean reading achievement of sixth-grade pupils in Midcity: samples selected with probabilities proportional to sixth-grade enrollments (PPS), single-stage cluster sampling of 10 schools from a population of 21 (raw scores on the Stanford Achievement Test).**

Most confidence intervals resulting from RSC-unbiased estimation included the population mean (Figure 7.5). Five of the 100 intervals failed to include the mean, and an additional six intervals barely included the mean. For these data then, simple random sampling with unbiased estimation provided accurate confidence statements; the population mean was contained in 95 percent of the confidence intervals. The confidence intervals resulting from RSC-unbiased estimation were extremely wide. The first confidence interval shown in Figure 7.5 ranges from 61 to 76, the fiftieth from 61 to 76, and the hundredth from 58 to 82. In corresponding grade equivalent units, the intervals range from 6.0 to 7.0, from 6.0 to 7.0, and from 5.9 to 7.4. Ranges of a whole grade equivalent unit probably are unacceptable when mean achievement is estimated for a school district.

Confidence intervals resulting from RSC-ratio estimation are shown in Figure 7.6. They were considerably narrower than intervals resulting from RSC-unbiased estimation. Nine of the 100 intervals failed to include the population mean, and an equal number barely included the mean. For these data, confidence statements for RSC-ratio estimation were highly optimistic. If the 100 confidence intervals shown here represented a long-run average, what were purportedly 95 percent confidence intervals were in fact 91 percent intervals. However, only one of the 100 confidence intervals failed by more than 2 raw score points to include the population mean. In fact, seven of the nine intervals which did not include the mean failed to do so by less than 1 raw score point.

Confidence intervals for the PPS method were narrower than those computed using RSC-unbiased estimation but were wider than those produced by RSC-ratio estimation. These data are consistent with the distributions of sample means and variances shown in Figures 7.1 and 7.4. Of the 100 confidence intervals produced by the PPS method, only three failed to include the population mean. Data in Figure 7.7 also show that 12 of the 100 intervals barely included the population mean. Assuming these 100 confidence intervals represented long-run results, intervals calculated to be 95 percent were really 97 percent confidence intervals.

One hundred confidence intervals resulting from the PPES method are shown in Figure 7.8. These confidence intervals were, on the average, narrower than those of the other three sampling and estimation methods. Results shown in Figure 7.8 were thus consistent with the distributions of sample means and variances presented in Figures 7.1 and 7.4. Six of the 100 confidence intervals failed to include the population mean. An additional 12 intervals barely included the mean. If we assume that these results represent a long-run average, 94 percent confidence intervals, rather than 95 percent intervals, were obtained from PPES sampling and estimation.

In summary, data from 100 independent samples support the conclusion that reasonably accurate confidence statements resulted from three of the four cluster sampling procedures applied to Midcity data. Simple random sampling of clusters with ratio estimation resulted in 91 percent confidence intervals for the Midcity population.

An alternative method of estimation is presented in the next section. This method resulted in somewhat wider confidence intervals for RSC-unbiased estimation and RSC-ratio estimation. Therefore, a higher proportion of confidence intervals included the population mean.

Jackknife Estimation of Population Means and Estimator Variances

ADVANTAGES OF JACKKNIFE PROCEDURES

The jackknife method of estimation was first introduced by Quenouille in 1956. The method was termed the *jackknife* in recognition of its wide applicability. Miller (1964) reviews the properties of jackknife estimators, and Mosteller

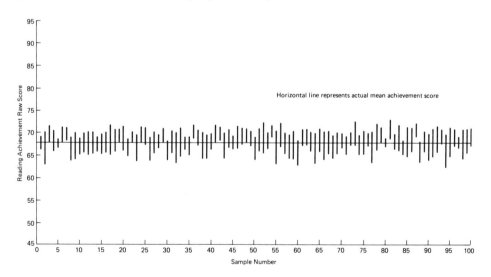

FIGURE 7.8. **Confidence intervals on the mean reading achievement of sixth-grade pupils in Midcity: samples selected with probabilities proportional to fifth-grade SCAT score totals (PPES), single-stage cluster sampling of 10 schools from a population of 21 (raw scores on the Stanford Achievement Test).**

and Tukey (1968) provide a particularly illuminating discussion of methods of application. The jackknife technique has not been used extensively in finite-population sampling problems and is not generally found in texts. Mickey (1959) used a variant of the method in conjunction with ratio estimation and regression estimation of means. Shoemaker (1973) applied the jackknife procedure to estimate the variances of estimators arising in multiple matrix sampling.

In situations where conventional estimators are biased, jackknife estimators frequently are unbiased. Also, when formulas for estimating estimator variances are not available or do not permit ready evaluation, the jackknife technique is a useful and practical alternative.

Since conventional variance estimation formulas are available for single-stage cluster sampling procedures, the results of conventional estimation methods and those of jackknifing can be compared. The accuracy of confidence statements computed from jackknife estimates can also be compared to those computed using standard methods. In the last section, we found that RSC-ratio estimation did not provide accurate confidence statements when applied to data from Midcity. Results presented in this section show that corresponding jackknife procedures did provide accurate confidence statements when used with these data.

APPLICATION OF THE JACKKNIFE PROCEDURE

The jackknife procedure requires eight steps. Whether the method is used with finite populations or with infinite populations, the general procedure is the same. Two steps involve the computation of conventional estimates; these estimates are then used in the rest of the computations. The detailed steps in applying the jackknife technique are given below, followed by an example.

1. A sample is selected just as for a conventional estimation procedure.
2. Sample data are divided into k nonoverlapping subgroups.
3. A conventional estimate (such as a ratio estimate of a population mean) is computed, using all the sample data. Call this estimate y_{all}.
4. k overlapping subsets of data are formed, each consisting of all sample data less one of the subgroups formed in step 2.
5. The same conventional estimator used in step 3 is applied to the data in each of the k subsets formed in step 4. The resulting estimate for the jth subset is called $y_{(j)}$.
6. Pseudo-values of estimates for each of the k subgroups formed in step 2 are computed by applying the formula

$$y_{*j} = ky_{\text{all}} - (k - 1)y_{(j)} \tag{7.25}$$

7. The jackknife estimate of a population parameter, y_*, is computed by forming the average of the pseudo-values computed in step 6:

$$y_* = \frac{1}{k} \sum_{j=1}^{k} y_{*j} \tag{7.26}$$

8. The jackknife estimate of estimator variance s_*^2 is computed by applying the formula

$$s_*^2 = \frac{\left(\sum_{j=1}^{k} y_{*j}^2\right) - ky_*^2}{k(k-1)} \tag{7.27}$$

An example should clarify the procedure.

Suppose we wanted to apply the jackknife technique in conjunction with single-stage cluster sampling and RSC-ratio estimation. We might select a sample of 10 classrooms and use their data to estimate the mean achievement of pupils in a population of 20 classrooms. Following the steps listed above, a sample of 10 classrooms would first be selected using simple random sampling without replacement. In step 2, the 10 sampled classrooms would be allocated to five nonoverlapping subgroups. If the classrooms were numbered from 1 to 10, the five subgroups might consist of classrooms 1 and 2, 3 and 4, 5 and 6, 7 and 8, and 9 and 10. In step 3, the RSC estimator given by Equation (7.9) would be applied to the mean achievement scores and enrollments of the 10 sampled classrooms. The value computed would be called y_{all}. Next, five overlapping subsets of eight classrooms each would be constructed using the subgroups formed in step 2. Each subset would be composed of all 10 classrooms less one of the subgroups. The five subsets would be as follows:

Subset	Classrooms in Subset
1	1 and 2, 3 and 4, 5 and 6, 7 and 8
2	1 and 2, 3 and 4, 5 and 6, 9 and 10
3	1 and 2, 3 and 4, 7 and 8, 9 and 10
4	1 and 2, 5 and 6, 7 and 8, 9 and 10
5	3 and 4, 5 and 6, 7 and 8, 9 and 10

Note that each subset would consist of all sampled classrooms less one of the subgroups formed in step 2.

In step 5, the RSC-ratio estimator [Equation (7.9)] would be applied, using the achievement means and enrollments of the classrooms in each of the five subsets formed in step 4. The result would be an estimate for each of the five subsets, labeled $y_{(1)}$, $y_{(2)}$, $y_{(3)}$, $y_{(4)}$, and $y_{(5)}$.

In step 6, Equation (7.25) would be applied to each estimate formed in step 5 and to the estimate formed in step 3. The following five pseudo-values would result:

$$y_{*1} = 5y_{\text{all}} - 4y_{(1)}$$

$$y_{*2} = 5y_{\text{all}} - 4y_{(2)}$$

$$y_{*3} = 5y_{\text{all}} - 4y_{(3)}$$

$$y_{*4} = 5y_{\text{all}} - 4y_{(4)}$$

$$y_{*5} = 5y_{\text{all}} - 4y_{(5)}$$

The jackknife estimate of mean achievement for pupils in the 20 classrooms of the population would be computed in step 7. The estimate would be given by Equation (7.26):

$$y_* = \frac{y_{*1} + y_{*2} + y_{*3} + y_{*4} + y_{*5}}{5}$$

The jackknife estimate of the estimator variance would be computed in step 8. The estimate [Equation (7.27)] would be given by

$$s_*^2 = \frac{(y_{*1}^2 + y_{*2}^2 + y_{*3}^2 + y_{*4}^2 + y_{*5}^2) - 5y_*^2}{5(4)}$$

FORMING CONFIDENCE INTERVALS

The jackknife estimator [Equation (7.26)] and the jackknife variance estimator [Equation (7.27)] can be used to compute confidence limits on population parameters. An approximate $100(1 - \alpha)$ percent upper confidence limit is given by

$$y_{*U} = y_* + ts_* \tag{7.28}$$

and an approximate $100(1 - \alpha)$ percent lower confidence limit is given by

$$y_{*L} = y_* - ts_* \tag{7.28}$$

Here, t is an abscissa on Student's t-distribution with $100(\alpha/2)$ percent of the distribution to its right. Usually α is set at 0.10, 0.05, or 0.01. The number of degrees of freedom used for t is one less than the number of different pseudovalues computed in step 6.

Mosteller and Tukey (1968) suggest that about 10 subgroups be used in applying the jackknife procedure. They consider ninety-fifth percentiles of Student's t-distribution in making that suggestion. With an infinite number of degrees of freedom, the t-distribution coincides with the normal distribution, and 95 percent of the distribution falls between the t values ± 1.96. For 10 degrees of freedom, ± 2.23 includes 95 percent of the t-distribution. Thus, for 10 degrees of freedom, t is 11 percent larger than for infinite degrees of freedom. Corresponding 95 percent confidence intervals are, on the average, about 22 percent wider.

Example 7.1. Application of Jackknife Estimation

In the section on empirical results, conventional cluster sampling procedures were used to estimate the mean achievement of sixth-grade pupils in Midcity. In this example, jackknife estimation procedures are applied to the same data, so results can be compared to those of conventional estimation. Jackknife estimators of means are the same as conventional estimators for the

RSC-unbiased, PPS, and PPES procedures. Jackknife RSC-ratio estimators differ slightly from conventional RSC-ratio estimators.

Jackknife and conventional estimators are identical whenever a conventional estimator is a simple average of observations or a constant multiple of a simple average. The conventional RSC-unbiased estimator [Equation (7.1)], the conventional PPS estimator [Equation (7.15)], and the conventional PPES estimator [Equation (7.19)] are simple averages, multiplied by a constant. In contrast, the conventional RSC-ratio estimator [Equation (7.9)] is a ratio of sums. No simple algebraic relationship to the corresponding jackknife estimator can be found.

For the data in this example, jackknife and conventional variance estimates differed only for RSC-unbiased estimation and RSC-ratio estimation. Therefore, the confidence intervals on the population mean computed with these methods also differ.

DISTRIBUTIONS OF JACKKNIFE ESTIMATES OF MEANS AND JACKKNIFE VARIANCE ESTIMATES

The CLUSAMP-II computer program (Appendix D-2) provides jackknife estimates of means and variances corresponding to those of conventional cluster sampling procedures. This program was used to compute jackknife estimates of mean reading achievement and estimates of estimator variances.

One hundred independent samples of 10 schools each, the data used in the section on empirical results, were also used in this example. Jackknife estimates of mean achievement and estimator variances were computed for each sample, using the conventional RSC-unbiased estimator and the conventional RSC-ratio estimator in steps 3 and 5 of the jackknife procedure. Sample statistics, computed from the distributions of jackknife estimates, are shown in Table 7.8.

Interpretation of the results shown in Table 7.8 requires consideration of data presented in earlier examples. The jackknife RSC-unbiased estimator had a larger range and a larger variance than did the jackknife RSC-ratio estimator.

TABLE 7.8

Two Jackknife Estimation Procedures Applied to Cluster-Sampled Midcity Data[a]

Procedure	Estimates of Mean			Estimates of Variance of Estimates of Mean		
	Largest of 100 Estimates	Mean of 100 Estimates	Smallest of 100 Estimates	Largest of 100 Estimates	Mean of 100 Estimates	Smallest of 100 Estimates
Jackknife RSC-unbiased estimation	80.88	68.39	60.48	72.04	42.06	12.37
Jackknife RSC-ratio estimation	71.38	68.06	64.32	6.76	3.39	0.79

[a] Ranges and means of 100 estimates of mean sixth-grade reading achievement and 100 estimates of estimator variance. Samples of 10 schools.

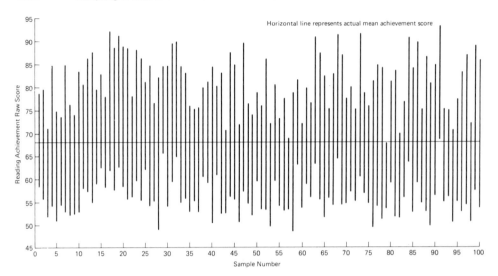

FIGURE 7.9. **Confidence intervals on the mean reading achievement of sixth-grade pupils in Midcity, computed using RSC single-stage cluster sampling of 10 schools from a population of 21, with jack-knife-unbiased estimation of the mean (raw scores on the Stanford Achievement Test).**

The ranges and means of these jackknife estimators were almost identical to those of corresponding conventional estimators (Table 7.6). The mean estimated variance of the jackknife RSC-unbiased estimator was almost twice as large as the mean estimated variance of the conventional RSC-unbiased estimator: 42 as compared to 22 (Table 7.7). A variance of 42 corresponds to a standard deviation of 6.5, and for nine degrees of freedom, resulting confidence intervals had an average width of 29 raw score points. Thus, the jackknife RSC-unbiased estimator exhibited very large sampling fluctuations, as did the conventional RSC-unbiased estimator.

The mean estimated variance of the jackknife RSC-ratio estimator was 3.39, compared to a mean estimated variance of 1.80 for the conventional RSC-ratio estimator. Proportional to their mean values, the fluctuations of the jackknife RSC-ratio variance estimator and those of the conventional RSC-ratio variance estimator were nearly equal (compare the largest and smallest observed values in Tables 7.7 and 7.8).

ACCURACY OF CONFIDENCE STATEMENTS

Distributions of confidence intervals were shown for each of four conventional estimators in Figures 7.5 through 7.8. From these data, we concluded that all conventional methods except the RSC-ratio estimator provided accurate confidence statements. Distributions of 95 percent confidence intervals for the jackknife RSC-unbiased estimator and the jackknife RSC-ratio estimator are shown in Figures 7.9 and 7.10. The large confidence intervals resulting from jackknife RSC-unbiased estimation can be seen in Figure 7.9. Although only two of these 100 intervals failed to include the population mean, they were so wide that jackknife-unbiased estimation was not practical for the Midcity population.

Ninety-five percent confidence intervals resulting from jackknife RSC-ratio estimation had an average width of 8.3 raw score points. This corresponds to an

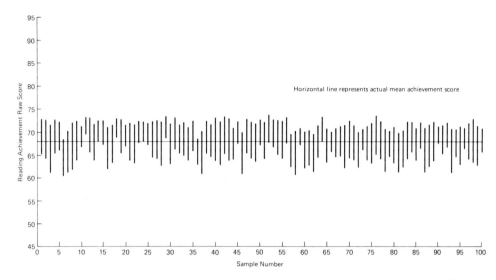

FIGURE 7.10. **Confidence intervals on the mean reading achievement of sixth-grade pupils in Midcity, computed using RSC single-stage cluster sampling of 10 schools from a population of 21, with jack-knife-ratio estimation of the mean (raw scores on the Stanford Achievement Test).**

average width of 0.5 grade equivalent units. Confidence intervals of this width might make the jackknife RSC-ratio estimator useful for Midcity. Estimation of mean reading achievement in a school district within ± 0.25 grade equivalent units would surely be useful for some purposes.

Conventional RSC-ratio estimation resulted in confidence intervals with an average width of 6.1 raw score points (Figure 7.6). However, in a sample of 100 intervals, nine intervals failed to include the population mean and an additional nine intervals barely included the population mean. Although the jackknife RSC-ratio estimator resulted in wider confidence intervals, it provided more accurate confidence statements than did the conventional RSC-ratio estimator. The population mean was covered by 96 of the 100 confidence intervals for jackknife RSC-ratio estimation (Figure 7.10). If these data were assumed to be representative of a long-run average, we could conclude that the jackknife RSC-ratio estimator provided accurate 95 percent confidence statements for Midcity data. These confidence statements were considerably more accurate than those resulting from conventional RSC-ratio estimation applied to the same data.

Comparison of Jackknife Estimates and Conventional Estimates

Four conventional estimation procedures and two jackknife procedures were applied to the same data. Findings are shown in the section on empirical results and in Example 7.1. These estimation procedures may be compared by considering mean values of estimated mean achievement, average widths of confidence intervals, and percentages of confidence intervals that include the population mean. These statistics are shown in Table 7.9.

TABLE 7.9
Comparative Statistics for Six Methods of Estimating the Mean Reading Achievement of Sixth-Grade Pupils in Midcity[a]

	Results from 100 Applications of Procedure		
Procedure	Average Value of Estimated Mean Achievement (Raw Score Points)	Average Width of Confidence Intervals (Raw Score Points)	Percentage of 95% Confidence Intervals That Include Population Mean
Simple random sampling of clusters with unbiased estimation	68.39	21.2	95
Simple random sampling of clusters with ratio estimation	68.03	6.1	91
Sampling proportional to sixth-grade enrollments (PPS)	67.99	8.7	97
Sampling proportional to fifth-grade SCAT score totals (PPES)	68.09	5.2	94
Simple random sampling of clusters with jackknife-unbiased estimation	68.39	29.3	98
Simple random sampling of clusters with jackknife-ratio estimation	68.06	8.3	96

[a] Statistics are based on 100 independent samples of 10 schools from a population of 21.

For samples of 10 schools from Midcity, the PPES procedure appears to be preferable to all other methods. The mean of 100 estimates of mean achievement was only one-tenth of a raw score point above the population mean. Ninety-four percent of 100 confidence intervals included the population mean. Confidence intervals were narrower than those found with any other procedure—over 60 percent narrower than those resulting from jackknife RSC-ratio estimation, which afforded confidence statements of comparable accuracy. All other methods provided confidence statements that were less accurate or resulted in wider confidence intervals.

Choosing a Sampling Unit

One of the first issues that arises when cluster sampling is used concerns the most appropriate units to use as clusters. In some surveys the choice will be obvious and limited. In others, however, particularly surveys in education and the social sciences, there will be several candidates. For example, in a survey of all adult residents of a city, cluster sampling of city blocks or cluster sampling of households might be feasible. In a statewide survey of public school students, potential clusters would include school districts, schools, or classrooms.

In these situations, an immediate question is what is the best choice of units to use as clusters. As in other choices among alternative sampling procedures, the major considerations are the precision afforded by each alternative, the administrative convenience or inconvenience associated with each alternative, and the overall cost of the survey resulting from each choice.

Hansen et al. (1953, pp. 270–284 and 306–311) and Cochran [1977, pp. 234–237, and based on the work of Jessen (1942, pp. 244–246)] provide discussions of the costs associated with the choice of sampling units in cluster sampling. Hansen et al. also provide excellent illustrations of cost functions for cluster sampling, when data-collection procedures include extensive travel and interviewing.

Estimating the costs of conducting a survey is a complex procedure. Some costs, such as the cost of purchasing survey materials and costs associated with reducing and analyzing survey data, require out-of-pocket expenditures. These costs can easily be itemized and are usually directly proportional to the number of elements sampled. Other costs, such as opportunity costs associated with survey administration, are not recognized so easily and may have very complicated relationships with the number of elements sampled.

We will illustrate the analysis of survey costs and the relationship between survey costs and the choice of a cluster sampling unit by returning to an example we have discussed many times: estimating the mean achievement of sixth-grade pupils in Midcity. In this example, we consider the relative costs of using schools or classrooms as clusters when achievement test data are collected through a cluster sampling design.

Example 7.2. Cost Analysis of the Choice of Cluster Sampling Units in a School Testing Program

Central school district offices incur some expenses in planning and administering testing programs, regardless of the number of pupils tested. If a director of testing is employed by a school district, his or her salary is an expense that is unrelated to the size of the school testing program. Whenever testing takes place in a school, administrative expenses are incurred, whether 100 pupils or 300 pupils are tested. When pupils in a classroom are tested, it is reasonable to consider a "classroom cost" in addition to the direct expenses of testing selected pupils. Teachers usually act as proctors when group achievement tests are given; their salaries must be prorated as expenses of testing. Disruption of regular classroom activity results in some inherent costs, whether the whole class or half the class is tested.

We will use a simple cost function to illustrate the choice between schools and classrooms as clusters. Define the following costs of testing:

C_0 is a constant "school district cost," independent of the choice of sampling unit.

C_1 is the cost of sampling in a school; it represents administrative expenses of testing incurred in a school in which testing takes place; C_1 is assumed to be independent of the number of pupils tested in a school.

C_2 is the cost of testing in a classroom; it represents expenses incurred as a result of class disruption and the cost of using a teacher as a test proctor.

C_3 is the cost of testing an individual pupil; it represents the purchase price of test materials and charges for scoring and data analysis for one pupil.

Using these definitions, the total cost of a testing program can be represented by the function

$$C = C_0 + MC_1 + NC_2 + LC_3 \qquad (7.29)$$

where

M = number of schools in which testing takes place
N = number of classrooms in which testing takes place
L = total number of pupils tested

This cost function permits comparison of the costs of testing for alternative choices of sampling units. Specific relationships will be assumed to hold among the cost coefficients C_0, C_1, C_2, and C_3. If a population mean is to be estimated at a specified level of precision, the number of clusters sampled to achieve this precision will be calculated for each potential cluster size. Estimators will be assumed to have normal distributions, and the relationship between sample size and variance will be used to determine required sample sizes. Values of M, N, and L will be calculated from required sample sizes, and these values will be used in Equation (7.29) to calculate the costs of sampling. In this way, we will determine the type of cluster that yields a specified level of precision at minimum cost.

If schools were used as clusters, the number of schools M in which testing took place would equal the number of clusters that would have to be sampled to afford a specified level of precision. The number of classrooms N in which testing would take place could be approximated by multiplying the average number of class sections per school for the grade being tested by the number of sampled schools M. The average enrollment per school in the grade being tested would then be multiplied by M to solve for L, the total number of pupils to be tested. These values would then be used in Equation (7.29) to solve for C, the cost of testing.

If classrooms were used as clusters, the values of M, N, and L would be determined similarly. The number of classrooms N in which testing took place would equal the sample size required to achieve a specified level of precision. The number of schools in which testing occurred would vary from sample to sample. If, on the average, there were k class sections per school in the grade tested, the average number of schools in which testing occurred would be N/k. We could assume conservatively that each tested classroom would be in a different school. If so, the number of schools M would equal the number of classrooms N. The number of pupils L would be approximated by the product of N and the average number of pupils per class. The computed values of M, N, and L would be used in Equation (7.29) to solve for the total cost of testing.

Testing would take place in fewer schools when schools, rather than class-rooms, were used as clusters. This would have administrative advantages, and if the school costs C_1 were high compared to other costs, substantial savings would result. Conversely, the number of pupils tested would be smaller when class-rooms, rather than schools, were used as clusters. If the costs of testing each pupil, C_3, were relatively high, reduction of the number of pupils tested might result in greater economy. If so, classrooms would be preferable to schools as clusters.

In the balance of this example, we determine the required sample sizes empirically, using sixth-grade reading achievement data from Midcity. Separate analyses are conducted using schools and classrooms as clusters. We assume several different relationships between the cost coefficients C_0, C_1, C_2, C_3. When classrooms are used as clusters, we assume that each tested classroom is in a different school. This is a conservative assumption, since it maximizes total school costs MC_1.

The mean reading achievement of sixth-grade pupils in Midcity was esti-mated in examples in Chapters 3 and 4. In those examples, allowable estimation error was specified as 0.2 grade equivalent units with 95 percent confidence. The same error limits are used in this example.

As in Example 3.1, a 95 percent confidence interval on the population mean with an average width of ± 0.2 grade equivalent units is assumed to correspond to an estimator variance of 4 raw score points squared. Four methods of estima-tion (RSC unbiased, RSC ratio, PPS, and PPES) and two cluster choices (schools and classrooms) are compared.

REQUIRED SAMPLE SIZES

Sample sizes that provide estimator variances ≤4 raw score points squared were determined for four sampling and estimation procedures and both types of clusters. Figure 7.3 shows estimator variance as a function of sample size when schools were used as clusters. The numbers of schools that must be sampled to provide estimator variances ≤4 raw score points squared are shown in Table 7.10.

There are 45 sixth-grade class sections in the 21 elementary schools in Midcity. To find the number of class sections that must be sampled, estimator variances must be calculated as a function of sample size. The CLUSAMP-I computer program, described in Appendix D-1, provides the needed calcula-tions. Results for Midcity are shown in Figure 7.11.

When sixth-grade class sections are used as sampling units, a different auxil-iary variable must be used for PPES sampling. Obviously, sixth-grade classrooms cannot be sampled with probabilities proportional to fifth-grade SCAT score totals for schools. Totals of scores on the Lorge-Thorndike ability test were available for pupils in each of the sixth-grade class sections in Midcity. PPES sampling involves selection of classrooms with probabilities proportional to totals of Lorge-Thorndike scores for pupils in these classrooms. Therefore, results for PPES sampling using classrooms as clusters are not directly comparable to results for PPES sampling using schools as clusters.

The number of classrooms that must be sampled to achieve estimator vari-ances not exceeding 4 raw score points squared are shown in Table 7.11.

TABLE 7.10

Four Cluster Sampling and Estimation Procedures: Sample Sizes (Numbers of Schools) Required for Estimator Variances of Four or Less[a]

Procedure	Required Sample Size
Simple random sampling of clusters with unbiased estimation	18 schools
Simple random sampling of clusters with ratio estimation	7 schools
Sampling with probabilities proportional to sixth-grade enrollment (PPS)	10 schools
Sampling with probabilities proportional to fifth-grade SCAT score totals (PPES)	4 schools

[a] Calculated for a population of 21 elementary schools in Midcity.

COSTS OF SAMPLING AND TESTING

Values of M, N, and L were calculated for each estimation method and both types of clusters, using data from Tables 7.10 and 7.11. Making the conservative assumption that each classroom used as a cluster was in a different school, values of N, M, and L were as shown in Table 7.12.

For each sampling and estimation method, the use of classrooms as clusters resulted in fewer pupils being tested than did the use of schools as clusters. PPES estimation with classrooms as clusters resulted in the fewest pupils being tested.

TABLE 7.11

Four Cluster Sampling and Estimation Procedures: Sample Sizes (Numbers of Classrooms) Required for Estimator Variances of Four or Less[a]

Procedure	Required Sample Size
Simple random sampling of clusters with unbiased estimation	33 classrooms
Simple random sampling of clusters with ratio estimation	10 classrooms
Sampling with probabilities proportional to class enrollments (PPS)	12 classrooms
Sampling with probabilities proportional to Lorge-Thorndike score totals for class sections (PPES)	2 classrooms

[a] Calculated for a population of 45 sixth-grade class sections in Midcity.

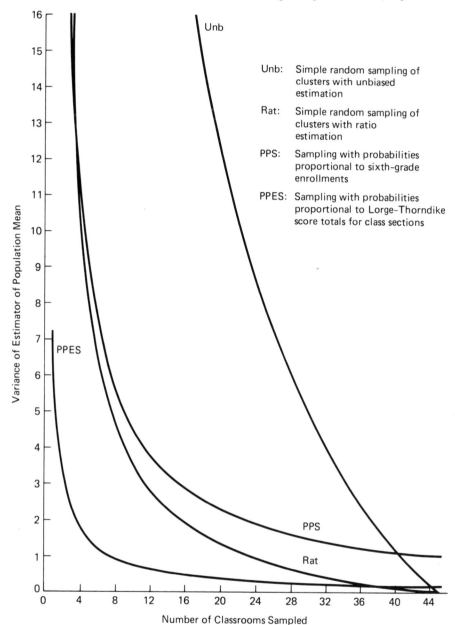

FIGURE 7.11. **Variances of estimators of mean achievement versus number of classrooms sampled (classrooms in Midcity used as clusters).**

When the total cost of testing is to be minimized, Equation (7.29) can be used to determine the most economical method of estimation. However, some assumptions on the relationships between cost coefficients are necessary.

The cost function [Equation (7.29)] can be simplified by assuming the school district cost C_0 to be independent of the choice of sampling unit. If C_0 is equal to zero, testing costs for each procedure are reduced by a constant amount. If the

TABLE 7.12

Number of Schools N, Classrooms M, and Pupils L Required to Estimate Mean Sixth-Grade Reading Achievement for Pupils in Midcity with Estimator Variances of Four or Less

Sampling Unit and Procedure	Required Sample Sizes		
	Number of Schools (N)	Number of Classrooms (M)	Number of Pupils (L)
Schools as clusters, RSC-unbiased estimation	18	39	1041
Schools as clusters, RSC-ratio estimation	7	15	394
Schools as clusters, PPS estimation	10	22	577
Schools as clusters, PPES estimation	4	9	236
Classrooms as clusters, RSC-unbiased estimation	21[a]	33	865
Classrooms as clusters, RSC-ratio estimation	10	10	262
Classrooms as clusters, PPS estimation	12	12	314
Classrooms as clusters, PPES estimation	2	2	53

[a] Maximum possible value.

costs of testing in a classroom, C_2, and the costs of testing an individual pupil, C_3, are assumed to be multiples of the costs of testing in a school, C_1, the total costs of all four sampling and estimation procedures, with either choice of clusters, can be compared directly. It is not necessary to assume a specific value for C_1 since the total costs for each method can be expressed as a multiple of C_1.

The simplest assumption, without ignoring any costs, is that all cost coefficients are equal; that is,

$$C_1 = C_2 = C_3$$

For this assumption, the total cost of using schools as clusters with RSC-unbiased estimation was found to be

$$C = 18C_1 + 39C_1 + 1041C_1 = 1098C_1$$

The corresponding cost using schools as clusters with RSC-ratio estimation was found to be

$$C = 7C_1 + 15C_1 + 394C_1 = 416C_1$$

As expected, RSC-ratio estimation was found to be far less costly than RSC-unbiased estimation.

In reality, it is unlikely that administrative costs incurred at the school level would be the same as those incurred at the classroom level. It is even less likely that the cost of testing one pupil would equal the administrative costs at either the school or the classroom level. Classroom administrative costs would probably be less than those incurred at the school level, amounting to perhaps one-half to one-fifth as much. The cost of testing one pupil might be only one-tenth to one-twenty-fifth the administrative costs at the classroom level.

Costs of testing using each sampling and estimation procedure and both choices of sampling unit are shown in Figures 7.12 and 7.13. The cost functions

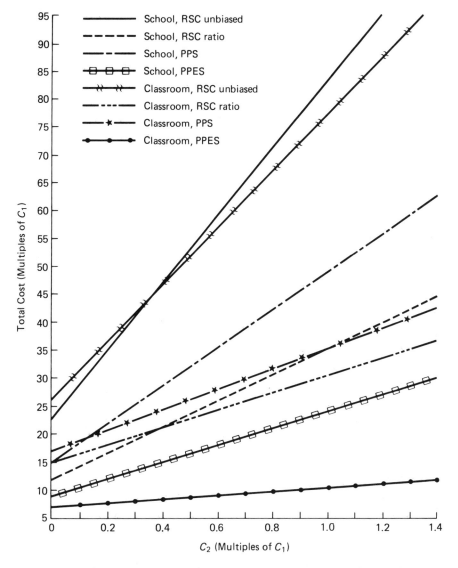

FIGURE 7.12. **Total testing cost as a function of costs of testing in a school C_2 (data from Midcity with a linear cost function: C_1 = school testing costs; C_2 = classroom testing costs; C_3 = cost of testing one pupil, assumed to equal 0.02 C_2).**

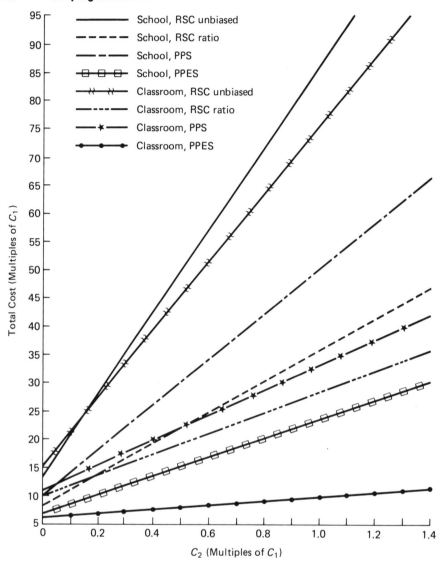

FIGURE 7.13. **Total testing cost as a function of costs of testing in a school C_2 (data for Midcity with a linear cost function: C_1 = school testing costs; C_2 = classroom testing costs; C_3 = cost of testing one pupil, assumed to equal 0.10 C_2).**

shown in Figure 7.12 were based on the assumption that the cost of testing one pupil, C_3, was one-fiftieth the administrative costs in a classroom, C_2. For the cost functions in Figure 7.13, C_3 was assumed to be one-tenth the size of C_2.

Comparison of the cost functions in Figures 7.12 and 7.13 showed the total cost of testing to be lowest when classrooms were used as clusters and PPES estimation was used. This combination of cluster size and estimation method provided the lowest total cost over the entire range of classroom costs, C_2, and for both choices of pupil costs, C_3. The next lowest total cost was provided by PPES estimation with schools as sampling units. This combination was second best over the entire range of cost coefficients considered. Different auxiliary variables were used with the PPES procedure for the two choices of cluster size (schools or classrooms). Had the same variable been used, it is likely that using schools as clusters would have provided the lowest total costs when C_2 was very small. This pattern was exhibited for all other methods of sampling and estimation.

When RSC-unbiased estimation was used with these data, total costs were much higher than those incurred using any other procedure. When pupil cost was assumed to be one-fiftieth the classroom cost, RSC-unbiased estimation with schools as clusters was less expensive than RSC-unbiased estimation with classrooms as clusters, provided the classroom cost C_2 was no more than one-half the school cost C_1. When pupil cost was assumed to be one-tenth the classroom cost, RSC-unbiased estimation with schools as clusters was less expensive than RSC-unbiased estimation with classrooms as clusters, provided C_2 as no more than one-eighth C_1. The same kinds of relationships held for RSC-ratio estimation and PPS estimation. School clusters provided lower total cost than did classroom clusters when C_2 was a small fraction of C_1. For larger values of C_2, classroom clusters were more economical. As the ratio C_3/C_2 became larger, classrooms as clusters provided lower total costs than did schools as clusters, for smaller values of C_2.

The cost function [Equation (7.29)] and the cost coefficients used in Example 7.2 were intended to be illustrative. No doubt, different cost functions will be appropriate in different situations, even if they involve school testing programs. For other applications, a totally different cost analysis would likely be required. However, in most applications, it will probably be necessary to associate sets of costs with each of the administrative levels involved in the survey, just as we did in Example 7.2. We would therefore expect our basic approach to cost analysis to generalize well beyond the specific example.

Estimation of Population Totals

All of the single-stage cluster sampling methods we have discussed for estimation of population means also apply to estimation of population totals. Although the estimators of population totals differ somewhat from corresponding estimators of population means, the differences are not substantial. We will therefore list appropriate estimators and estimator variances with little elaboration.

THE BASIC ESTIMATORS

RSC-UNBIASED ESTIMATION

When a sample of n clusters is selected from a population of N using simple random sampling and unbiased estimation, a consistent and unbiased estimator of the population total is given by

$$\hat{Y} = \frac{N}{n} \sum_{i=1}^{n} M_i \bar{y}_i \qquad (7.30)$$

where M_i denotes the number of elements in the ith sampled cluster and $\bar{y}_i$ denotes the mean of the sampling variable for elements in the ith sampled cluster.

RSC-RATIO ESTIMATION

When a sample of n clusters is selected from a population of N using simple random sampling and ratio estimation, a consistent estimator of the population total is given by

$$\hat{Y}_R = M_0 \frac{\sum_{i=1}^{n} M_i \bar{y}_i}{\sum_{i=1}^{n} M_i} \tag{7.31}$$

where M_0 denotes the total number of elements in the population. All other terms are as defined previously.

SAMPLING WITH PROBABILITIES PROPORTIONAL TO CLUSTER SIZES (PPS)

When clusters are selected with probabilities proportional to the number of elements they contain and standard PPS estimation is used, an unbiased estimator of the population total is given by

$$\hat{Y}_{\text{pps}} = \frac{M_0}{n} \sum_{i=1}^{n} \bar{y}_i \tag{7.32}$$

All terms are as defined above.

SAMPLING WITH PROBABILITIES PROPORTIONAL TO TOTALS OF AN AUXILIARY VARIABLE (PPES)

When clusters are sampled with probabilities proportional to the totals of an auxiliary variable over all elements they contain and standard PPES estimation is used, an unbiased estimator of the population total is given by

$$\hat{Y}_{\text{ppes}} = \frac{1}{n} \sum_{i=1}^{n} \frac{M_i \bar{y}_i}{z_i} \tag{7.33}$$

where z_i denotes the probability of selection for the ith cluster. All other terms are as defined previously.

ESTIMATORS OF ESTIMATOR VARIANCE

RSC-UNBIASED PROCEDURE

When clusters are selected through simple random sampling and the unbiased estimator of the population total is used, the following equation provides an unbiased estimator of the variance of the estimated population total:

$$v(\hat{Y}) = \frac{N^2(1-f)}{(n-1)n} \sum_{i=1}^{n} (y_i - \bar{y})^2 \tag{7.34}$$

where f denotes the sampling fraction for clusters, n/N, and $\bar{y}$ denotes the average of the totals in sampled clusters, and equals

$$\frac{1}{n} \sum_{i=1}^{n} M_i \bar{y}_i$$

All other terms have been defined previously.

RSC-RATIO PROCEDURE

When clusters are selected through simple random sampling and the ratio estimator of the population total is used, the following equation provides a slightly biased estimator of the variance of the estimated population total. The degree of bias is proportional to the inverse of the sample size of clusters.

$$v(\hat{Y}_R) = \frac{N^2(1 - f)}{(n - 1)n} \sum_{i=1}^{n} M_i^2(\bar{y}_i - \hat{\bar{Y}}_R)^2 \tag{7.35}$$

where $\hat{\bar{Y}}_R$ denotes the RSC-ratio estimator of the population mean and is given by Equation (7.9). Other terms have been defined previously.

PPS SAMPLING AND ESTIMATION

When clusters are selected with probabilities proportional to the number of elements they contain and standard PPS estimation is used, an unbiased estimator of the variance of the estimated population total is given by

$$v(\hat{Y}_{\text{pps}}) = \frac{M_0^2}{(n - 1)n} \sum_{i=1}^{n} (\bar{y}_i - \hat{\bar{Y}}_{\text{pps}})^2 \tag{7.36}$$

where $\hat{\bar{Y}}_{\text{pps}}$ denotes the PPS estimator of the population mean, defined by Equation (7.15). All other terms are as defined above.

PPES SAMPLING AND ESTIMATION

When clusters are sampled with probabilities proportional to the totals of an auxiliary variable over all elements they contain and standard PPES estimation is used, an unbiased estimator of the variance of the estimated population total is given by

$$v(\hat{Y}_{\text{ppes}}) = \frac{1}{(n - 1)n} \sum_{i=1}^{n} \left(\frac{M_i \bar{y}_i}{z_i} - \hat{Y}_{\text{ppes}} \right)^2 \tag{7.37}$$

where $\hat{Y}_{\text{ppes}}$ denotes the PPES estimator of the population total, defined by Equation (7.33). All other terms are as defined above.

CONFIDENCE INTERVALS

As was the case in formulating confidence intervals on the population mean, an approximate $100(1 - \alpha)$ percent confidence interval on the population total can be formed for any of the cluster sampling and estimation methods by using the general formulas

$$\hat{Y}_U = \hat{Y} + t\sqrt{v(\hat{Y})}$$

$$\hat{Y}_L = \hat{Y} - t\sqrt{v(\hat{Y})}$$

(7.38)

where

> t = abscissa on Student's t-distribution with degrees of freedom equal to $n - 1$ and $100(\alpha/2)$ percent of the distribution to its right
>
> $\hat{Y}$ = estimate of the population total associated with a cluster sampling procedure
>
> $v(\hat{Y})$ = estimated variance of estimate of the population total associated with a cluster sampling procedure.

Use of the t-distribution with Equation (7.38) assumes that the estimator of the population total follows a normal distribution and that the variance estimator has a chi-square distribution with $n - 1$ degrees of freedom. Violation of either of these assumptions might result in inaccurate confidence statements. The empirical analyses of confidence intervals on the population mean discussed earlier in this chapter, although limited in generalizability, lend some support to the expectation that these approximate confidence intervals are likely to be reasonably accurate in the long run.

Estimation of Population Proportions

The transformations of the sampling variable defined in earlier chapters for estimation of population proportions can also be used with single-stage cluster sampling and estimation procedures. Suppose we wish to estimate the proportion of elements in the population that have values of the sampling variable within some interval, say between C_1 and C_2, or that have some attribute.

If y_{ij} denotes the value of the sampling variable for the jth element in the ith cluster in the population, the transformation

$$\delta(ij) = \begin{cases} 1 & \text{if } C_1 \leq y_{ij} \leq C_2, \text{ or if the element has the attribute of interest} \\ 0 & \text{if } y_{ij} < C_1 \text{ or } y_{ij} > C_2, \text{ or if the element does not have the attribute of interest} \end{cases}$$

(7.39)

will allow all the formulas for estimation of the population mean, as well as those for estimation of the variances of the estimators, to be used directly with the transformed values. That is, the population proportion and the variances of estimators of the population proportion can be computed and estimated using the equations given above for computing and estimating the population mean and the variances of estimators of the population mean, merely by using transformed values of the sampling variable, as defined in Equation (7.39).

Using the resulting estimators in Equation (7.38), approximate confidence intervals on the population proportion can also be found.

Summary

Single-stage cluster sampling offers great administrative convenience and, in some situations, will be the only feasible approach to sampling and estimation. One advantage of single-stage cluster sampling is its requirement that the sampling frame contain an exhaustive listing of clusters, rather than an exhaustive listing of population elements. In many applications, a sampling frame of elements will not be available, and this alone will preclude the use of a number of other sampling and estimation methods.

When survey data must be collected from individuals in a geographically diverse region, single-stage cluster sampling can have the advantage of concentrating groups of potential respondents in a relatively small number of contiguous locations. When used in an interview survey, the savings resulting from cluster sampling can be substantial if interviewers' travel costs are high.

The advantages of single-stage cluster sampling may, in some applications, be outweighed by the disadvantage of low statistical efficiency. Depending on the population being sampled and the specific sampling and estimation procedure employed, single-stage cluster sampling might require a far larger sample of elements than alternative procedures (such as stratified random sampling) in order to realize the same degree of estimation precision. Whether low statistical efficiency results from the use of single-stage sampling depends on such factors as the choice of sampling units (clusters), the variability of elements within clusters, and the kinds of information available to influence the selection of clusters. In general, the more heterogeneous the elements within clusters, the higher the efficiency of single-stage sampling and estimation procedures. Since, the precision of estimators associated with single-stage procedures depends more on the number of clusters sampled than on the number of elements sampled, it is typically the case that smaller cluster units will lead to higher estimation precision than will larger ones.

The empirical results presented in this chapter, although limited in their generalizability, suggest that some cluster sampling and estimation procedures can provide a high degree of statistical efficiency, at least when applied to some populations. In particular, the PPES sampling and estimation method was found to be highly efficient when applied to the problem of estimating the mean reading achievement of a population of sixth-grade pupils in a medium-sized city.

The CLUSAMP-I computer program described in Appendix D-1 and the CLUSAMP-II computer program described in Appendix D-2 should facilitate the evaluation of single-stage cluster sampling and estimation procedures in specific applications. Among other statistics, these programs compute the variances of estimators resulting from single-stage cluster sampling, when data for an entire population are used as input. Statistics that permits a comparative analysis of the efficiencies of all the single-stage cluster sampling and estimation procedures discussed in this chapter are provided by these computer programs.

Exercises

In the exercises that follow, you will use data for an entire population to evaluate the efficiencies of various single-stage cluster sampling and estimation procedures.

Suppose you want to estimate the mean weekly income of female heads of households who reside in 20 public housing projects in the city of Chicago. Various single-stage cluster sampling and estimation procedures are to be considered, with housing projects used as clusters.

Relevant data for all 20 housing projects in Chicago are as follows:

Housing Project	Number of Female Residents[a]	Mean Age of Female Residents	Mean Weekly Income of Female Residents, $
1	56	34	66
2	65	33	67
3	71	38	71
4	58	34	56
5	47	34	65
6	66	38	71
7	55	37	75
8	99	38	71
9	57	39	75
10	40	37	68
11	59	36	70
12	72	40	73
13	43	35	59
14	63	36	66
15	38	37	71
16	51	34	66
17	51	35	71
18	41	34	68
19	29	41	82
20	74	35	66

[a] Assume that all female residents are heads of households.

Use the data in this table to solve the following problems:

1. Suppose you were to select a simple random sample of 10 housing projects from the population of 20 (*Note:* Do *not* draw a sample at this point), and you used an unbiased estimator of the mean weekly income of female heads of households in all 20 housing projects. What would be the population value of the variance of your estimator?

2. Suppose you were to use the same sampling procedure and sample size as in Exercise 1, but you used a ratio estimator of the mean weekly income of female heads of households in all 20 housing projects. What would be the approximate population value of the mean square error (MSE) of your estimator? *Note:* MSE = variance + (bias)^2.

3. Suppose you were to select a sample (sampling with replacement) of 10 of the 20 housing projects by sampling projects with probabilities proportional to their total number of female residents. If you then used conventional PPS estimation, what would be the population value of the variance of the PPS estimator of the mean weekly income of female heads of households in the 20 housing projects?

4. Suppose you were to select a sample (sampling with replacement) of 10 of the 20 housing projects by sampling projects with probabilities proportional to their totals of the *ages* of female residents. What would be the population variance of the conventional PPES estimator of the mean weekly income of female heads of households in the 20 housing projects?

5. For the one sampling and estimation procedure that affords the greatest estimation precision (based on the results of Exercises 1 through 4), determine the number of housing projects you would have to sample in order to estimate the mean weekly income of female heads of households within $5 per week, with 95 percent confidence.

6. Suppose that your simple random sample of housing projects had resulted in the following projects being selected: numbers 2, 4, 6, 8, 10, 12, 14, 16, 18, and 20. For this sample, use ratio estimation (RSC-ratio) to estimate:

   A. The mean weekly income of female heads of households in all 20 housing projects
   B. The variance of your estimator (ignore the bias)

Compute a 90 percent confidence interval on the population mean weekly income of female heads of households.

8

Implications for Practice: Summing Up

Why Use Sampling?

Generalization is an essential component of research in education and the social sciences. We can think of no practical case, with the possible exception of a health professional studying an individual patient or client, where a researcher is interested in results that apply solely to the observed objects of inquiry. It is far more common for researchers to collect data for some units and then assert that their findings apply to some larger population, of which the observed units are a part. So researchers typically treat their observed units as a sample, regardless of the process used to select them. Sampling, whether formal and planned or informal and haphazard, is an essential part of research in education and the social sciences.

In Chapter 1 we discussed the forms of generalization that occur when research results are interpreted. One form, statistical generalization, depends largely on the quality of the plans and procedures used to select units from the population. Two types of errors invalidate statistical generalizations: bias errors and random errors. Bias errors occur when the observed units are not representative of the population to which research results are generalized. Lack of representativeness can be caused by a faulty design for sampling or by inadequate care in fulfilling the requirements of a sound design. Random errors are inevitable in empirical research, unless data are collected from an entire population. A major virtue of a sound design for sampling is the opportunity it affords for controlling the magnitude of random errors.

Since, in most sampling designs, the level of random error is reduced when sample size is increased, it might seem best to avoid sampling altogether by collecting data from the entire population. This procedure is usually impractical.

In addition, conducting a "census," that is, collecting data from or for every unit in the population, is not always the best procedure.

Even in those situations where measurements or observations can be obtained for the entire population of interest, it will often be beneficial to limit data collection to a sample. When research funds are constrained, as is almost always the case, collection of data from an entire population may limit the types of data collected, the quality of data collected, or the depth and quality of data analysis. Funds expended for data collection cannot be used for research planning, data reduction, data analysis, or reporting of results. Within a fixed budget, it may be possible to monitor carefully the quality of data collection, if data are collected from a small percentage of a population, but impossible to monitor the quality of data collected from the entire population. When experts have considered both the statistical and nonstatistical errors that arise in survey research, they often conclude that total error can actually be reduced through the use of sampling.

In earlier chapters, we reviewed a substantial number of sampling and estimation procedures. Although these procedures differ in their assumptions, mechanics, and information requirements, they share some essential characteristics. All are *probability sampling procedures*. Effective statistical generalization requires the use of a probability sampling procedure, since only these procedures ensure the long-term representativeness of resulting samples. When ad hoc sampling procedures are used, for example, purposeful selection of survey respondents or "snowball sampling" in which one respondent suggests others who are similar in some characteristics, bias inevitably results. We can never be sure that any collection of ad hoc samples, regardless of the number of samples drawn, will adequately represent the population of interest.

Another important consequence of probability sampling is the opportunity it affords to estimate the magnitude of likely estimation errors. The sampling distributions of most estimators associated with probability sampling procedures are known. It is therefore possible to compute the probability of encountering estimation errors in excess of some tolerable bound and to design probability sampling plans that have sample sizes sufficient to limit estimation errors to acceptable levels. In addition, probability sampling procedures allow us to compute confidence limits on the population parameters of interest. Nonprobability sampling procedures do not offer these advantages and opportunities.

Considerations in Selecting a Sampling Procedure

Once you are convinced that sampling is a good thing and that a probability sampling procedure ought to be used, the next question is which sampling procedure to choose. Economy is a clear and obvious virtue and certainly should be considered when choosing a sampling procedure. But economy comes in many forms, and the sampling procedure that minimizes the sample size for a fixed level of estimation precision might not be the most economical. Feasibility is an even more obvious virtue. An economical sampling procedure is worthless if it cannot be applied in the situation at hand. We will consider these factors and a few others in greater detail.

Before a sampling procedure is selected, the population of interest and the parameters to be estimated must be clearly defined. In Chapter 1, we illustrated the critical need for a precise operational definition of the population to which generalizations are to be made. One defining property of a probability sampling procedure is that it be applied in a situation where any potential sampling unit can be unequivocally classified either as a member of the population of interest or outside the population of interest.

It is impossible to make a reasoned choice of the best sampling procedure unless the population parameters to be estimated have been clearly defined. Although this admonition might seem obvious, distinctions among population parameters are sometimes quite subtle. For example, estimating the average annual income per household and estimating the average annual income per wage earner present very different problems. In the first problem, households are the units of analysis and, at some point in a sampling design, will have to be used as sampling units. In the second problem, individual wage earners are the units of analysis. Although households might be used as sampling units in this problem also, it is theoretically possible to sample wage earners, either directly or ultimately, in a cluster sampling design. In addition, the estimators used in the second problem would not be the same as those used in the first. So specifying what is to be estimated and the population for which estimations are desired are essential first steps in selecting a sampling procedure.

Feasibility must be considered before addressing economy. If a sampling procedure cannot be applied in the situation at hand, its economy is no longer an issue.

Often, the lack of an appropriate sampling frame is the greatest impediment to applying a desired sampling procedure. Many of the sampling procedures reviewed in earlier chapters require a complete listing of all elements in the population of interest. When populations are large or cover a large political unit such as a state or the nation, a listing of individuals with specified characteristics will rarely be available. In some situations, sampling frames can be constructed for use in a research study, but more often, the cost and time required will be prohibitive. In such situations, the only feasible sampling procedures are those that involve cluster sampling, perhaps with sampling of elements at the final stage.

Even when existing sampling frames permit the use of procedures that sample elements directly, inconvenience and administrative complexity might make such procedures undesirable. An example considered at some length in earlier chapters, estimating the mean achievement of pupils in a school system, is a good case in point. School systems always have a complete list of the pupils they enroll. However, selection of a simple random sample of pupils or stratified sampling of pupils would almost certainly mean that some pupils would be selected from each classroom in the school system. The administrative costs of sampling pupils from every classroom would be high. All classrooms would be disrupted by having some pupils leave while others remained. The sampled pupils would be deprived of instruction that the other pupils obtained and would have to be given remedial or supplementary instruction at a later time. Special facilities, outside regular classrooms, would have to be used for data collection. A cluster sampling procedure that used classrooms as clusters would have avoided these problems.

Another consideration in selecting a sampling procedure is the form in which sampling frames are available. If the sampling frame exists in a form that can be read directly by a computer (for example, on magnetic tape or punched cards), selection of a simple random sample or a stratified random sample could be completed readily. However, if the sampling frame consists of physical records in individual file folders and the population is large, selection of a simple random sample would be very tedious and time-consuming. All the records would have to be counted and numbered, or a computerized listing would have to be constructed. An obvious sampling alternative that would not require the numbering of records would be linear systematic sampling; folders could be selected from the files at a fixed physical interval.

Choosing the sampling procedure that maximizes estimation precision for a fixed sample size would seem to be an appropriate selection strategy. Certainly, statistical precision is a major consideration in designing a sampling plan. However, the sampling procedure that uses the fewest elements to obtain a specified level of precision might not result in the most economical sampling design. Other costs must be considered as well.

In a survey that involves face-to-face interviews, a major expense is the transportation of interviewers. If potential respondents are scattered over a wide area, the cost of reaching them may be prohibitive. Although a sampling plan that calls for direct selection of respondents might require a small number of respondents in order to realize the desired level of estimation precision, such a plan might cost more than a cluster sampling procedure that requires a larger sample of respondents. With a judicious choice of cluster sampling units, potential respondents can be grouped in contiguous locations, with a consequent reduction in travel distance and time. Even though a larger number of interviews might be required in a cluster sampling design, the reduction in travel cost might minimize overall research costs. Clearly, a careful cost analysis is essential in selecting the best sampling procedure for a given situation.

Summary of Empirical Results

Many of the examples given in earlier chapters had as an objective the estimation of the same population parameter: the mean reading achievement of sixth-grade pupils in a medium-sized city (Midcity). We applied 17 different sampling and estimation procedures to this problem, including the use of two types of cluster sampling units. Although these examples were intended to be illustrative and the generalizability of their statistical results is limited, it is of some interest to examine the relative efficiencies of these procedures in this single application. Such an examination will show, among other things, that sample size requirements vary substantially, depending on the choice of sampling and estimation procedure.

One sampling and estimation procedure is said to be more efficient than another if, using the same sample size, it provides more precise estimation of a population parameter. One procedure is also said to be more efficient than another if, while affording the same level of estimation precision, it requires a smaller sample size. We will examine the relative efficiencies of the 17 sampling and estimation procedures used with Midcity data by comparing the sample sizes required to estimate the mean reading achievement of sixth-grade pupils within ±0.2 grade equivalent units of the actual population mean with 95 percent confidence. To be concise, we will use the following abbreviations when referring to the sampling and estimation procedures:

Sampling and Estimation Procedure	Abbreviation
Simple random sampling with unbiased estimation	SRS
Stratified random sampling with unbiased estimation and proportional allocation of sample sizes to strata	Strat-prop
Stratified random sampling with unbiased estimation and Neyman optimal allocation of sample sizes to strata	Strat-opt
Linear systematic sampling from a population arranged in alphabetic order	LSS-alpha
Linear systematic sampling from a population arranged in increasing order of an auxiliary variable	LSS-inc
Linear systematic sampling from a population arranged in increasing order of an auxiliary variable, with end corrections	LSS-E.C.
Linear systematic sampling from a population arranged in increasing order of an auxiliary variable, with order reversed in alternate strata	LSS-O.R.
Centrally located systematic sampling	CSS
Balanced systematic sampling	BSS
Single-stage cluster sampling with simple random sampling of clusters, schools used as clusters, and unbiased estimation	RSC-schools-unb
Single-stage cluster sampling with simple random sampling of clusters, schools used as clusters, and ratio estimation	RSC-schools-rat
Single-stage cluster sampling with schools used as clusters; clusters selected with probabilities proportional to cluster sizes and PPS estimation	PPS-schools
Single-stage cluster sampling with schools used as clusters; clusters selected with probabilities proportional to values of a school-related auxiliary variable and PPES estimation	PPES-schools
Single-stage cluster sampling with simple random sampling of clusters, classrooms used as clusters, and unbiased estimation	RSC-class-unb

Sampling and Estimation Procedure	Abbreviation
Single-stage cluster sampling with simple random sampling of clusters, classrooms used as clusters, and ratio estimation	RSC-class-rat
Single-stage cluster sampling with classrooms used as clusters; clusters selected with probabilities proportional to cluster sizes and PPS estimation	PPS-class
Single-stage cluster sampling with classrooms used as clusters; clusters selected with probabilities proportional to values of a classroom-related auxiliary variable and PPES estimation	PPES-class

These 17 sampling and estimation procedures required the sample sizes shown in Table 8.1 in order to estimate the mean reading achievement of Midcity's sixth-graders within ±0.2 grade equivalent units with 95 percent confidence.

The results in Table 8.1 show a large variation in required sample sizes. Strat-prop and Strat-opt using Lorge-Thorndike scores for stratification of pupils were the most efficient procedures. Using these procedures, desired estimation precision was obtained with a sampling fraction smaller than 3 percent. All the systematic sampling procedures except BSS provided acceptable precision with a sampling fraction of 5 percent. Most notable was the efficiency of LSS-alpha, which combined the advantages of simplicity and administrative convenience. The mean square errors associated with LSS-alpha showed an orderly progression as a function of sampling fraction, which was not shown by several other systematic sampling procedures (see Table 6.2). The mean square error for LSS-alpha was 2.49 raw score points squared when the sampling fraction was 5 percent, suggesting that our precision requirements might be realized with an even smaller sampling fraction. PPES-class also required a sampling fraction of less than 5 percent to meet our precision requirements. Unlike several of the systematic sampling procedures, PPES-class provided consistent estimation of the population mean. It also afforded the administrative convenience of testing all pupils in selected classrooms.

Some sampling and estimation procedures required fewer pupils, while others required fewer classrooms. Reducing the number of pupils selected and reducing the number of classrooms in which testing must take place are both forms of economy. School administrators may ascribe different values to each, with some administrators giving greater weight to reduction of the number of classrooms used for testing.

When SRS is used, the number of classrooms from which sampled pupils are selected is a random variable. Data in Table 8.1 show that with SRS, 106 pupils had to be tested to achieve the desired level of precision. Since Midcity had 43 classrooms with an average enrollment of 25 pupils per classroom, the 106 pupils required by SRS might be selected from all 43 classrooms or from as few as four or five. The smaller numbers are unlikely. With SRS, the expected number of pupils that would be selected from each classroom is about 2.5.

TABLE 8.1

Sample Size Required to Estimate Mean Reading Achievement within ±0.2 Grade Equivalent Units with 95 Percent Confidence[a]

Sampling Procedure	Required Sample Size
Simple random sampling (SRS)	106 pupils
Stratified sampling by Lorge-Thorndike ability test scores—six strata:	
Proportional allocation (Strat-prop)	26 pupils
Optimal allocation (Strat-opt)	25 pupils
Linear systematic sampling:	
Alphabetic order (LSS-alpha)	≤ 59 pupils[b]
Increasing order of Lorge-Thorndike scores (LSS-inc)	≤ 59 pupils[b]
Increasing order of Lorge-Thorndike scores; end corrections used (LSS-E.C.)	≤ 59 pupils[b]
Order reversed in alternate strata (LSS-O.R.)	≤ 59 pupils[b]
Centrally located systematic samples (CSS)	≤ 59 pupils[b]
Balanced systematic sampling (BSS)	118 pupils
Single-stage cluster sampling:	
Unbiased estimation, schools used as clusters (RSC-schools-unb)	1041 pupils
Ratio estimation, schools used as clusters (RSC-schools-rat)	394 pupils
Probabilities proportional to school enrollments, schools used as clusters (PPS-schools)	577 pupils
Probabilities proportional to fifth-grade SCAT score totals, schools used as clusters (PPES-schools)	236 pupils
Unbiased estimation, classrooms used as clusters (RSC-class-unb)	865 pupils
Ratio estimation, classrooms used as clusters (RSC-class-rat)	262 pupils
Probabilities proportional to classroom enrollments, classrooms used as clusters (PPS-class)	314 pupils
Probabilities proportional to Lorge-Thorndike score totals, classrooms used as clusters (PPES-class)	53 pupils

[a] Midcity data, population size = 1180 sixth-grade pupils.
[b] Five percent is the smallest sampling fraction investigated. Smaller sampling fractions might provide acceptable precision for these sampling methods.

Therefore, in most samples, at least one pupil probably would be selected from each classroom.

For systematic sampling procedures also, the number of pupils selected from each classroom would be a random variable. The arrangement of pupils' names in the sampling frame and the criteria for assigning pupils to classrooms would affect the number of classrooms from which pupils would be selected. With LSS-alpha, since the sample size needed to realize our precision requirements was 59 pupils (see Table 8.1), pupils probably would be selected from every classroom in Midcity. With LSS-inc, BSS, or CSS, if we assume that pupils were arranged in the sampling frame in increasing order of their Lorge-Thorndike ability test scores, then determining the number of classrooms from which pupils would be selected is more complex. Mean Lorge-Thorndike scores differed substantially among the schools in Midcity. It is therefore likely that Lorge-Thorndike means differed substantially among classrooms as well. If pupils were assigned to classrooms on the basis of their ability test scores, the homogeneity of their Lorge-Thorndike scores would be maximized within classrooms. With required sample sizes of 59 pupils (see Table 8.1), it is likely that LSS-inc, BSS, and CSS would require sampling pupils from every classroom in Midcity. However, no more than two pupils each would be selected from most classrooms.

The results shown for Strat-opt and Strat-prop in Table 8.1 suggest that pupils would be sampled from no more than 25 or 26 classrooms if these procedures were used, even if each selected pupil came from a different classroom.

Required sample sizes of classrooms were shown for the single-stage cluster sampling procedures in Table 7.12. Among these procedures, greatest economy was realized with the PPES-class method, since 53 pupils would be drawn from two classrooms. If the PPES-schools method were used, pupils would be selected from only nine classrooms.

These results indicate that the sampling and estimation method that requires the smallest number of pupils is not the same as the procedure that requires the smallest number of classrooms, in order to realize the specified level of estimation precision.

The arguments we have advanced with regard to the numbers of classrooms sampled also apply to the numbers of schools sampled. From Table 7.12, we can see that the PPES-class method required sampling pupils from only two schools. If SRS, Strat-opt, Strat-prop, or any systematic sampling procedure was to be used, we would expect pupils to be sampled from all or nearly all 21 elementary schools in Midcity.

To summarize, PPES-class was most economical of classrooms and schools from which sampled pupils were drawn. Strat-prop and Strat-opt required sampling the fewest pupils in order to satisfy our precision requirements. Some of the systematic sampling procedures required testing only a few more pupils than did the stratified sampling procedures.

In several examples in earlier chapters, we estimated the proportion of sixth-graders in Midcity who had reading achievement scores at least 1 grade equivalent unit below the national norm. We used SRS and all the systematic sampling procedures to estimate this parameter. Although we did not evaluate the efficiency of stratified sampling for this problem, we would not expect great economies to be realized. Since proportions in the range 0.2 to 0.8 have nearly identical variances, stratified sampling usually offers little advantage over SRS when estimating population proportions. The exception is the rare case where proportions within strata are extremely disparate.

Although we would expect the efficiencies of PPES-class and PPES-school to be similar for estimation of population proportions and population means, we have not verified this expectation empirically.

Estimation of the proportion of low-achieving sixth-graders in Midcity was examined in Chapters 3, 5, and 6. The results of these analyses are summarized in Table 8.2. The proportion of Midcity sixth-graders with reading achievement

TABLE 8.2

Sampling Fractions Required for Estimation of Proportion of Low-Achieving Pupils[a] for Various Sampling and Estimation Procedures

Sampling and Estimation Procedure	Required Sampling Fractions for Estimation	
	Within ±0.02 at 95% Confidence, %	Within ±0.05 at 95% Confidence, %
Simple random sampling (SRS)	60	20
Linear systematic sampling:		
Population in alphabetic order (LSS-alpha)	25	20
Population in increasing order of Lorge-Thorndike test scores (LSS-inc)	25	10
Population in increasing order of Lorge-Thorndike test scores, end corrections used (LSS-E.C.)	25	10
Population in increasing order of Lorge-Thorndike scores, order reversed in alternate strata (LSS-O.R.)	25	10
Balanced systematic sampling, population in increasing order of Lorge-Thorndike scores (BSS)	25	10
Centrally located systematic samples, population in increasing order of Lorge-Thorndike test scores (CSS)	25	10

[a] Proportion of 1180 sixth-grade pupils in Midcity with reading achievement scores at least 1 grade equivalent unit below national norm.

scores at least 1 grade equivalent unit below the national norm was 0.194. What constitutes an acceptable difference between an actual proportion and an estimated proportion must be determined subjectively. In earlier examples we adopted precision criteria of errors no larger than ±0.02 or ±0.05 with 95 percent confidence. The sampling fractions required to meet these levels of precision are shown in Table 8.2.

Results presented in Table 8.2 show that all systematic sampling procedures were considerably more efficient than SRS for estimating proportions of low-achieving sixth-graders. For a 95 percent confidence interval with a width of ±0.02, SRS required sampling 60 percent of the population, whereas the systematic sampling procedures required sampling fractions of 25 percent. With a 25 percent sampling fraction, all the systematic samples resulted in sample proportions that were within ±0.02 of the population proportion, so the assumption of a 95 percent confidence level appears to be conservative.

When 95 percent confidence intervals with an average width of ± 0.05 were required, the systematic sampling procedures were, with one exception, about twice as efficient as SRS. The exception was LSS-alpha, which was expected to be about as efficient as SRS. SRS and LSS-alpha required sampling fractions of 20 percent; all other systematic sampling procedures required sampling fractions of 10 percent.

If pupils were selected through a systematic sampling procedure with a sampling fraction of 10 percent, they would probably be drawn from every classroom in Midcity. If SRS were used with a sampling fraction of 20 percent, the expected number of pupils selected from a classroom with an enrollment of 25 would be five. It is therefore very likely that pupils would be sampled from every classroom in Midcity in almost all applications of this procedure.

Single-stage cluster sampling would probably be most economical of classrooms and schools when proportions of low achievers are to be estimated. A sampling fraction of 25 percent, which would be five times as large as that required to estimate pupils' mean achievement with acceptable precision, would result in sampling pupils from 11 of Midcity's 43 sixth-grade classes and from no more than six of Midcity's 21 elementary schools.

In summary, when estimating the proportion of low-achieving pupils, almost all systematic sampling procedures required sampling fewer pupils than did SRS. Single-stage cluster sampling, with classrooms as clusters, would be expected to require sampling of pupils from the fewest classrooms and schools.

Computer Programs for Sampling and Estimation

When sampling procedures are used, estimation of population parameters often requires complex and tedious computations. These computations inhibit the use of sampling procedures in research in education and the social sciences.

Four computer programs, of a sort previously unavailable, have been developed to perform the calculations associated with the sampling and estimation procedures discussed in this book. These programs will eliminate the need for hand calculations in evaluating simple random sampling, stratified sampling, systematic sampling, and single-stage cluster sampling.

The STRASAMP-I computer program (Appendix B) provides statistics for evaluating the Strat-prop and Strat-opt sampling and estimation procedures. It accepts data on an entire population and permits evaluation of up to six alternative stratification variables. The program calculates stratum boundaries, sample sizes required to meet specified levels of precision, and sample sizes to be drawn from each stratum. It also calculates the variances of estimators of population means and proportions. With these data, efficiencies of stratified sampling procedures can be compared to those of other methods.

The SYSAMP-I computer program (Appendix C) provides statistics for evaluation and comparison of eight systematic sampling procedures. It accepts data for an entire population. For specified sampling intervals, it calculates mean square errors of estimation for LSS, applied with and without use of end corrections, to populations in the order read on input, to populations that it arranges in increasing order of an auxiliary variable (LSS-inc), and to populations that it first arranges in increasing order of an auxiliary variable and then arranges in reverse order in alternate strata (LSS-O.R.). It also computes mean square errors for BSS and CSS. Additionally, the program computes efficiencies for each systematic sampling procedure, relative to SRS and stratified sampling with one element per stratum.

The CLUSAMP-I computer program (Appendix D-1) provides estimator variances as a function of sample size for four single-stage cluster sampling and estimation procedures: RSC-unb, RSC-rat, PPS, and PPES. The program also computes estimator bias and mean square errors as a function of sample size for the RSC-rat procedure. With these statistics, the efficiencies of single-stage cluster sampling procedures can be compared to those of other procedures.

The CLUSAMP-II computer program (Appendix D-2) provides data for evaluating six single-stage cluster sampling and estimation procedures. The program computes estimated population means and estimates of estimator variances for RSC-unb, RSC-rat, RSC with jackknife-unbiased estimation, and RSC with jackknife-ratio estimation. It also computes these statistics for the PPS and PPES sampling and estimation methods. The program accepts data for an entire population as input. Using the Monte Carlo method, the program selects a specified number of independent cluster samples and computes estimates from the data in each sample. Program output can be used to analyze the empirical distributions of estimators.

All these computer programs are written in the FORTRAN language and require less than 1 minute of central processor time on a mainframe computer when used with a population of 1200 elements. With the exception of CLUSAMP-I, each program requires data for every element or every cluster in a population. To use these programs effectively then, a researcher must have access to data for an entire population that is similar to the population of research interest. Since data archives are common in education and the social sciences, this requirement will not often pose an insurmountable problem. For example, since school systems collect and store data on every student routinely, the information needed to evaluate the use of alternative sampling and estimation procedures in educational research settings is generally available. The data archives of social welfare agencies and research laboratories can provide the information needed to evaluate the use of sampling procedures in many social and behavioral science applications.

The Use of Random Number Tables

Sampling procedures are based on probabilistic models that assume sampled elements are selected "at random." All potential samples are assumed to have prescribed probabilities of selection, and the statistical theories associated with sampling methods are valid only when these probabilities are realized in practice.

Experiments have shown that people are generally not random in their choices of numbers or objects. For example, if a person is asked to repeatedly "randomly choose one of the digits between zero and nine," some digits will be chosen far more frequently than others. Tables of random numbers have been compiled to guard against this sampling bias and to facilitate experiments that require sampling. Three of the most extensive tables and the use of these tables are discussed here.

In 1969, the RAND Corporation published a random number table entitled *A Million Random Digits with 100,000 Normal Deviates.* As the title suggests, the table contains 1 million digits, with each of the numbers 0 through 9 appearing equally often, without a prescribed order. Digits in the table are arranged in groups of five, with 10 groups printed per line. The lines of the table are numbered from 00000 to 19999.

When using a random number table, one must guard against selecting the same digits repeatedly in such a way that randomness is destroyed. For example, reading a column of numbers from the top of a page to the bottom and then reading the same column from the middle of the page to the bottom would result in a correlated set of numbers. To avoid such problems, a random number table should be used with a procedure that ensures randomness. For the RAND table, the following procedure is appropriate.

1. Open the book to any page, alternating between the first third, second third, and last third of the book in repeated experiments.
2. With eyes closed, point to a group of five digits.
3. If the first digit of the group is even, change it to a zero; if it is odd, change it to a one.
4. Use the group of five digits (with the first digit changed to zero or one) to designate the line of the table where reading of random numbers will start.
5. Use the last digit of the five-digit group to designate the column where reading of random numbers will start (if the digit is zero, start in column 10).
6. Starting with the line and column determined above, read as many groups of random numbers as are needed, going from left to right in the table and starting at the left-hand group of digits in each successive row (just as in reading a book).

Groups of five adjacent digits will be sufficient for sampling in populations of 100,000 or less. For larger populations, sampling by hand is impractical and a pseudo-random number generator (computer program) should be used.

Simple random sampling requires that each element in the population have an equal chance of selection and that sampling be conducted without replacement. That is, once an element has been chosen for the sample, it cannot be chosen again. The RAND table of random numbers can be used for simple random sampling. The instructions given above for reading groups of five digits should be followed, and additional steps should be used to meet the assumptions of simple random sampling.

1. Number the elements in the sampling frame sequentially.
2. Use as many digits in each group of five as the exponent of the smallest power of 10 that exceeds the population size. For example, if the population size is 79, it is exceeded by 100, or 10^2; 10,000, or 10^4, is the smallest power of 10 that exceeds a population of size 8274. In the first case, use the last two digits of each group of five; in the second case, use the last four.
3. Select as many elements as desired for the sample, reading the numbers of sampled elements from the RAND table in accordance with the instructions given above. Two classes of random numbers should be ignored: those of elements already sampled and those larger than the size of the population. For example, if the population contains 82 elements, random numbers between 83 and 99 would be ignored as would the numbers of any elements previously designated in the sample.

The process of sampling can be made more efficient by modifying step 3. Each of the random numbers read from the table is multiplied by a fraction equal to the population size divided by the smallest power of 10 that exceeds the population size. The resulting products are rounded to the nearest integer. Then only the numbers of any elements previously designated for the sample are ignored. For example, suppose the population size is 232. The smallest power of 10 that exceeds 232 is 10^3, or 1000. Thus each random number is multiplied by 232/1000, or 0.232. The only random numbers that are ignored are those of elements already in the sample.

A second widely used random number table, compiled by M. G. Kendall and B. B. Smith, is entitled *Tracts for Computers XXIV, Tables of Random Sampling Numbers*. It was published in 1960 and contains 100,000 random digits. The Kendall and Smith table is divided into numbered blocks of 1000 digits. Within each block, lines of the table are numbered from 1 to 25. Within lines, digits are grouped into blocks of four. A randomized procedure for selecting numbers from the Kendall and Smith table is as follows.

1. Open the book to any page, alternating between the first third, second third, and last third of the book in repeated experiments.
2. With eyes closed, point to a group of four digits.
3. Add one to the last two digits of the four-digit group, and use the resulting value to designate the number of a group of 1000 digits.
4. Use the leftmost digit of the block of four to designate the number of a row within the designated thousand digits.
5. Starting with the leftmost block of four digits within the group of 1000 digits and row determined above, read as many consecutive blocks of four digits as are needed, going from left to right in each row (just as in reading a book).

Groups of four adjacent random digits will be sufficient for sampling in populations of 10,000 or less. For larger populations, five or six adjacent digits can be treated as a unit. Once the starting group of digits has been determined, the same procedure for simple random sampling is used for the Kendall and Smith table as for the RAND table.

Since in simple random sampling an element can enter the sample only once, many of the digits read from a random number table will have to be ignored when they appear a second time. With large sampling fractions, many more numbers will be ignored than used, when most of the sample has been selected.

A table that is ideally suited to simple random sampling was compiled by Moses and Oakford (1963). It is entitled *Tables of Random Permutations* and should be useful in sampling problems with populations of 1000 or less. The book contains a series of tables, each table containing random arrangements of numbers within a specified range. The ranges of numbers in the tables are 1–9, 1–16, 1–20, 1–30, 1–50, 1–100, 1–200, 1–500, and 1–1000. Within each table, all numbers within the specified range are printed in randomly arranged blocks. Thus in the table with range 1–20, the numbers from 1 to 20 are randomly arranged in each block, and there are a large number of blocks.

To select a simple random sample from the Moses and Oakford tables, the procedure is as follows.

1. Find the pages that contain the table with the smallest specified range larger than the population size. For example, with a population size of 165, the smallest specified range larger than the population size is 200.

2. Open the book to any page of the table determined in step 1.
3. With eyes closed, point to a block of numbers on the page.
4. Reading from left to right, starting at the upper left-hand corner of the selected block, read the numbers of the elements to be included in the sample. Ignore any numbers larger than the population size.

The Moses and Oakford tables can only be used with populations smaller than 1000.

The STRASAMP-I Computer Program

Program Description

The procedures required for application of stratified sampling involve long and tedious calculations. If several alternative stratification variables are considered to the point of calculating sample size requirements under optimal and proportional allocation, many hours of hand calculation would be necessary to derive the desired results.

A computer program that applies the Dalenius and Hodges procedure to determine stratum boundaries, computes total sample sizes, and determines sample sizes for each stratum under proportional allocation has been developed. The program allows the simultaneous assessment of the sampling efficiency afforded by six or fewer potential stratification variables. The program is written in Fortran IV. Stratum boundaries and sample sizes for a problem involving a population of 1180 pupils were computed for four alternative stratification variables in 4.05 seconds using an IBM 360/67 computer.

PROGRAM INPUT

As input, the STRASAMP-I computer program requires six values entered on a single card and observations on all potential stratification variables for all elements in the population. The six values required are

1. Size of population (N)
2. Desired number of strata (K)
3. Number of potential stratification variables (to be used one at a time) (L)
4. Acceptable limit of estimation error (D)—corresponds to ε in Examples 3.1 through 4.2

5. An abscissa of the normal distribution corresponding to desired level of confidence (T)
6. The index number of the stratification variable that is to be used to estimate variances within strata, called a *proxy variable* (MM)

These values are punched on a single card in the following format:

N—a 4-digit integer value
K—a 2-digit integer value
L—a 2-digit integer value
D—a 10-digit number with decimal punched
T—a 10-digit number with decimal punched
MM—a 2-digit integer value

The next cards should contain the data to be used for stratification. A single card should be punched for each unit in the population. Each data card should contain a value for every potential stratification variable, with a consistent format and order on every card. The data format card is numbered 200, and a card consistent with the user's data should replace the "200 format" card in the listed program.

PROGRAM OUTPUT

Potential stratification variables are evaluated in order, as arranged in the input data. Output consists of a set of data indicating stratum limits and other statistics for each stratification variable. The index number of the stratification variable is listed first:

STRATIFICATION VARIABLE =

Seven columns of data are printed for each stratification variable. STRATUM indicates stratum number. LOWER LIMIT indicates the minimum value of the stratification variable in each stratum. UPPER LIMIT indicates the maximum value of the stratification variable in each stratum. SIZE indicates the size of the population in each stratum. MEAN indicates the mean value of the proxy variable in each stratum. STD. DEVIATION indicates the standard deviation of the proxy variable in each stratum. SAMPLE SIZE PROP. indicates the sample size required in each stratum, if proportional allocation is used.

Total sample sizes required using optimal allocation and proportional allocation, respectively, are printed to the right of the headings

SAMPLE SIZE FOR OPTIMAL ALLOCATION =
SAMPLE SIZE FOR PROPORTIONAL ALLOCATION =

The index number of the variable used as a proxy for the sampling variable when computing within-stratum variances is printed to the right of the heading

INDEX OF PROXY VARIABLE =

Following the printing of the data described above for each stratification variable, the program provides a number of statistics used in the Dalenius and Hodges' procedure for computing stratum boundaries. Output is in columnar form, in the order of input of the stratification variables. The first three headings MINIMUM VALUES OF STRATIFICATION VARIABLES, MAXIMUM VALUES OF STRATIFICATION VARIABLES, and RANGES OF STRATIFICATION VARIABLES are self-explanatory. Under the heading TOP LIMITS OF 20 INTERVAL STRATIFICATIONS are printed the upper limits of the small intervals used in the Dalenius and Hodges procedure for establishing stratum boundaries. Corresponding lower limits of these intervals are next printed under the heading BOTTOM LIMITS OF 20 INTERVAL STRATIFICATIONS. The frequencies or numbers of units that fall in each of the intervals are printed for each stratification variable under the heading POPULATION SIZES WITHIN 20 INTERVAL STRATIFICATIONS. Values of the cumulative square root frequencies are printed under the heading CUMULATIVE SQRT. OF POP. SIZE FOR 20 INTERVALS.

USE OF THE PROGRAM

With the labor of computing stratum boundaries and sample sizes reduced to preparing a card deck for computer input, it becomes feasible to consider a number of alternative stratification variables. Minimization of required sample size is an obvious criterion in the selection of stratification variables, but it is certainly not the only factor to be considered. A set of variables with data already on punched cards would be preferred to variables that required additional tabulation and keypunching. In some cases, the costs of preparing data for use in stratification might offset the savings to be realized through reduction in sample size. Variables with data more readily available would then be preferred for stratification even though they did not lead to minimum sample sizes.

The STRASAMP-I computer program can be used to gain practical insight into the efficiencies afforded by different stratification variables for an array of sampling variables and different numbers of strata. The theoretical results discussed above that relate sampling efficiency to number of strata and the correlation between sampling variables and stratification variables depend on a number of assumptions that may not be met in practice. The stratified sampling program permits these relationships to be verified in practical problems involving real data.

The stratified sampling program is thus a tool to aid in the design of efficient sampling procedures. If used by school systems, it would provide some of the data necessary to develop effective complementary testing programs for individual and institutional analyses.

(Program listing and sample output follow.)

Program Listing

```
// EXEC FORTGCLG
//SYSIN DD *
C THIS PROGRAM IS USEFUL FOR THE EVALUATION OF STRATIFIED SAMPLING
C FOR A PARTICULAR POPULATION.  IT ALLOWS THE  IDENTIFICATION OF A
CNUMBER OF ALTERNATIVE STRATIFICATION VARIABLES, AND  WHEN THE
C NUMBER OF DESIRED STRATA IS INDICATED, CALCULATES OPTIMUM BOUNDARIES
C FOR STRATA ANDTHE SAMPLE SIZES REQUIRED FOR EACH. WHEN AN
C ACCEPTABLE ERROR TOLERANCE AND A CONFIDENCE LEVEL ARE SPECIFIED,
C SAMPLE SIZE REQUIRED FOR STRATIFIED SAMPLING IS COMPUTED FOR
C THE NEYMAN ALLOCATION OF SAMPLES TO STRATA AND FOR PROPORTIONAL
C ALLOCATIONOF SAMPLES TO STRATA. OUTPUT INCLUDES TOTAL SAMPLE
C SIZE REQUIRED AND SAMPLE SIZES FOR EACH STRATUM. STRATUM BOUNDARIES
C ARE ALSO PROVIDED AS OUTPUT.
      INTEGER BK,BSTAR,XN(10,6)
      REAL Y(1500,6),YMIN(6),YMAX(6),RANGE(6),PINT(6),TLIM(20,6),
     CBLIM(20,6),NN(20,6),CUMN(20,6),ZINT(6),SVLIM(10,6),SLIM(10,6),
     CSUMV(10,6),SUMSQV(10,6),XBAR(10,6),S(10,6),SUMSQ(6),
     CSUMPR(6),W(10,6),A(10,6),B(10,6),SSOPT(6),SS(10,6),SSPROP(6)
  100 FORMAT(I4,I2,I2,F10.5,F10.5,I2)
  200 FORMAT(F3.0,1X,F3.0)
C N IS POPULATION SIZE, K IS NO. OF STRATA, L IS NO. OF STRATIFICATION
C VARIABLES, D IS ERROR LIMIT, T IS NORMAL ABSCISSA, MM IS INDEX OF
C PROXY CHARACTERISTIC
C Y(I,J) IS I-TH OBSERVATION FOR THE J-TH STRATIFICATION VARIABLE
      READ (5,100) N,K,L,D,T,MM
      LL=L+1
      DO 300 I=1,N
      READ (5,200) (Y(I,J),J=1,LL)
  300 CONTINUE
      DO 1000 J=1,LL
      YMIN(J)=Y(1,J)
      YMAX(J)=Y(1,J)
      DO 1000 I=2,N
      IF(Y(I,J).LE.YMIN(J)) YMIN(J)=Y(I,J)
      IF(Y(I,J).GE.YMAX(J)) YMAX(J)=Y(I,J)
 1000 CONTINUE
      DO 1200 J=1,LL
      RANGE(J)=YMAX(J)-YMIN(J)
      PINT(J)=RANGE(J)/20.0
      DO 1200 KK=1,20
      TLIM(KK,J)=YMIN(J)+KK*PINT(J)
      BLIM(KK,J)=YMIN(J)+(KK-1)*PINT(J)
      NN(KK,J)=0.0
 1200 CONTINUE
      DO 5100 I=1,N
      DO 5100 J=1,LL
      DO 5000 KK=1,20
      IF(Y(I,J).GT.BLIM(KK,J).AND.Y(I,J).LE.TLIM(KK,J)) GO TO 4500
      GO TO 5000
 4500 NN(KK,J)=NN(KK,J) +1
 5000 CONTINUE
 5100 CONTINUE
      DO 6000 J=1,LL
      DO 6000 KK=1,20
      CUMN(KK,J)=0.0
```

```
6000 CONTINUE
     DO 7000 J=1,LL
     CUMN(1,J)=SQRT(NN(1,J))
     DO 7000 KK=2,20
     AK=KK-1
     CUMN(KK,J)=CUMN(AK,J)+SQRT(NN(KK,J))
7000 CONTINUE
     DO 8000 J=1,LL
     ZINT(J)=CUMN(20,J)/K
8000 CONTINUE
     K2=K-1
     DO 8100 J=1,LL
     DO 8100 II=1,K2
     SVLIM(II,J)=II*ZINT(J)
8100 CONTINUE
     DO 8220 J=1,LL
     KK=1
     DO 8220 II=1,K2
     SLIM(II,J)=TLIM(KK,J)
     DO 8200 KK=1,19
     BK=KK+1
     IF(ABS(CUMN(BK,J)-SVLIM(II,J)).LE.ABS(CUMN(KK,J)-SVLIM(II,J)))
    C GO TO 8150
     GO TO 8220
8150 SLIM(II,J)=TLIM(BK,J)
8200 CONTINUE
8220 CONTINUE
     DO 8300 J=1,LL
     XN(1,J)=0
     SUMV(1,J)=0.0
     SUMSQV(1,J)=0.0
     DO 8300 I=1,N
     IF(Y(I,J).LE.SLIM(1,J)) GO TO 8250
     GO TO 8300
8250 XN(1,J)=XN(1,J) +1
     SUMV(1,J)=SUMV(1,J)+Y(I,MM)
     SUMSQV(1,J)=SUMSQV(1,J)+(Y(I,MM))**2
8300 CONTINUE
     DO 9999 J=1,LL
     SLIM(K,J)=YMAX(J)
9999 CONTINUE
     DO 8400 J=1,LL
     DO 8400 II=2,K
     XN(II,J)=0
     SUMV(II,J)=0.0
     SUMSQV(II,J)=0.0
     DO 8400 I=1,N
     I2=II-1
     IF(Y(I,J).GT.SLIM(I2,J).AND.Y(I,J).LE.SLIM(II,J)) GO TO 8350
     GO TO 8400
8350 XN(II,J)=XN(II,J)+1
     SUMV(II,J)=SUMV(II,J)+Y(I,MM)
     SUMSQV(II,J)=SUMSQV(II,J)+(Y(I,MM))**2
8400 CONTINUE
     DO 8500 J=1,LL
```

```
      DO 8500 II=1,K
      IF(XN(II,J).GT.0)  GO TO 8410
      XBAR(II,J)=0.0
      GO TO 8411
8410  XBAR(II,J)=SUMV(II,J)/XN(II,J)
8411  IF(XN(II,J).GE.2) GO TO 8420
      S(II,J)=0.0
      GO TO 8500
8420  S(II,J)=(SUMSQV(II,J)-(XN(II,J))*(XBAR(II,J))**2)/(XN(II,J)-1)
      S(II,J)=SQRT(S(II,J))
8500  CONTINUE
      DO 350 J=1,LL
      SUMSQ(J)=0.0
      SUMPR(J)=0.0
 350  CONTINUE
      DO 400 J=1,LL
      DO 400 I=1,K
      W(I,J)=FLOAT(XN(I,J))/FLOAT(N)
      A(I,J)=(W(I,J))*(S(I,J))
      SUMPR(J)=SUMPR(J)+A(I,J)
      B(I,J)=(A(I,J))*(S(I,J))
      SUMSQ(J)=SUMSQ(J)+B(I,J)
 400  CONTINUE
      V=(D/T)**2
      DO 500 J=1,LL
      SSOPT(J)=((SUMPR(J))**2)/(V+(1.0/N)*SUMSQ(J))
      SSPROP(J)=(SUMSQ(J)/V)/(1.0+SUMSQ(J)/(V*N))
      DO 500 I=1,K
      SS(I,J)=(SSPROP(J))*(W(I,J))
 500  CONTINUE
      DO 90 J=1,LL
      WRITE (6,2) J
      WRITE (6,3)
      I=1
      WRITE (6,10) I,YMIN(J),SLIM(I,J),XN(I,J),XBAR(I,J),S(I,J),SS(I,J)
      DO 80 I=2,K
      IJ=I-1
      WRITE (6,10)I,SLIM(IJ,J),SLIM(I,J),XN(I,J),XBAR(I,J),S(I,J),SS(I,J
     1)
  80  CONTINUE
      WRITE (6,5) SSOPT(J)
      WRITE (6,6) SSPROP(J)
      WRITE (6,7) MM
  90  CONTINUE
   2  FORMAT('0','STRATIFICATION VARIABLE=',I4)
   3  FORMAT('0','STRATUM LOWER LIMIT UPPER LIMIT  SIZE  MEAN STD. DEVIA
     CTION SAMPLE SIZE PROP.')
   4  FORMAT('0','DESIRED VARIANCE=',F10.5)
   5  FORMAT('0','SAMPLE SIZE FOR OPTIMAL ALLOCATION=',E14.5)
   6  FORMAT('0','SAMPLE SIZE FOR PROPORTIONAL ALLOCATION=',E14.5)
   7  FORMAT('0','INDEX OF PROXY VARIABLE=',I4)
  10  FORMAT(' ',I4,5X,F10.4,2X,F10.4,I6,F10.4,2X,F10.4,12X,F10.4)
      WRITE (6,11)
  11  FORMAT('0','MINIMUM VALUES OF STRATIFICATION VARIABLES')
      WRITE (6,12) (YMIN(J),J=1,LL)
```

```
 12 FORMAT('0',6F12.4)
    WRITE (6,13)
 13 FORMAT('0','MAX VALUES OF STRATIFICATION VRIABLES')
    WRITE (6,14) (YMAX(J),J=1,LL)
 14 FORMAT('0',6F12.4)
    WRITE (6,15)
 15 FORMAT('0','RANGES OF STRATIFICATION VARIABLES')
    WRITE (6,16) (RANGE(J),J=1,LL)
 16 FORMAT('0',6F12.4)
    WRITE (6,17)
 17 FORMAT('0','TOP LIMITS OF 20 INTERVAL STRATIFICATIONS')
    DO 18 KK=1,20
    WRITE (6,19) (TLIM(KK,J),J=1,LL)
 18 CONTINUE
 19 FORMAT(' ',6F12.4)
    WRITE (6,20)
 20 FORMAT('0','BOTTOM LIMITS OF 20 INTERVAL STRATIFICATIONS')
    DO 21 KK=1,20
    WRITE (6,19) (BLIM(KK,J),J=1,LL)
 21 CONTINUE
    WRITE (6,22)
 22 FORMAT('0','POPULATION SIZES WITHIN 20-INTERVAL STRATIFICATIONS')
    DO 23 KK=1,20
    WRITE (6,19) (NN(KK,J),J=1,LL)
 23 CONTINUE
    WRITE (6,24)
 24 FORMAT('0','CUMULATIVE SQRT OF POP. SIZE FOR 20-INTERVALS')
    DO 25 KK=1,20
    WRITE (6,19) (CUMN(KK,J),J=1,LL)
 25 CONTINUE
    WRITE (6,26)
 26 FORMAT('0','VALUES OF SVLIM(II,J)')
    DO 27 II=1,K2
    WRITE (6,19) (SVLIM(II,J),J=1,LL)
 27 CONTINUE
    WRITE (6,28)
 28 FORMAT('0','STRATUM BOUNDARIES ALL VARIABLES')
    DO 29 II=1,K2
    WRITE (6,19) (SLIM(II,J),J=1,LL)
 29 CONTINUE
    WRITE (6,30)
 30 FORMAT('0','A(I,J),B(I,J),W(I,J),PRINTED IN ORDER')
    DO 31 I=1,K
    WRITE (6,19) (A(I,J),J=1,LL)
    WRITE (6,19) (B(I,J),J=1,LL)
    WRITE (6,19) (W(I,J),J=1,LL)
    WRITE (6,32)
 31 CONTINUE
 32 FORMAT(' ','          ')
    WRITE (6,33)
 33 FORMAT('0','SUMPR(J)')
    WRITE (6,19) (SUMPR(J),J=1,LL)
    WRITE (6,34)
 34 FORMAT('0','SUMSQ(J)')
    WRITE (6,19) (SUMSQ(J),J=1,LL)
    RETURN
    END
//GO.SYSIN DD *
```

Sample Output

$DATA

STRATIFICATION VARIABLE = 1

STRATUM	LOWER LIMIT	UPPER LIMIT	SIZE	MEAN	STD. DEVIATION	SAMPLE SIZE PROP.
1	3.1000	4.1800	12	3.7500	0.3425	0.9083
2	4.1800	4.9000	31	4.5129	0.2291	2.3465
3	4.9000	5.4400	15	5.2133	0.1357	1.1354
4	5.4400	6.7000	20	5.8700	0.3063	1.5139

SAMPLE SIZE FOR OPTIMAL ALLOCATION = 0.54770E 01

SAMPLE SIZE FOR PROPORTIONAL ALLOCATION = 0.59041E 01

INDEX OF PROXY VARIABLE = 1

STRATIFICATION VARIABLE = 2

STRATUM	LOWER LIMIT	UPPER LIMIT	SIZE	MEAN	STD. DEVIATION	SAMPLE SIZE PROP.
1	3.6000	4.4500	23	4.0956	0.5023	4.5264
2	4.4500	4.9600	17	4.6588	0.3430	3.3456
3	4.9600	5.6400	19	5.1421	0.4073	3.7392
4	5.6400	7.0000	19	5.7579	0.4903	3.7392

SAMPLE SIZE FOR OPTIMAL ALLOCATION = 0.15036E 02

SAMPLE SIZE FOR PROPORTIONAL ALLOCATION = 0.15351E 02

INDEX OF PROXY VARIABLE = 1

STRATIFICATION VARIABLE = 3

STRATUM	LOWER LIMIT	UPPER LIMIT	SIZE	MEAN	STD. DEVIATION	SAMPLE SIZE PROP.
1	0.0000	90.4000	16	4.0250	0.5745	2.8587
2	90.4000	96.0500	21	4.4333	0.2310	3.7521
3	96.0500	101.7000	18	5.0889	0.3969	3.2161
4	101.7000	113.0000	23	5.7130	0.4445	4.1094

SAMPLE SIZE FOR OPTIMAL ALLOCATION = 0.12792E 02

SAMPLE SIZE FOR PROPORTIONAL ALLOCATION = 0.13936E 02

INDEX OF PROXY VARIABLE = 1

STRATIFICATION VARIABLE = 4

STRATUM	LOWER LIMIT	UPPER LIMIT	SIZE	MEAN	STD. DEVIATION	SAMPLE SIZE PROP.
1	14.0000	35.5000	21	5.5857	0.5425	6.2438
2	35.5000	57.0000	18	5.0389	0.4984	5.3518
3	57.0000	82.8000	16	4.6312	0.6194	4.7572
4	82.8000	100.0000	23	4.2783	0.6599	6.8384

SAMPLE SIZE FOR OPTIMAL ALLOCATION = 0.22914E 02

SAMPLE SIZE FOR PROPORTIONAL ALLOCATION = 0.23191E 02

INDEX OF PROXY VARIABLE = 1

MINIMUM VALUES OF STRATIFICATION VARIABLES

    3.1000   3.6000   0.0000   14.0000

MAX VALUES OF STRATIFICATION VARIABLES

    6.7000   7.0000   113.0000   100.0000

RANGES OF STRATIFICATION VARIABLES

    3.6000   3.4000   113.0000   86.0000

TOP LIMITS OF 20-INTERVAL STRATIFICATIONS

3.2800	3.7700	5.6500	18.3000
3.4600	3.9400	11.3000	22.6000
3.6400	4.1100	16.9500	26.9000
3.8200	4.2800	22.6000	31.2000
4.0000	4.4500	28.2500	35.5000
4.1800	4.6200	33.9000	39.8000
4.3600	4.7900	30.5500	44.1000
4.5400	4.9600	45.2000	48.4000
4.7200	5.1300	50.8500	52.7000
4.9000	5.3000	56.5000	57.0000
5.0800	5.4700	62.1500	61.3000
5.2600	5.6400	67.8000	65.6000
5.4400	5.8100	73.4500	69.9000
5.6200	5.9800	79.1000	74.2000
5.8000	6.1500	84.7500	78.5000
5.9800	6.3200	90.4000	82.8000
6.1600	6.4900	96.0500	87.1000
6.3400	6.6600	101.7000	91.4000
6.5200	6.8300	107.3500	95.7000
6.7000	7.0000	113.0000	100.0000

BOTTOM LIMITS OF 20-INTERVAL STRATIFICATIONS

3.1000	3.6000	0.0000	14.0000
3.2800	3.7700	5.6500	18.3000
3.4600	3.9400	11.3000	22.6000
3.6400	4.1100	16.9500	26.9000
3.8200	4.2800	22.6000	31.2000
4.0000	4.4500	28.2500	35.5000
4.1800	4.6200	33.9000	39.8000
4.3600	4.7900	39.5500	44.1000
4.5400	4.9600	45.2000	48.4000
4.7200	5.1300	50.8500	52.7000
4.9000	5.3000	56.5000	57.0000
5.0800	5.4700	62.1500	61.3000
5.2600	5.6400	67.8000	65.6000
5.4400	5.8100	73.4500	69.9000
5.6200	5.9800	79.1000	74.2000
5.8000	6.1500	84.7500	78.5000
5.9800	6.3200	90.4000	82.8000
6.1600	6.4900	96.0500	87.1000
6.3400	6.6600	101.7000	91.4000
6.5200	6.8300	107.3500	95.7000

POPULATION SIZES WITHIN 20-INTERVAL STRATIFICATIONS

1.0000	0.0000	0.0000	0.0000
1.0000	6.0000	0.0000	2.0000
0.0000	5.0000	0.0000	3.0000
2.0000	4.0000	0.0000	8.0000
5.0000	6.0000	0.0000	7.0000
2.0000	7.0000	0.0000	5.0000
10.0000	3.0000	0.0000	2.0000
5.0000	7.0000	0.0000	5.0000
11.0000	6.0000	0.0000	4.0000
5.0000	5.0000	0.0000	2.0000
2.0000	2.0000	0.0000	8.0000
5.0000	6.0000	0.0000	1.0000
8.0000	5.0000	1.0000	1.0000
6.0000	0.0000	0.0000	2.0000
3.0000	6.0000	2.0000	3.0000
3.0000	6.0000	12.0000	1.0000
5.0000	1.0000	21.0000	4.0000
2.0000	0.0000	18.0000	3.0000
0.0000	0.0000	16.0000	2.0000
0.0000	0.0000	4.0000	13.0000

CUMULATIVE SQRT OF POP. SIZE FOR 20-INTERVALS

1.0000	0.0000	0.0000	0.0000
2.0000	2.4495	0.0000	1.4142
2.0000	4.6856	0.0000	3.1463
3.4142	6.6856	0.0000	5.9747
5.6503	9.1350	0.0000	8.6204
7.0645	11.7808	0.0000	10.8565
10.2268	13.5128	0.0000	12.2707
12.4628	16.1586	0.0000	14.5068
15.7795	18.6081	0.0000	16.5068
18.0155	20.8441	0.0000	17.9210
19.4297	22.2583	0.0000	20.7494
21.6658	24.7078	0.0000	21.7494
24.4942	26.9438	1.0000	22.7494
26.9437	26.9438	1.0000	24.1636
28.6757	29.3933	2.4142	25.8956
30.4077	31.8428	5.8783	26.8956
32.6438	32.8428	10.4609	28.8956
34.0580	32.8428	14.7035	30.6277
34.0580	32.8428	18.7035	32.0419
34.0580	32.8428	20.7035	35.6474

VALUES OF SVLIM(II,J)

8.5145	8.2107	5.1759	8.9119
17.0290	16.4214	10.3518	17.8237
25.5435	24.6321	15.5276	26.7355

STRATUM BOUNDARIES ALL VARIABLES

4.1800	4.4500	90.4000	35.5000
4.9000	4.9600	96.0500	57.0000
5.4400	5.6400	101.7000	82.8000

A(I,J),B(I,J),W(I,J), PRINTED IN ORDER

0.0527	0.1481	0.1178	0.1461
0.0180	0.0744	0.0677	0.0792
0.1538	0.2949	0.2051	0.2692
0.0911	0.0747	0.0622	0.1150
0.0209	0.0256	0.0144	0.0573
0.3974	0.2179	0.2692	0.2308
0.0261	0.0992	0.0916	0.1271
0.0035	0.0404	0.0364	0.0787
0.1923	0.2436	0.2308	0.2051
0.0785	0.1194	0.1311	0.1946
0.0241	0.0586	0.0583	0.1284
0.2564	0.2436	0.2949	0.2949

SUMPR(J)

0.2484	0.4415	0.4027	0.5827

SUMSQ(J)

0.0665	0.1990	0.1767	0.3436

Appendix **C**

The SYSAMP-I Computer Program

Program Description

The SYSAMP-I computer program computes statistics useful in evaluating a number of systematic sampling procedures. It accepts as input values of a sampling variable and an auxiliary variable for an entire population. The program selects all systematic samples, computes statistics for each, and computes mean square errors of estimation and efficiencies relative to simple random sampling and stratified sampling.

The sample mean, variance, and percentage of sample elements with values less than or equal to a specified cutoff are computed for each systematic sample selected, using the following sampling procedures:

1. Linear systematic sampling from an unordered population
2. Linear systematic sampling from an unordered population with use of end corrections
3. Linear systematic sampling from a population ordered by values of an auxiliary variable
4. Linear systematic sampling from a population ordered by values of an auxiliary variable with use of end corrections
5. Balanced systematic sampling from a population ordered by values of an auxiliary variable
6. Centrally located systematic samples from a population ordered by values of an auxiliary variable
7. Linear systematic sampling from a population ordered by values of an auxiliary variable with order reversed in alternate strata
8. Linear systematic sampling from a population ordered by values of an

auxiliary variable with order reversed in alternate strata and with use of end corrections

Additionally, the mean square error of estimates of the population mean and efficiencies relative to simple random sampling and stratified sampling are computed for each of the eight systematic sampling procedures.

SYSAMP-I computes the overall population mean, variance, and the percentage of elements in the population with values less than or equal to a specified cutoff. Four statistics useful in comparing systematic sampling to stratified sampling and simple random sampling are also computed. They are the pooled variance within systematic samples, the pooled variance within strata, the within-stratum correlation coefficient, and the within-sample intraclass correlation coefficient.

The program utilizes 12 subroutines. In order of appearance in the program listing they are

> REV: a subroutine that reverses the order of elements in every other stratum (group of k elements) of the population.
> EFFIC: a subroutine that computes the efficiency of a systematic sampling procedure relative to simple random sampling and stratified sampling with one element per stratum.
> STAT: a subroutine that computes the mean, variance, and percent of elements with values less than or equal to a specified cutoff within each systematic sample.
> RHO: a subroutine that computes the within-stratum intraclass correlation coefficient and the within-sample intraclass correlation coefficient.
> SEL: a subroutine that selects all linear systematic samples with sampling interval k from a population read as program input.
> SYSVAR: a subroutine that computes the mean square error among the means of systematic samples.
> ENDCOR: a subroutine that computes end corrections for linear systematic samples.
> UNCOR: a subroutine that removes end corrections from linear systematic samples.
> s2: a subroutine that computes the pooled variance within systematic samples, the pooled variance within strata, the means within systematic samples, and the means within strata.
> CLS: a subroutine that computes sample means, percentages of sample elements with values less than or equal to a specified cutoff, and the mean square error of estimate for centrally located systematic samples.
> BAL: a subroutine that selects all balanced systematic samples from a population ordered by values of an auxiliary variable and computes means, variances, and percentages of elements with values less than or equal to a specified cutoff for each balanced systematic sample. The mean square error of estimate is also computed for balanced systematic samples.
> QSORT: a subroutine that arranges the elements of a one-dimensional array in increasing order of the values of another one-dimensional array.

PROGRAM INPUT

Input to SYSAMP-I consists of a parameter card, followed by a data card for every element in the population. The parameter card contains three values: the number of elements in the population, the desired sampling interval, and a cutoff value used to compute population and sample percentages. The percentage of elements with values less than or equal to the cutoff value is computed for each systematic sample and for the entire population. The format for the parameter card is F4.0,F2.0,F6.2—a four-digit integer field, followed by a two-digit integer field, followed by a six-digit floating-point field. The format number for the parameter card is 1100.

The remaining program input consists of a data card for each element in the population. Each data card contains two values: the first is the value of the sampling variable, and the second is the value of an auxiliary variable used to order the population elements for some systematic sampling procedures. The format for data cards is an 11-digit blank field, followed by a three-digit integer field, followed by a six-digit blank field, followed by a three-digit integer field—11X,F3.0,6X,F3.0. The format number for data cards is 1200.

PROGRAM OUTPUT

Output for SYSAMP-I is printed in eight consecutive blocks; one for each of the eight systematic sampling procedures. These blocks of printout follow printing of the sampling interval, to the right of the heading SAMPLING INTERVAL =, and the population size to the right of the heading POPULATION SIZE =.

The first block of printout contains results for linear systematic sampling from a population in the order read on program input. The heading RESULTS FOR UNORDERED POPULATION WITHOUT END CORRECTIONS identifies the block.

The statistics s_{wsy}^2, s_{wst}^2, ρ_w, and ρ_{wst} are printed to the right of the respective headings

POOLED VARIANCE WITHIN SYSTEMATIC SAMPLES =
POOLED VARIANCE WITHIN STRATA =
WITHIN-SAMPLE INTRACLASS CORRELATION COEFFICIENT =
WITHIN-STRATUM CORRELATION COEFFICIENT =

Under the general heading WITHIN-SAMPLE STATISTICS, statistics for each systematic sample are printed in columnar form. The self-explanatory column headings are SAMPLE NUMBER, MEAN, VARIANCE, and PERCENT LE xxxx. The percentage of sample elements with values less than or equal to the cutoff value (xxxx) is printed for each systematic sample, under the last heading described.

Population parameter values are printed next, to the right of the headings

POPULATION MEAN =
POPULATION VARIANCE =
PERCENT OF POPULATION LE XXXX =

Population values of mean square errors are printed for systematic sampling, simple random sampling, and stratified sampling with one element per stratum, under the general heading VARIANCE OF ESTIMATED MEAN and the respective column headings SYSTEMATIC SAMPLING, SIMPLE RANDOM SAMPLING, and STRATIFIED SAMPLING.

The efficiency of the systematic sampling procedure relative to simple random sampling is printed next, to the right of the heading EFFICIENCY OF SYST. SAMPLING RELATIVE TO SRS = PERCENT.

Finally, the efficiency of the systematic sampling procedure relative to stratified sampling with one element per stratum is printed to the right of the heading EFFICIENCY OF SYST. SAMPLING RELATIVE TO STRAT. SAMPLING = PERCENT.

The remaining blocks of printout are consistent with the first block in arrangement of output. The initial statistics printed in the first block are not repeated when their values are invariant under changes in sampling procedure. Thus the values of s_{wsy}^2, s_{wst}^2, ρ_w, and ρ_{wst} are not repeated when end corrections are used with linear systematic sampling from the population arranged as read on input.

General headings for the remaining blocks of printout and corresponding sampling procedures are as follows:

RESULTS FOR UNORDERED POPULATION WITH END CORRECTIONS heads results for linear systematic sampling from the population read on input but with use of end corrections.

RESULTS FOR ORDERED POPULATION WITHOUT END CORRECTIONS heads results for linear systematic sampling from a population arranged in increasing order of the auxiliary variable read as input. End corrections are not used.

RESULTS FOR BALANCED SYSTEMATIC SAMPLING WITHOUT END CORRECTIONS heads results for balanced systematic sampling applied to the population arranged in increasing order of the auxiliary variable read as input. End corrections are not used.

RESULTS FOR CENTRALLY LOCATED SYSTEMATIC SAMPLES WITHOUT END CORRECTIONS heads results for centrally located sampling applied to the population arranged in increasing order of the auxiliary variable read on input. End corrections are not used.

RESULTS FOR ORDERED POPULATION WITH END CORRECTIONS heads results for the population arranged in increasing order of the auxiliary variable read on input. End corrections are used.

RESULTS FOR POPULATION WITH ORDER REVERSED IN ALTERNATE STRATA WITHOUT END CORRECTIONS heads results for the population arranged as follows: Elements are first arranged in increasing order of the auxiliary variable read on input. Elements within every other stratum (group of k elements) are then arranged in reverse order. End corrections are not used.

RESULTS FOR POPULATION WITH ORDER REVERSED IN ALTERNATE STRATA WITH END CORRECTIONS heads results for the population arranged as described above. End corrections are used.

Two additional statistics are printed in the block of output for balanced systematic sampling. For some population sizes, least squares linear regression is used to extend the population in an attempt to reduce estimator bias. When the population extension is not used, the values 0.0 are printed to the right of the headings REGRESSION INTERCEPT = and REGRESSION SLOPE =. When the population extension is used, values printed to the right of these headings are the intercept and slope of the least squares regression line that best fits the population arranged in increasing order of the auxiliary variable read on input. The sampling variable is used as a dependent variable, and the index of population elements is used as an independent variable.

Decision rules and formulas for extension of the population are as follows: A sample size n is first computed as the largest integer less than or equal to the quotient N/k, where N denotes the population size and k denotes the sampling interval. If n is an even integer, the population is not extended. If n is an odd integer, the population is extended by creating new elements with indices between $nk + R + 1$ and $nk + R + k$, where $R = N - nk$. The value of the element with index $nk + R + i$ is equal to $a + b(nk + R + i)$, where

$$a = \frac{(4N + 2) \sum_{i=1}^{N} y_i - 6 \sum_{i=1}^{N} i y_i}{N(N - 1)} \tag{C.1}$$

and

$$b = \frac{12 \sum_{i=1}^{N} i y_i}{N(N^2 - 1)} - \frac{6 \sum_{i=1}^{N} y_i}{N(N - 1)} \tag{C.2}$$

(Program listing and sample output follow.)

Program Listing

```
C        THIS PROGRAM EVALUATES THE EFFICIENCY OF ALTERNATIVE LINEAR
C        SYSTEMATIC SAMPLING PROCEDURES RELATIVE TO SIMPLE RANDOM SAMPLING
C        AND STRATIFIED SAMPLING. ALL SYSTEMATIC SAMPLES ARE DRAWN FROM A
C        POPULATION IN NATURAL ORDER, INCREASING ORDER OF AN AUXILIARY
C        VARIABLE, AND WITH ORDER IN EVERY STRATUM REVERSED. THE EFFECT OF
C        END CORRECTIONS IS EVALUATED IN EACH CASE.
C        Y IS THE SAMPLING VARIABLE, X IS THE AUXILIARY VARIABLE, NN IS
C        POPULATION SIZE, K IS SAMPLING INTERVAL, XX IS PCT. CUTOFF.
         COMMON Z(10,750)
         REAL MI,MJ
         INTEGER R
         DIMENSION Y(1500),X(1500),MI(20),VI(20),SDI(20)
         DIMENSION PCI(20),MJ(750),VJ(750)
         READ (*,1100) NN,K,XX
         READ (*,1200) (Y(I),X(I),I=1,NN)
         PRINT 1,K
    1    FORMAT('1','SAMPLING INTERVAL=',I3)
         N=NN/K
         NN1=N*K
         R=NN-NN1
         PCT=0.0
         GM=0.0
         DO 10 I=1,NN
         GM=GM+Y(I)
         IF(Y(I).LE.XX) PCT=PCT+1.0
   10    CONTINUE
         PCT=(PCT/NN)*100.0
         GM=GM/NN
         GV=0.0
         DO 20 I=1,NN
         GV=GV+(Y(I)-GM)**2
   20    CONTINUE
         GV=GV/NN
         VSRS=GV*(NN-N)/FLOAT((NN-1)*N)
         IF(R.EQ.0) N=N-1
         IF(R.EQ.0) NN1=NN1-K
         IF(R.EQ.0) R=K
 1100    FORMAT(I4,I2,F6.2)
 1200    FORMAT(F3.0,1X,F3.0)
C        FORMAT 1100 IS FOR READING POP SIZE, INTERVAL AND PCT. CUTOFF.
C        FORMAT 1200 IS FOR READING SAMPLING VARIABLE AND AUXILIARY VAR.
C        THIS BLOCK FORMS SYSTEMATIC SAMPLES FROM UNORDERED POPULATION
         CALL SEL(NN1,N,K,R,Y)
C        THIS BLOCK COMPUTES VAR.FOR STRAT. AND SRS COMPARISONS
         CALL S2(N,R,K,S2WSY,S2WST,MI,MJ)
         PRINT 501
  501    FORMAT('1','RESULTS FOR UNORDERED POPULATION WITHOUT END CORRECTI
        CONS')
         PRINT 502,S2WSY
  502    FORMAT('0','POOLED VARIANCE WITHIN SYSTEMATIC SAMPLES=',2X,E14.5)
         PRINT 503,S2WST
  503    FORMAT('0','POOLED VARIANCE WITHIN STRATA=',2X,E14.5)
C        THIS BLOCK COMPUTES INTRACLASS CORRELATION COEFFIECIENTS
         CALL RHO(N,R,K,GM,S2WST,MJ,PW,PWST)
         PRINT 499,PW
```

```
499 FORMAT('0','WITHIN-SAMPLE INTRACLASS CORRELATION COEFF=',2X,E14.5)
    PRINT 504,PWST
504 FORMAT('0','WITHIN-STRATUM CORRELATION COEFFICIENT=',2X,E14.5)
    CALL STAT(N,K,R,XX,MI,PCI,VI)
    PRINT 505
505 FORMAT('0',10X,'WITHIN-SAMPLE STATISTICS')
    PRINT 506,XX
506 FORMAT('0','SAMPLE NUMBER',6X,'MEAN',2X,'VARIANCE',2X,'PERCENT LE'
   C,2X,F6.2)
    PRINT 507,(I,MI(I),VI(I),PCI(I),I=1,K)
507 FORMAT(' ',5X,I3,7X,F8.2,F10.2,2X,F10.2)
    PRINT 508,GM
508 FORMAT('0','POPULATION MEAN=',F12.4)
    PRINT 509,GV
509 FORMAT('0','POPULATION VARIANCE=',F12.4)
    PRINT 511,XX,PCT
511 FORMAT('0','PERCENT OF POPULATION LE',2X,F6.2,'=',F7.2)
    CALL SYSVAR(K,GM,MI,VV)
    CALL EFFIC(R,N,K,NN,VSRS,S2WST,VV,VSTRT,ESRS,ESTRT)
    PRINT 512
512 FORMAT('0','VARIANCE OF ESTIMATED MEAN')
    PRINT 513
513 FORMAT('0','SYSTEMATIC SAMPLING  SIMPLE RANDOM SAMPLING  STRATIF
   CIED SAMPLING')
    PRINT 514,VV,VSRS,VSTRT
514 FORMAT(' ',3X,F12.4,4X,7X,F12.4,5X,6X,F12.4)
    PRINT 515,ESRS
515 FORMAT('0','EFFICIENCY OF SYST. SAMPLING RELATIVE TO SRS='2X,F7.2,
   C2X,'PERCENT')
    PRINT 516,ESTRT
516 FORMAT('0','EFFICIENCY OF SYST. SAMPLING RELATIVE TO STRAT. SAMPLI
   CNG='2X,F7.2,2X,'PERCENT')
    CALL ENDCOR(NN,K)
    CALL STAT(N,K,R,XX,MI,PCI,VI)
    CALL SYSVAR(K,GM,MI,VV)
    PRINT 518
518 FORMAT('1','RESULTS FOR UNORDERED POPULATION WITH END CORRECTIONS'
   C)
    PRINT 505
    PRINT 506,XX
    PRINT 507,(I,MI(I),VI(I),PCI(I),I=1,K)
    PRINT 508,GM
    PRINT 509,GV
    PRINT 511,XX,PCT
    CALL EFFIC(R,N,K,NN,VSRS,S2WST,VV,VSTRT,ESRS,ESTRT)
    PRINT 512
    PRINT 513
    PRINT 514,VV,VSRS,VSTRT
    PRINT 515,ESRS
    PRINT 516,ESTRT
    THIS BLOCK ARANGES OBSERVATIONS IN INCREASING ORDER OF AN
    AUXILIARY VARIABLE
    CALL QSORT(X,Y,NN)
998 FORMAT('0','A LISTING OF THE REORDERED POPULATION')
999 FORMAT(' ',4X,F10.5,4X,F10.5)
```

```
      PRINT 998
      PRINT 999,(Y(I),X(I),I=1,NN)
      CALL SEL(NN1,N,K,R,Y)
      CALL S2(N,R,K,S2WSY,S2WST,MI,MJ)
      CALL RHO(N,R,K,GM,S2WST,MJ,PW,PWST)
      PRINT 517
517   FORMAT('1','RESULTS FOR ORDERED POPULATION WITHOUT END CORRECTIONS
     C')
      PRINT 502,S2WSY
      PRINT 503,S2WST
      PRINT 499,PW
      PRINT 504,PWST
      CALL STAT(N,K,R,XX,MI,PCI,VI)
      PRINT 505
      PRINT 506,XX
      PRINT 507,(I,MI(I),VI(I),PCI(I),I=1,K)
      PRINT 508,GM
      PRINT 509,GV
      PRINT 511,XX,PCT
      CALL SYSVAR(K,GM,MI,VV)
      CALL EFFIC(R,N,K,NN,VSRS,S2WST,VV,VSTRT,ESRS,ESTRT)
      PRINT 512
      PRINT 513
      PRINT 514,VV,VSRS,VSTRT
      PRINT 515,ESRS
      PRINT 516,ESTRT
      CALL BAL(NN,K,XX,GM,Y,A,B,MI,VI,PCI,VV)
      PRINT 619
619   FORMAT('1','RESULTS FOR BALANCED SYSTEMATIC SAMPLING WITHOUT END C
     CORRECTIONS')
      PRINT 502,S2WSY
      PRINT 503,S2WST
      PRINT 499,PW
      PRINT 504,PWST
      PRINT 505
      PRINT 506,XX
      PRINT 507,(I,MI(I),VI(I),PCI(I),I=1,K)
      PRINT 508,GM
      PRINT 509,GV
      PRINT 511,XX,PCT
      CALL EFFIC(R,N,K,NN,VSRS,S2WST,VV,VSTRT,ESRS,ESTRT)
      PRINT 512
      PRINT 514,VV,VSRS,VSTRT
      PRINT 515,ESRS
      PRINT 516,ESTRT
      PRINT 711,A,B
711   FORMAT('0','REGRESSION INTERCEPT=',F9.2,2X,'REGRESSION SLOPE=',F9.
     C2)
      CALL CLS(NN,K,XX,GM,Y,YBAR,YBAR1,YBAR2,PC,PC1,PC2,EMSE)
      PRINT 621
621   FORMAT('1','RESULTS FOR CENTRALLY LOCATED SYSTEMATIC SAMPLES WITHO
     CUT END CORRECTIONS')
      PRINT 502,S2WSY
      PRINT 503,S2WST
      PRINT 499,PW
```

```
      PRINT 504,PWST
      PRINT 505
      PRINT 506,XX
      L=K/2
      L=2*L
      IF (L.EQ.K) GO TO 333
      PRINT 622
622   FORMAT('0','WITHIN-SAMPLE STATISTICS')
      PRINT 623,XX
623   FORMAT('0','MEAN',2X,'PERCENT LE',2X,F6.2)
      PRINT 624,YBAR,PC
624   FORMAT(' ',F8.2,2X,F10.2)
      GO TO 334
333   PRINT 622
      PRINT 625,XX
625   FORMAT('0','SAMPLE NUMBER',6X,'MEAN',2X,'PERCENT LE',2X,F6.2)
      PRINT 626,YBAR1,PC1
626   FORMAT(' ',6X,'1',8X,F8.2,2X,F10.2)
      PRINT 627,YBAR2,PC2
627   FORMAT(' ',6X,'2',8X,F8.2,2X,F10.2)
334   VV=EMSE
      CALL EFFIC(R,N,K,NN,VSRS,S2WST,VV,VSTRT,ESRS,ESTRT)
      PRINT 512
      PRINT 513
      PRINT 514,VV,VSRS,VSTRT
      PRINT 515,ESRS
      PRINT 516,ESTRT
      CALL ENDCOR(NN,K)
      CALL STAT(N,K,R,XX,MI,PCI,VI)
      CALL SYSVAR(K,GM,MI,VV)
      PRINT 519
519   FORMAT('1','RESULTS FOR ORDERED POPULATION WITH END CORRECTIONS')
      PRINT 505
      PRINT 506,XX
      PRINT 507,(I,MI(I),VI(I),PCI(I),I=1,K)
      PRINT 508,GM
      PRINT 509,GV
      PRINT 511,XX,PCT
      CALL EFFIC(R,N,K,NN,VSRS,S2WST,VV,VSTRT,ESRS,ESTRT)
      PRINT 512
      PRINT 513
      PRINT 514,VV,VSRS,VSTRT
      PRINT 515,ESRS
      PRINT 516,ESTRT
      CALL UNCOR(NN,K)
      CALL REV(N,R,K)
      CALL S2(N,R,K,S2WSY,S2WST,MI,MJ)
      CALL RHO(N,R,K,GM,S2WST,MJ,PW,PWST)
      PRINT 531
531   FORMAT('1','RESULTS FOR POPULATION WITH ORDER REVERSED IN ALTERNAT
     CE STRATA')
      PRINT 532
532   FORMAT('0','WITHOUT END CORRECTIONS')
      PRINT 502,S2WSY
      PRINT 503,S2WST
```

```
      PRINT 499,PW
      PRINT 504,PWST
      CALL STAT(N,K,R,XX,MI,PCI,VI)
      PRINT 505
      PRINT 506,XX
      PRINT 507,(I,MI(I),VI(I),PCI(I),I=1,K)
      PRINT 508,GM
      PRINT 509,GV
      PRINT 511,XX,PCT
      CALL SYSVAR(K,GM,MI,VV)
      CALL EFFIC(R,N,K,NN,VSRS,S2WST,VV,VSTRT,ESRS,ESTRT)
      PRINT 512
      PRINT 513
      PRINT 514,VV,VSRS,VSTRT
      PRINT 515,ESRS
      PRINT 516,ESTRT
      CALL ENDCOR(NN,K)
      CALL STAT(N,K,R,XX,MI,PCI,VI)
      CALL SYSVAR(K,GM,MI,VV)
      PRINT 531
      PRINT 541
  541 FORMAT('0','WITH END CORRECTIONS')
      PRINT 505
      PRINT 506,XX
      PRINT 507,(I,MI(I),VI(I),PCI(I),I=1,K)
      PRINT 508,GM
      PRINT 509,GV
      PRINT 511,XX,PCT
      CALL EFFIC(R,N,K,NN,VSRS,S2WST,VV,VSTRT,ESRS,ESTRT)
      PRINT 512
      PRINT 513
      PRINT 514,VV,VSRS,VSTRT
      PRINT 515,ESRS
      PRINT 516,ESTRT
      END
```

```
      SUBROUTINE REV(N,R,K)
      COMMON Z(10,750)
      REAL TEMP(10)
      INTEGER R
      M=N
      IF(R.EQ.K) M=M+1
      DO 30 J=2,M,2
      DO 10 I=1,K
      L=K-I+1
      TEMP(L)=Z(I,J)
10    CONTINUE
      DO 20 I=1,K
      Z(I,J)=TEMP(I)
20    CONTINUE
30    CONTINUE
      N3=M/2
      N3=2*N3
      IF(M.EQ.N3) GO TO 60
      IF(R.EQ.K) GO TO 60
      N1=M+1
      DO 40 I=1,R
      L=R-I+1
      TEMP(L)=Z(I,N1)
40    CONTINUE
      DO 50 I=1,R
      Z(I,N1)=TEMP(I)
50    CONTINUE
60    RETURN
      END
```

```
SUBROUTINE EFFIC(R,N,K,NN,VSRS,S2WST,VV,VSTRT,ESRS,ESTRT)
   INTEGER R
   M=N
   IF(R.EQ.K) M=M+1
   VSTRT=((NN-M)*S2WST)/FLOAT(NN*M)
   IF(VV.EQ.0.0) GO TO 10
   ESRS=(VSRS/VV)*100.0
   ESTRT=(VSTRT/VV)*100.0
   GO TO 20
10 ESRS=9999.0
   ESTRT=9999.0
20 RETURN
   END
```

```
SUBROUTINE STAT(N,K,R,XX,MI,PCI,VI)
   COMMON Z(10,750)
   REAL MI(10),PCI(10),VI(10)
   INTEGER R
   INTEGER R1
   M=N
   NP=M+1
   IF(R.EQ.K) M=M+1
   DO 10 I=1,K
   MI(I)=0.0
   PCI(I)=0.0
   VI(I)=0.0
10 CONTINUE
   IF (R.EQ.K) NP=M
   DO 20 I=1,K
   DO 19 J=1,NP
   IF(I.GT.R.AND.J.EQ.NP) GO TO 20
   IF(Z(I,J).LE.XX) PCI(I)=PCI(I)+1.0
   MI(I)=MI(I)+Z(I,J)
19 CONTINUE
20 CONTINUE
   DO 30 I=1,R
   PCI(I)=(PCI(I)/NP)*100.0
   MI(I)=MI(I)/NP
30 CONTINUE
   IF(R.EQ.K) GO TO 41
   R1=R+1
   DO 40 I=R1,K
   PCI(I)=(PCI(I)/M)*100.0
   MI(I)=MI(I)/M
40 CONTINUE
   IF(R.EQ.K) NP=M
41 DO 50 I=1,K
   DO 49 J=1,NP
   IF(I.GT.R.AND.J.EQ.NP) GO TO 50
   VI(I)=VI(I)+(Z(I,J)-MI(I))**2
49 CONTINUE
50 CONTINUE
   DO 60 I=1,R
   VI(I)=VI(I)/(NP-1)
60 CONTINUE
   IF(R.EQ.K) GO TO 71
   DO 70 I=R1,K
   VI(I)=VI(I)/(M-1)
70 CONTINUE
71 RETURN
   END
```

```
SUBROUTINE RHO(N,R,K,GM,S2WST,MJ,PW,PWST)
   COMMON Z(10,750)
   REAL MJ(750)
   INTEGER R
   M=N
   IF(R.EQ.K) M=M+1
   NZ=M-1
   CP1=0.0
   CP2=0.0
   SS=0.0
   DO 10 I=1,K
   DO 10 J=1,NZ
   J1=J+1
   DO 10 L=J1,M
   CP1=CP1+((Z(I,J)-GM)*(Z(I,L)-GM))
   CP2=CP2+((Z(I,J)-MJ(J))*(Z(I,L)-MJ(L)))
10 CONTINUE
   DO 20 I=1,K
   DO 20 J=1,M
   SS=SS+(Z(I,J)-GM)**2
20 CONTINUE
   SS=SS/FLOAT(M*K-1)
   PW=2.0*CP1/(SS*(M-1)*(M*K-1))
   PWST=2.0*CP2/(S2WST*M*(M-1)*(K-1))
   RETURN
   END
```

```
SUBROUTINE SEL(NN1,N,K,R,Y)
   COMMON Z(10,750)
   REAL Y(1500)
   INTEGER R
   L1=0
   DO 100 I=1,NN1,K
   L1=L1+1
   DO 100 J=1,K
   IJ=I+J-1
100 Z(J,L1)=Y(IJ)
   NP=N+1
   DO 101 I=1,R
   NJ=NN1+I
101 Z(I,NP)=Y(NJ)
   RETURN
   END
```

```
SUBROUTINE SYSVAR(K,GM,MI,VV)
    REAL MI(10)
    VV=0.0
    DO 10 I=1,K
    VV=VV+(MI(I)-GM)**2
10  CONTINUE
    VV=VV/K
    RETURN
    END
```

```
SUBROUTINE ENDCOR(NN,K)
   COMMON Z(10,750)
   INTEGER R
   INTEGER R1
   N=NN/K
   R=NN-N*K
   IF(R.EQ.0) GO TO 100
   N1=N+1
   DO 10 I=1,R
   Z(I,1)=(Z(I,1))*(1.0+((N+1)*(2*I-R-1)/FLOAT(2*N*K)))
   Z(I,N1)=(Z(I,N1))*(1.0-((N+1)*(2*I-R-1)/FLOAT(2*N*K)))
10 CONTINUE
   R1=R+1
   DO 20 I=R1,K
   Z(I,1)=(Z(I,1))*(1.0+(N*(2*I-K-R-1)/FLOAT(2*(N-1)*K)))
   Z(I,N)=(Z(I,N))*(1.0-(N*(2*I-K-R-1)/FLOAT(2*(N-1)*K)))
20 CONTINUE
100 J=K
   DO 110 I=1,K
   Z(I,1)=(Z(I,1))*(1.0+(N*(2*I-K-1)/FLOAT(2*(N-1)*K)))
   Z(I,N)=(Z(I,N))*(1.0-(N*(2*I-K-1)/FLOAT(2*(N-1)*K)))
110 CONTINUE
   RETURN
   END
```

```
SUBROUTINE UNCOR(NN,K)
   COMMON Z(10,750)
   INTEGER R
   INTEGER R1
   N=NN/K
   R=NN-N*K
   IF(R.EQ.0) GO TO 100
   N1=N+1
   DO 10 I=1,R
   Z(I,1)=(Z(I,1))/(1.0+((N+1)*(2*I-R-1)/FLOAT(2*N*K)))
   Z(I,N1)=(Z(I,N1))/(1.0-((N+1)*(2*I-R-1)/FLOAT(2*N*K)))
10 CONTINUE
   R1=R+1
   DO 20 I=R1,K
   Z(I,1)=(Z(I,1))/(1.0+(N*(2*I-K-R-1)/FLOAT(2*(N-1)*K)))
   Z(I,N)=(Z(I,N))/(1.0-(N*(2*I-K-R-1)/FLOAT(2*(N-1)*K)))
20 CONTINUE
100 J=K
   DO 110 I=1,K
   Z(I,1)=(Z(I,1))/(1.0+(N*(2*I-K-1)/FLOAT(2*(N-1)*K)))
   Z(I,N)=(Z(I,N))/(1.0-(N*(2*I-K-1)/FLOAT(2*(N-1)*K)))
110 CONTINUE
   RETURN
   END
```

```
SUBROUTINE S2(N,R,K,S2WSY,S2WST,MI,MJ)
   COMMON Z(10,750)
   REAL MI(10),MJ(750),VI(10),VJ(750)
   INTEGER R
   M=N
   IF(R.EQ.K) M=M+1
   DO 10 I=1,K
   VI(I)=0.0
10 MI(I)=0.0
   DO 20 I=1,M
   VJ(I)=0.0
20 MJ(I)=0.0
   DO 30 I=1,K
   DO 30 L=1,M
   MI(I)=MI(I)+Z(I,L)
30 CONTINUE
   DO 40 J=1,M
   DO 40 L=1,K
   MJ(J)=MJ(J)+Z(L,J)
40 CONTINUE
   DO 50 I=1,K
50 MI(I)=MI(I)/FLOAT(M)
   DO 60 J=1,M
60 MJ(J)=MJ(J)/FLOAT(K)
   DO 70 I=1,K
   DO 70 L=1,M
   VI(I)=VI(I)+(Z(I,L)-MI(I))**2
70 CONTINUE
   S2WSY=0.0
   DO 80 I=1,K
   S2WSY=S2WSY+VI(I)
80 CONTINUE
   S2WSY=S2WSY/FLOAT(K*(M-1))
   DO 90 J=1,M
   DO 90 L=1,K
   VJ(J)=VJ(J)+(Z(L,J)-MJ(J))**2
90 CONTINUE
   S2WST=0.0
   DO 100 J=1,M
   S2WST=S2WST+VJ(J)
100 CONTINUE
   S2WST=S2WST/FLOAT(M*(K-1))
   RETURN
   END
```

```
      SUBROUTINE CLS(NN,K,XX,GM,Y,YBAR,YBAR1,YBAR2,PC,PC1,PC2,EMSE)
         REAL Y(1500)
         INTEGER R
         YBAR=0.0
         YBAR1=0.0
         YBAR2=0.0
         PC=0.0
         PC1=0.0
         PC2=0.0
         N=NN/K
         L=K/2
         L=2*L
         IF(L.EQ.K) GO TO 100
         DO 10 I=1,N
         J=(I-1)*K+(K+1)/2
         YBAR=YBAR+Y(J)
         IF(Y(J).LE.XX) PC=PC+1.0
   10 CONTINUE
         R=NN-N*K
         KK=(K+1)/2
         IF(R.LT.KK) GO TO 40
         J=N*K+(K+1)/2
         IF(Y(J).LE.XX) PC=PC+1.0
         YBAR=YBAR+Y(J)
         PC=(PC/FLOAT(N+1))*100.0
         YBAR=YBAR/FLOAT(N+1)
         GO TO 50
   40 YBAR=YBAR/FLOAT(N)
         PC=(PC/FLOAT(N))*100.0
   50 EMSE=(YBAR-GM)**2
         GO TO 200
  100 YBAR1=0.0
         DO 110 I=1,N
         J=(I-1)*K+K/2
         L=(I-1)*K+(K+2)/2
         YBAR1=YBAR1+Y(J)
         IF(Y(J).LE.XX) PC1=PC1+1.0
         YBAR2=YBAR2+Y(L)
         IF(Y(L).LE.XX) PC2=PC2+1.0
  110 CONTINUE
         R=NN-N*K
         KK=(K+2)/2
         IF(R.LT.KK) GO TO 130
         J=N*K+K/2
         L=N*K+(K+2)/2
         YBAR1=(YBAR1+Y(J))/FLOAT(N+1)
         IF(Y(J).LE.XX) PC1=PC1+1.0
         PC1=(PC1/FLOAT(N+1))*100.0
         YBAR2=(YBAR2+Y(L))/FLOAT(N+1)
         IF(Y(L).LE.XX) PC2=PC2+1.0
         PC2=(PC2/FLOAT(N+1))*100.0
         GO TO 150
  130 YBAR1=YBAR1/FLOAT(N)
         YBAR2=YBAR2/FLOAT(N)
         PC1=(PC1/FLOAT(N))*100.0
         PC2=(PC2/FLOAT(N))*100.0
  150 EMSE=0.5*(((YBAR1-GM)**2)+(YBAR2-GM)**2)
  200 RETURN
         END
```

```
      SUBROUTINE BAL(NN,K,XX,GM,Y,A,B,MI,VI,PCI,VV)
      REAL Y(1500),U(10,400),W(10,400),MI(10),VI(10),PCI(10)
      INTEGER Q,P
      N=NN/K
      M=N/2
      L=2*M
      A=0.0
      B=0.0
      IF(N.EQ.L) GO TO 100
      N=N+1
      M=N/2
      SUM1=0.0
      SUM2=0.0
      DO 10 I=1,NN
      SUM1=SUM1+Y(I)
      SUM2=SUM2+I*Y(I)
 10   CONTINUE
      A=((4.0*NN+2.0)*SUM1-6.0*SUM2)/FLOAT(NN*(NN-1))
      B=12.0*SUM2/FLOAT(NN*((NN**2)-1))-6.0*SUM1/FLOAT(NN*(NN-1))
      Q=(N+1)*K
      P=Q-NN
      IF(P.EQ.0) GO TO 100
      DO 20 I=1,P
      J=NN+I
      Y(J)=A+B*J
 20   CONTINUE
100   DO 120 I=1,K
      DO 120 J=1,M
      IJ=I+2*(J-1)*K
      U(I,J)=Y(IJ)
      JI=2*J*K-I+1
      W(I,J)=Y(JI)
120   CONTINUE
      DO 125 I=1,K
      VI(I)=0.0
      MI(I)=0.0
      PCI(I)=0.0
125   CONTINUE
      DO 130 I=1,K
      DO 130 J=1,M
      MI(I)=MI(I)+U(I,J)+W(I,J)
      IF(U(I,J).LE.XX) PCI(I)=PCI(I)+1.0
      IF(W(I,J).LE.XX) PCI(I)=PCI(I)+1.0
130   CONTINUE
      DO 140 I=1,K
      MI(I)=MI(I)/N
140   CONTINUE
      DO 150 I=1,K
      DO 150 J=1,M
      VI(I)=VI(I)+((U(I,J)-MI(I))**2)+((W(I,J)-MI(I))**2)
150   CONTINUE
      DO 160 I=1,K
      VI(I)=VI(I)/(N-1)
      PCI(I)=(PCI(I)/N)*100.0
160   CONTINUE
```

```
      VV=0.0
      DO 170 I=1,K
      VV=VV+(MI(I)-GM)**2
170   CONTINUE
      VV=VV/K
      RETURN
      END
```

```
      SUBROUTINE QSORT(A,B,N)
         REAL A(1),T,X,B(1),Z
         INTEGER LT(17),UT(17),I,J,K,M,N,P,Q
C        THE DIMENSIONS FOR LT AND UT HAVE TO BE AT LEAST LOG   N.
C                                                            2
C        17 WAS CHOSEN TO HANDLE N<131,073
C
         J=N
         M=1
         I=1
C
C          IF THIS SEGMENT HAS MORE THAN TWO ELEMENTS WE SPLIT IT
   10 IF (J-I-1) 100,90,15
C
C        P IS THE POSITION OF AN ARBITARY ELEMENT IN THE SEGMENT
C        WE CHOOSE THE MIDDLE ELEMENT. UNDER CERTIAN CIRCUMSTANCES
C        IT MAY BE ADVANTAGEOUS TO CHOOSE P AT RANDOM.
   15 P=(J+I)/2
      T=A(P)
      TT=B(P)
      A(P)=A(I)
      B(P)=B(I)
C        STARTING AT THE BEGINNING OF THE SEGMENT, SEARCH FOR K
C        SUCH THAT A(K)>T
      Q=J
      K=I
   20 K=K+1
      IF (K.GT.Q) GO TO 60
      IF (A(K).LE.T) GO TO 20
C
C        SUCH AN ELEMENT HAS NOW BEEN FOUND
C        NOW SEARCH FOR A Q SUCH THAT A(Q)<T STARTING AT THE END OF THE
C        SEGMENT.
   30 IF (A(Q).LT.T) GO TO 40
      Q=Q-1
      IF (Q.GT.K) GO TO 30
      GO TO 50
C
C        A(Q) HAS NOW BEEN FOUND. WE INTERCHANGE A(Q) AND A(K)
   40 X=A(K)
      A(K)=A(Q)
      A(Q)=X
      Z=B(K)
      B(K)=B(Q)
      B(Q)=Z
C
C        UPDATE Q AND SEARCH FOR ANOTHER PAIR TO INTERCHANGE
      Q=Q-1
      GO TO 20
   50 Q=K-1
   60 CONTINUE
C
C        THE UPWARDS SEARCH HAS NOW MET THE DOWNWARDS SEARCH
      A(I)=A(Q)
      B(I)=B(Q)
```

```
      A(Q)=T
      B(Q)=TT
C
C     THE SEGMENT IS NOW DIVIDED IN THREE PARTS: (I,Q-1),(Q),(Q+1,J)
C
C     STORE THE POSITION OF THE LARGEST SEGMENT IN LT AND UT
      IF (2*Q.LE.I+J) GO TO 70
      LT(M)=I
      UT(M)=Q-1
      I=Q+1
      GO TO 80
   70 LT(M)=Q+1
      UT(M)=J
      J=Q-1
C
C     UPDATE M AND SPLIT THE NEW SMALLER SEGMENT
   80 M=M+1
      GO TO 10
C
C     WE ARRIVE HERE IF THE SEGMENT HAS TWO ELEMENTS
C     WE TEST TO SEE IF THE SEGMENT IS PROPERLY ORDERED
C     IF NOT, WE PERFORM AN INTERCHANGE
   90 IF (A(I).LE.A(J)) GO TO 100
      X=A(I)
      A(I)=A(J)
      A(J)=X
      Z=B(I)
      B(I)=B(J)
      B(J)=Z
C
C     IF LT AND UT CONTAIN MORE SEGMENTS TO BE SORTED REPEAT PROCESS
  100 M=M-1
      IF (M.LE.0) RETURN
      I=LT(M)
      J=UT(M)
      GO TO 10
      END
```

Sample Output

SAMPLING INTERVAL = 10

RESULTS FOR UNORDERED POPULATION WITHOUT END CORRECTIONS

POOLED VARIANCE WITHIN SYSTEMATIC SAMPLES = 0.45673E 03

POOLED VARIANCE WITHIN STRATA = 0.44020E 03

WITHIN-SAMPLE INTRACLASS CORRELATION COEFF = -0.65428E-02

WITHIN-STRATUM CORRELATION COEFFICIENT = -0.62634E-02

WITHIN-SAMPLE STATISTICS

SAMPLE NUMBER	MEAN	VARIANCE	PERCENT LE	46.00
1	68.83	5.85	16.95	
2	66.42	6.98	25.42	
3	69.14	1.26	19.49	
4	68.54	5.98	22.03	
5	66.86	0.88	22.03	
6	69.08	4.14	18.80	
7	67.40	10.32	15.38	
8	67.26	9.24	14.53	
9	68.41	1.16	20.51	
10	68.22	1.29	18.80	

POPULATION MEAN = 68.0153

POPULATION VARIANCE = 453.8213

PERCENT OF POPULATION LE 46.00 = 19.40

VARIANCE OF ESTIMATED MEAN

SYSTEMATIC SAMPLING	SIMPLE RANDOM SAMPLING	STRATIFIED SAMPLING
0.8367	3.4956	3.3877

EFFICIENCY OF SYST. SAMPLING RELATIVE TO SRS = 417.78 PERCENT

EFFICIENCY OF SYST. SAMPLING RELATIVE TO STRAT. SAMPLING = 404.89 PERCENT

RESULTS FOR UNORDERED POPULATION WITH END CORRECTIONS

WITHIN-SAMPLE STATISTICS

SAMPLE NUMBER	MEAN	VARIANCE	PERCENT LE	46.00
1	69.03	17.41	17.80	
2	66.60	12.33	26.27	
3	69.13	1.26	19.49	
4	68.53	2.44	22.03	
5	66.81	0.24	22.03	
6	69.08	10.42	18.80	
7	67.35	6.73	15.38	
8	67.23	0.49	14.53	
9	68.56	4.18	20.51	
10	68.42	16.69	19.66	

POPULATION MEAN = 68.0153

POPULATION VARIANCE = 453.8213

PERCENT OF POPULATION LE   46.00 = 19.40

VARIANCE OF ESTIMATED MEAN

SYSTEMATIC SAMPLING	SIMPLE RANDOM SAMPLING	STRATIFIED SAMPLING
0.8612	3.4956	3.3877

EFFICIENCY OF SYST. SAMPLING RELATIVE TO SRS = 405.88 PERCENT

EFFICIENCY OF SYST. SAMPLING RELATIVE TO STRAT. SAMPLING = 393.36 PERCENT

RESULTS FOR ORDERED POPULATION WITHOUT END CORRECTIONS

POOLED VARIANCE WITHIN SYSTEMATIC SAMPLES = 0.45341E 03

POOLED VARIANCE WITHIN STRATA = 0.88957E 02

WITHIN-SAMPLE INTRACLASS CORRELATION COEFF = -0.64821E-02

WITHIN-STRATUM CORRELATION COEFFICIENT = 0.27730E-02

      WITHIN-SAMPLE STATISTICS

SAMPLE NUMBER	MEAN	VARIANCE	PERCENT LE	46.00
1	66.62	6.88	18.64	
2	67.78	11.84	19.49	
3	68.21	9.19	22.03	
4	67.76	13.16	19.49	
5	66.97	13.70	22.03	
6	68.96	10.59	15.38	
7	68.62	9.61	20.51	
8	68.62	12.04	22.22	
9	67.27	13.61	17.09	
10	69.37	6.11	17.09	

POPULATION MEAN = 68.0153

POPULATION VARIANCE = 453.8213

PERCENT OF POPULATION LE 46.00 = 19.40

VARIANCE OF ESTIMATED MEAN

SYSTEMATIC SAMPLING	SIMPLE RANDOM SAMPLING	STRATIFIED SAMPLING
0.7206	3.4956	0.6846

EFFICIENCY OF SYST. SAMPLING RELATIVE TO SRS = 485.09 PERCENT

EFFICIENCY OF SYST. SAMPLING RELATIVE TO STRAT. SAMPLING = 95.01 PERCENT

RESULTS FOR BALANCED SYSTEMATIC SAMPLING WITHOUT END CORRECTIONS

POOLED VARIANCE WITHIN SYSTEMATIC SAMPLES = 0.45341E 03

POOLED VARIANCE WITHIN STRATA = 0.88957E 02

WITHIN-SAMPLE INTRACLASS CORRELATION COEFF = -0.64821E-02

WITHIN STRATUM CORRELATION COEFFICIENT = 0.27730E-02

                  WITHIN-SAMPLE STATISTICS

SAMPLE NUMBER	MEAN	VARIANCE	PERCENT LE	46.00
1	68.32	479.98	17.80	
2	66.93	442.31	18.64	
3	67.41	491.02	23.73	
4	67.96	504.46	20.34	
5	66.72	446.85	19.49	
6	69.47	426.18	17.80	
7	68.69	458.13	19.49	
8	69.70	480.88	20.34	
9	68.41	432.33	17.80	
10	67.93	433.28	17.80	

POPULATION MEAN = 68.0153

POPULATION VARIANCE = 453.8213

PERCENT OF POPULATION LE   46.00 = 19.40

VARIANCE OF ESTIMATED MEAN

SYSTEMATIC SAMPLING	SIMPLE RANDOM SAMPLING	STRATIFIED SAMPLING
0.8920	3.4956	0.6846

EFFICIENCY OF SYST. SAMPLING RELATIVE TO SRS = 391.87 PERCENT

EFFICIENCY OF SYST. SAMPLING RELATIVE TO STRAT. SAMPLING = 76.75 PERCENT

REGRESSION INTERCEPT = 35.13   REGRESSION SLOPE = 0.06

RESULTS FOR CENTRALLY LOCATED SYSTEMATIC SAMPLES WITHOUT END CORRECTIONS

POOLED VARIANCE WITHIN SYSTEMATIC SAMPLES = 0.45341E 03

POOLED VARIANCE WITHIN STRATA = 0.88957E 02

WITHIN-SAMPLE INTRACLASS CORRELATION COEFF = -0.64821E-02

WITHIN-STRATUM CORRELATION COEFFICIENT = 0.27730E-02

WITHIN-SAMPLE STATISTICS

SAMPLE NUMBER   MEAN   VARIANCE   PERCENT LE   46.00

WITHIN-SAMPLE STATISTICS

SAMPLE NUMBER	MEAN	PERCENT LE	46.00
1	66.62	22.22	
2	68.96	15.38	

VARIANCE OF ESTIMATED MEAN

SYSTEMATIC SAMPLING	SIMPLE RANDOM SAMPLING	STRATIFIED SAMPLING
1.4116	3.4956	0.6846

EFFICIENCY OF SYST. SAMPLING RELATIVE TO SRS = 247.63 PERCENT

EFFICIENCY OF SYST. SAMPLING RELATIVE TO STRAT. SAMPLING = 48.50 PERCENT

RESULTS FOR ORDERED POPULATION WITH END CORRECTIONS

WITHIN-SAMPLE STATISTICS

SAMPLE NUMBER	MEAN	VARIANCE	PERCENT LE	46.00
1	67.11	18.92	18.64	
2	68.15	19.24	19.49	
3	68.38	9.09	22.03	
4	67.79	6.90	19.49	
5	66.85	2.95	22.03	
6	69.02	21.26	15.38	
7	68.57	6.16	20.51	
8	68.45	1.01	22.22	
9	66.99	0.19	17.09	
10	68.99	6.35	17.95	

POPULATION MEAN = 68.0153

POPULATION VARIANCE = 453.8213

PERCENT OF POPULATION LE   46.00 = 19.40

VARIANCE OF ESTIMATED MEAN

SYSTEMATIC SAMPLING	SIMPLE RANDOM SAMPLING	STRATIFIED SAMPLING
0.5892	3.4956	0.6846

EFFICIENCY OF SYST. SAMPLING RELATIVE TO SRS = 593.27 PERCENT

EFFICIENCY OF SYST. SAMPLING RELATIVE TO STRAT. SAMPLING = 116.19 PERCENT

RESULTS FOR POPULATION WITH ORDER REVERSED IN ALTERNATE STRATA

WITHOUT END CORRECTIONS

POOLED VARIANCE WITHIN SYSTEMATIC SAMPLES = 0.45345E 03

POOLED VARIANCE WITHIN STRATA = 0.88957E 02

WITHIN-SAMPLE INTRACLASS CORRELATION COEFF = -0.65727E-02

WITHIN STRATUM CORRELATION COEFFICIENT = 0.22658E-02

$\qquad$ WITHIN-SAMPLE STATISTICS

SAMPLE NUMBER	MEAN	VARIANCE	PERCENT LE	46.00
1	68.37	12.75	17.80	
2	66.98	13.69	18.64	
3	67.41	9.65	23.73	
4	67.99	11.71	20.34	
5	66.67	6.86	19.49	
6	69.15	10.47	17.95	
7	68.37	9.75	19.66	
8	69.44	11.53	20.51	
9	68.09	13.05	17.95	
10	67.70	6.90	17.95	

POPULATION MEAN = 68.0153

POPULATION VARIANCE = 453.8213

PERCENT OF POPULATION LE   46.00 = 19.40

VARIANCE OF ESTIMATED MEAN

SYSTEMATIC SAMPLING	SIMPLE RANDOM SAMPLING	STRATIFIED SAMPLING
0.6919	3.4956	0.6846

EFFICIENCY OF SYST. SAMPLING RELATIVE TO SRS = 505.21 PERCENT

EFFICIENCY OF SYST. SAMPLING RELATIVE TO STRAT. SAMPLING = 98.95 PERCENT

RESULTS FOR POPULATION WITH ORDER REVERSED IN ALTERNATE STRATA

WITH END CORRECTIONS

WITHIN-SAMPLE STATISTICS

SAMPLE NUMBER	MEAN	VARIANCE	PERCENT LE	46.00
1	68.89	30.46	17.80	
2	67.35	21.74	18.64	
3	67.58	9.55	23.73	
4	68.02	5.95	20.34	
5	66.57	0.73	19.49	
6	69.22	21.09	17.95	
7	68.32	6.27	19.66	
8	69.26	0.86	20.51	
9	67.81	0.27	17.95	
10	67.32	5.59	18.80	

POPULATION MEAN = 68.0153

POPULATION VARIANCE = 453.8213

PERCENT OF POPULATION LE   46.00 = 19.40

VARIANCE OF ESTIMATED MEAN

SYSTEMATIC SAMPLING	SIMPLE RANDOM SAMPLING	STRATIFIED SAMPLING
0.7097	3.4956	0.6846

EFFICIENCY OF SYST. SAMPLING RELATIVE TO SRS = 492.55 PERCENT

EFFICIENCY OF SYST. SAMPLING RELATIVE TO STRAT. SAMPLING = 96.47 PERCENT

The CLUSAMP-I Computer Program

Program Description

The CLUSAMP-I computer program provides population values for estimator variances. It computes variances as a function of sample size for single-stage cluster sampling using four methods of estimation. It also computes approximate estimator bias and mean square error as a function of sample size for single-stage cluster sampling with ratio estimation.

The methods of sampling and estimation for which variances of the estimated population mean and other parameters are provided are as follows:

1. Simple random sampling of clusters with unbiased estimation
2. Simple random sampling of clusters with ratio estimation
3. Sampling with probabilities proportional to cluster sizes (PPS) and conventional PPS estimation
4. Sampling with probabilities proportional to cluster totals for an auxiliary variable (PPES) and conventional PPES estimation

In addition to computing variances of estimators, the program provides population values of the standard errors of estimated means for specified sample sizes. Other parameters computed include the covariance of cluster size and the within-cluster average of the variable for which estimation is desired, the correlation of these two variables, the average cluster size, the standard deviation of cluster size, and the coefficient of variation of cluster size.

PROGRAM INPUT

Input to CLUSAMP-I consists of a parameter card, followed by a data card for every cluster in the population. The parameter card contains three values:

the number of clusters in the population, an index indicating the sample size interval for which variances are desired, and an indicator of the presence or absence of PPES estimation. The format number for the parameter card is 100. The format is I4,I2,I2: a four-digit integer field followed by 2 two-digit integer fields. The second value on the parameter card is an index that indicates the sample size interval for which variances are desired. It can be given any integer value between one and the number of clusters in the population. If the index is set at one, variances are computed for every value of sample size up to and including the number of clusters in the population. If the index is set at two, variances are computed for samples of size two, four, six, etc. The third value on the parameter card is an indicator variable. If it is set at one, the program will read an auxiliary variable for computation of PPES estimator variances. If the indicator variable equals two, an auxiliary variable is not read and variances are not computed for the PPES estimator.

The remaining program input consists of a data card for each cluster in the population. Each data card must contain three values if PPES estimation is desired. If PPES estimation is not desired, the third value is omitted. Each data card contains, in order, the within-cluster average of the variable for which estimation is desired, the number of elements in the cluster, and the within-cluster average of the auxiliary variable used in PPES sampling. The input format is F8.5,F2.0,F8.5; an eight-digit floating-point field, a two-digit floating-point field, and an eight-digit floating-point field. The input format number is 300.

PROGRAM OUTPUT

CLUSAMP-I first prints those parts of estimator variance formulas that do not depend upon sample size. For the unbiased estimator and the ratio estimator, the sample size is omitted in the denominator, and one minus the sampling fraction is omitted in the numerator. For the PPS and PPES estimators, only the sample size is omitted in the denominator. The values of these expressions are printed to the right of the following headings:

> VARIANCE OF UNBIASED ESTIMATOR =
> VARIANCE OF RATIO ESTIMATOR =
> VARIANCE OF PPS ESTIMATOR =
> VARIANCE OF OPTIMAL PPES ESTIMATOR =

Variances of estimators are printed in columns, below the general heading VARIANCES OF ESTIMATED MEANS FOR INDICATED SAMPLE SIZES AND ESTIMATION METHODS and below appropriate column headings. The leftmost column is headed SAMPLE SIZE, which is self-explanatory. Below the heading UNBIASED EST. are printed variances of the unbiased estimator for sample sizes printed in the first column. Variances of the ratio estimator, the PPS estimator, and the PPES estimator are printed below the headings RATIO EST., PPS ESTIMATOR, and OPTIMAL PPES EST., respectively. A single line of printout contains variances for all four estimation methods for the indicated sample size.

Standard deviations of estimators are printed with a format identical to that used for estimator variances. Standard deviations are printed below the heading

STANDARD DEVIATIONS OF ESTIMATED MEANS FOR INDICATED SAMPLE SIZES AND ESTIMATION METHODS.

Next, the covariance of cluster size and the within-cluster average of the variable for which estimation is desired is printed, to the right of the heading COVARIANCE OF PSU SIZE, ACHIEVEMENT AVG. =. The correlation of cluster size and the within-cluster average of the variable for which estimation is desired is printed to the right of the heading CORRELATION OF PSU SIZE, ACHIEVEMENT AVG. =. Average cluster size is next printed to the right of the heading AVERAGE PSU SIZE =. The standard deviation of cluster size is printed to the right of STD. DEVIATION OF PSU SIZE =.

The ratio estimator is the only biased estimator for which parameters are computed. The bias and mean square error of the ratio estimator are printed for desired sample sizes under the general heading BIAS AND MEAN SQUARE ERROR OF RATIO ESTIMATE OF MEAN. The values are printed in columnar form, in a format similar to that used for estimator variances. The column headings SAMPLE SIZE, BIAS, and MEAN SQUARE ERROR are self-explanatory.

The final value printed is the coefficient of variation of cluster size. It is printed to the right of the heading COEFFICIENT OF VARIATION OF PSU SIZE =.

(Program listing and sample output follow.)

Program Listing

```
$WATFOR
C THIS PROGRAM COMPUTES VARIANCES AND OTHER PARAMETERS FOR ESTIMATORS OF
C MEANS USING ALTERNATIVE SINGLE STAGE CLUSTER SAMPLING TECHNIQUES
      REAL Y(150), M(150),Z(150),YTOT(150),X(150),F(150),S2UNB(150),SUNB
     C(150),S2RAT(150),SRAT(150),S2PPS(150),SPPS(150),S2PPES(150),SPPES(
     C150),BIAS(150),MSE(150)
      READ 100,N,L,J
 100 FORMAT(I4,I2,I2)
C N IS NUMBER OF CLUSTERS IN POPULATION,L IS VARIANCE COMPUTATION INTERV.
C (VALUES OF SAMPLE SIZE), J=1 IF OPTIONAL PPS VARIABLE IS INPUT,J=2 IF N
      IF(J.EQ.2) GO TO 200
      READ 300,(Y(I),M(I),Z(I),I=1,N)
 300 FORMAT(F8.5,F2.0,F8.5)
      GO TO 400
 200 READ 300,(Y(I),M(I),I=1,N)
 400 DO 500 I=1,N
      YTOT(I)=Y(I)*M(I)
      Z(I)=Z(I)*M(I)
 500 CONTINUE
      SUMM=0.0
      SUM=0.0
      SUMZ=0.0
      DO 600 I=1,N
      SUM=SUM+YTOT(I)
      SUMSQ=SUMSQ+(YTOT(I))**2
      SUMZ=SUMZ+Z(I)
      SUMM=SUMM+M(I)
 600 CONTINUE
      XN=FLOAT(N)
      XN1=FLOAT(N-1)
      VUNB=((SUMSQ-((SUM/XN)**2)*XN)*((XN/SUMM)**2))/XN1
      ELAVG=SUM/SUMM
      SUMRAT=0.0
      DO 700 I=1,N
      SUMRAT=SUMRAT+(YTOT(I)-M(I)*ELAVG)**2
 700 CONTINUE
      VRAT=((XN/SUMM)**2)*SUMRAT/XN1
      SUMPPS=0.0
      DO 800 I=1,N
      SUMPPS=SUMPPS+(M(I))*(Y(I)-ELAVG)**2
 800 CONTINUE
      VPPS=SUMPPS/SUMM
      VPPES=0.0
      DO 900 I=1,N
      X(I)=(Z(I))/SUMZ
      VPPES=VPPES+(X(I))*((YTOT(I)/X(I))-SUM)**2
 900 CONTINUE
      VPPES=VPPES/SUMM**2
      PRINT 2000,VUNB
2000 FORMAT('1','VARIANCE OF UNBIASED ESTIMATOR=',E14.5)
      PRINT 3000,VRAT
3000 FORMAT ('0','VARIANCE OF RATIO ESTIMATOR=',E14.5)
      PRINT 4000,VPPS
4000 FORMAT('0','VARIANCE OF PPS ESTIMATOR=',E14.5)
      PRINT 5000,VPPES
```

```
5000 FORMAT('0','VARIANCE OF OPTIMAL PPES ESTIMATOR=','E14.5)
     DO 6000 I=1,N,L
     F(I)=FLOAT(I)/XN
     S2UNB(I)=((1.0-F(I))/FLOAT(I))*VUNB
     SUNB(I)=SQRT(S2UNB(I))
     S2RAT(I)=((1.0-F(I))/FLOAT(I))*VRAT
     SRAT(I)=SQRT(S2RAT(I))
     S2PPS(I)=(1.0/FLOAT(I))*VPPS
     SPPS(I)=SQRT(S2PPS(I))
     S2PPES(I)=(1.0/FLOAT(I))*VPPES
     SPPES(I)=SQRT(S2PPES(I))
6000 CONTINUE
     PRINT 6050
6100 FORMAT('0','SAMPLE SIZE UNBIASED EST.      RATIO EST. PPS ESTIMATOR
    C OPTIMAL PPES EST.')
6050 FORMAT('0','VARIANCES OF ESTIMATED MEANS FOR INDICATED SAMPLE SIZE
    CS AND ESTIMATION METHODS')
     PRINT 6100
     PRINT 6200,(I,S2UNB(I),S2RAT(I),S2PPS(I),S2PPES(I),I=1,N,L)
6200 FORMAT(' ',4X,I3,4X,E14.5,1X,E14.5,1X,E14.5,1X,E14.5)
     PRINT 6250
6250 FORMAT('1','STANDARD DEVIATIONS OF ESTIMATED MEANS FOR INDICATED S
    CAMPLE SIZES AND ESTIMATION METHODS')
     PRINT 6100
     PRINT 6200,(I,SUNB(I),SRAT(I),SPPS(I),SPPES(I),I=1,N,L)
     ZUM=0.0
     ZUMMSQ=0.0
     ZUMSQ=0.0
     ZMBAR=SUMM/XN
     ZUMPR=0.0
     DO 7000 I=1,N
     ZUMMSQ=ZUMMSQ+(M(I))**2
     ZUM=ZUM+Y(I)
     ZUMPR=ZUMPR+(Y(I)-ELAVG)*(M(I)-ZMBAR)
     ZUMSQ=ZUMSQ+(Y(I))**2
7000 CONTINUE
     SIGY=ZUMSQ-ELAVG*ZUM+XN*ELAVG**2
     SIGM=ZUMMSQ-XN*ZMBAR**2
     COV=ZUMPR/XN1
     RHO=XN1*COV/SQRT(SIGY*SIGM)
     SIGY=SQRT(SIGY/XN1)
     SIGM=SQRT(SIGM/XN1)
     CM=SIGM/ZMBAR
     PRINT 9000,COV
9000 FORMAT('0','COVARIANCE OF PSU SIZE,ACHIEVEMENT AVG.=',E14.5)
     PRINT 9050,RHO
9050 FORMAT('0','CORRELATION OF PSU SIZE,ACHIEVEMENT AVG.=',E14.5)
     PRINT 9100,ZMBAR
9100 FORMAT('0','AVERAGE PSU SIZE=',E14.5)
     PRINT 9200,SIGM
9200 FORMAT('0','STD. DEVIATION OF PSU SIZE=',E14.5)
     DO 9300 I=1,N,L
     BIAS(I)=(CM**3)*SIGY*RHO/FLOAT(I)
     MSE(I)=(BIAS(I))**2+S2RAT(I)
9300 CONTINUE
```

```
      PRINT 9400
9400  FORMAT('1','BIAS AND MEAN SQUARE ERROR OF RATIO ESTIMATE OF MEAN')
      PRINT 9500
9500  FORMAT('0','SAMPLE SIZE            BIAS   MEAN SQUARE ERROR')
      PRINT 9600,(I,BIAS(I),MSE(I),I=1,N,L)
9600  FORMAT(' ',4X,I3,6X,E14.5,3X,E14.5)
      PRINT 9700,CM
9700  FORMAT('0','COEFFICIENT OF VARIATION OF PSU SIZE=',E14.5)
      RETURN
      END
```

Sample Output

VARIANCE OF UNBIASED ESTIMATOR = 0.44601E 03

VARIANCE OF RATIO ESTIMATOR = 0.45656E 02

VARIANCE OF PPS ESTIMATOR = 0.44963E 02

VARIANCE OF OPTIMAL PPES ESTIMATOR = 0.71680E 01

VARIANCES OF ESTIMATED MEANS FOR INDICATED SAMPLE SIZES AND ESTIMATION METHODS

SAMPLE SIZE	UNBIASED EST.	RATIO EST.	PPS ESTIMATOR	OPTIMAL PPES EST
1	0.43610E 03	0.44642E 02	0.44963E 02	0.71680E 01
2	0.21309E 03	0.21813E 02	0.22482E 02	0.35840E 01
3	0.13876E 03	0.14204E 02	0.14988E 02	0.23893E 01
4	0.10159E 03	0.10399E 02	0.11241E 02	0.17920E 01
5	0.79290E 02	0.81166E 01	0.89926E 01	0.14336E 01
6	0.64423E 02	0.65948E 01	0.74939E 01	0.11947E 01
7	0.53804E 02	0.55077E 01	0.64233E 01	0.10240E 01
8	0.45840E 02	0.46924E 01	0.56204E 01	0.89601E 00
9	0.39645E 02	0.40583E 01	0.49959E 01	0.79645E 00
10	0.34689E 02	0.35510E 01	0.44963E 01	0.71680E 00
11	0.30635E 02	0.31360E 01	0.40876E 01	0.65164E 00
12	0.27256E 02	0.27901E 01	0.37469E 01	0.59734E 00
13	0.24397E 02	0.24974E 01	0.34587E 01	0 55139E 00
14	0.21946E 02	0.22466E 01	0.32117E 01	0.51200E 00
15	0.19823E 02	0.20292E 01	0.29975E 01	0.47787E 00
16	0.17964E 02	0.18389E 01	0.28102E 01	0.44800E 00
17	0.16324E 02	0.16711E 01	0.26449E 01	0.42165E 00
18	0.14867E 02	0.15219E 01	0.24980E 01	0.39822E 00
19	0.13563E 02	0.13884E 01	0.23665E 01	0.37727E 00
20	0.12389E 02	0.12682E 01	0.22482E 01	0.35840E 00
21	0.11327E 02	0.11595E 01	0.21411E 01	0.34134E 00
22	0.10362E 02	0.10607E 01	0.20438E 01	0.32582E 00
23	0.94803E 01	0.97047E 00	0.19549E 01	0.31165E 00
24	0.86724E 01	0.88776E 00	0.18735E 01	0.29867E 00
25	0.79290E 01	0.81166E 00	0.17985E 01	0.28672E 00
26	0.72428E 01	0.74142E 00	0.17294E 01	0.27569E 00
27	0.66075E 01	0.67639E 00	0.16653E 01	0.26548E 00
28	0.60176E 01	0.61600E 00	0.16058E 01	0.25600E 00
29	0.54683E 01	0.55977E 00	0.15505E 01	0.24717E 00
30	0.49556E 01	0.50729E 00	0.14988E 01	0.23893E 00
31	0.44761E 01	0.45820E 00	0.14504E 01	0.23123E 00
32	0.40265E 01	0.41217E 00	0.14051E 01	0.22400E 00
33	0.36041E 01	0.36894E 00	0.13625E 01	0.21721E 00
34	0.32066E 01	0.32825E 00	0.13224E 01	0.21082E 00
35	0.28318E 01	0.28988E 00	0.12847E 01	0.20480E 00
36	0.24778E 01	0.25365E 00	0.12490E 01	0.19911E 00
37	0.21430E 01	0.21937E 00	0.12152E 01	0.19373E 00
38	0.18258E 01	0.18690E 00	0.11832E 01	0.18863E 00
39	0.15248E 01	0.15609E 00	0.11529E 01	0.18380E 00
40	0.12389E 01	0.12682E 00	0.11241E 01	0.17920E 00
41	0.96695E 00	0.98984E-01	0.10967E 01	0.17483E 00
42	0.70795E 00	0.72470E-01	0.10705E 01	0.17067E 00
43	0.46099E 00	0.47190E-01	0.10457E 01	0.16670E 00
44	0.22526E 00	0.23059E-01	0.10219E 01	0.16291E 00
45	0.00000E 00	0.00000E 00	0.99918E 00	0.15929E 00

STANDARD DEVIATIONS OF ESTIMATED MEANS FOR INDICATED SAMPLE SIZES AND EST.

SAMPLE SIZE	UNBIASED EST.	RATIO EST.	PPS ESTIMATOR	OPTIMAL PPES EST.
1	0.20883E 02	0.66814E 01	0.67055E 01	0.28773E 01
2	0.14598E 02	0.46705E 01	0.47415E 01	0.18932E 01
3	0.11780E 02	0.37688E 01	0.38714E 01	0.15458E 01
4	0.10079E 02	0.32248E 01	0.33527E 01	0.13387E 01
5	0.89045E 01	0.28490E 01	0.29988E 01	0.11973E 01
6	0.80264E 01	0.25680E 01	0.27375E 01	0.10930E 01
7	0.73351E 01	0.23469E 01	0.25344E 01	0.10119E 01
8	0.67705E 01	0.21662E 01	0.23707E 01	0.94658E 00
9	0.62964E 01	0.20145E 01	0.22352E 01	0.89244E 00
10	0.58898E 01	0.18844E 01	0.21205E 01	0.84664E 00
11	0.55349E 01	0.17709E 01	0.20218E 01	0.80724E 00
12	0.52207E 01	0.16704E 01	0.19357E 01	0.77288E 00
13	0.49393E 01	0.15803E 01	0.18598E 01	0.74256E 00
14	0.46847E 01	0.14989E 01	0.17921E 01	0.71554E 00
15	0.44522E 01	0.14245E 01	0.17313E 01	0.69128E 00
16	0.42384E 01	0.13561E 01	0.16764E 01	0.66933E 00
17	0.40403E 01	0.12927E 01	0.16263E 01	0.64935E 00
18	0.38558E 01	0.12336E 01	0.15805E 01	0.63105E 00
19	0.36828E 01	0.11783E 01	0.15383E 01	0.61422E 00
20	0.35198E 01	0.11262E 01	0.14994E 01	0.59867E 00
21	0.33656E 01	0.10768E 01	0.14632E 01	0.58424E 00
22	0.32190E 01	0.10299E 01	0.14296E 01	0.57081E 00
23	0.30790E 01	0.98512E 00	0.13982E 01	0.55826E 00
24	0.29449E 01	0.94221E 00	0.13687E 01	0.54651E 00
25	0.28158E 01	0.90092E 00	0.13411E 01	0.53546E 00
26	0.26913E 01	0.86106E 00	0.13150E 01	0.52507E 00
27	0.25705E 01	0.82243E 00	0.12905E 01	0.51525E 00
28	0.24531E 01	0.78485E 00	0.12672E 01	0.50597E 00
29	0.23384E 01	0.74818E 00	0.12452E 01	0.49717E 00
30	0.22261E 01	0.71224E 00	0.12242E 01	0.48881E 00
31	0.21157E 01	0.67690E 00	0.12043E 01	0.48086E 00
32	0.20066E 01	0.64201E 00	0.11854E 01	0.47329E 00
33	0.18984E 01	0.60740E 00	0.11673E 01	0.46606E 00
34	0.17907E 01	0.57293E 00	0.11500E 01	0.45916E 00
35	0.16828E 01	0.53841E 00	0.11334E 01	0.45255E 00
36	0.15741E 01	0.50363E 00	0.11176E 01	0.44622E 00
37	0.14639E 01	0.46837E 00	0.11024E 01	0.44015E 00
38	0.13512E 01	0.43232E 00	0.10878E 01	0.43432E 00
39	0.12348E 01	0.39508E 00	0.10737E 01	0.42871E 00
40	0.11131E 01	0.35612E 00	0.10602E 01	0.42332E 00
41	0.98334E 00	0.31462E 00	0.10472E 01	0.41813E 00
42	0.84140E 00	0.26920E 00	0.10347E 01	0.41312E 00
43	0.67896E 00	0.21723E 00	0.10226E 01	0.40829E 00
44	0.47461E 00	0.15185E 00	0.10109E 01	0.40362E 00
45	0.00000E 00	0.00000E 00	0.99959E 00	0.39911E 00

COVARIANCE OF PSU SIZE, ACHIEVEMENT AVG. = 0.58215E 01

CORRELATION OF PSU SIZE, ACHIEVEMENT AVG. = 0.11518E-01

AVERAGE PSU SIZE = 0.26156E 02

STD. DEVIATION OF PSU SIZE = 0.73205E 01

BIAS AND MEAN SQUARE ERROR OF RATIO ESTIMATE OF MEAN

SAMPLE SIZE	BIAS	MEAN SQUARE ERROR
1	0.17435E-01	0.44642E 02
2	0.87175E-02	0.21814E 02
3	0.58117E-02	0.14204E 02
4	0.43588E-02	0.10399E 02
5	0.34870E-02	0.81167E 01
6	0.29058E-02	0.65948E 01
7	0.24907E-02	0.55077E 01
8	0.21794E-02	0.46924E 01
9	0.19372E-02	0.40583E 01
10	0.17435E-02	0.35510E 01
11	0.15850E-02	0.31360E 01
12	0.14529E-02	0.27901E 01
13	0.13412E-02	0.24974E 01
14	0.12454E-02	0.22466E 01
15	0.11623E-02	0.20292E 01
16	0.10897E-02	0.18389E 01
17	0.10256E-02	0.16711E 01
18	0.96862E-03	0.15219E 01
19	0.91764E-03	0.13884E 01
20	0.87175E-03	0.12682E 01
21	0.83024E-03	0.11595E 01
22	0.79250E-03	0.10607E 01
23	0.75805E-03	0.97047E 00
24	0.72646E-03	0.88776E 00
25	0.69740E-03	0.81167E 00
26	0.67058E-03	0.74142E 00
27	0.64574E-03	0.67639E 00
28	0.62268E-03	0.61600E 00
29	0.60121E-03	0.55977E 00
30	0.58117E-03	0.50729E 00
31	0.56242E-03	0.45820E 00
32	0.54485E-03	0.41217E 00
33	0.52834E-03	0.36894E 00
34	0.51280E-03	0.32825E 00
35	0.49815E-03	0.28988E 00
36	0.48431E-03	0.25365E 00
37	0.47122E-03	0.21937E 00
38	0.45882E-03	0.18690E 00
39	0.44705E-03	0.15609E 00
40	0.43588E-03	0.12682E 00
41	0.42525E-03	0.98984E-01
42	0.41512E-03	0.72470E-01
43	0.40547E-03	0.47190E-01
44	0.39625E-03	0.23059E-01
45	0.38745E-03	0.15011E-06

COEFFICIENT OF VARIATION OF PSU SIZE = 0.27988E 00

COMPILE TIME = 0.74 SEC, EXECUTION TIME = 0.63 S

$STOP

The CLUSAMP-II Computer Program

Program Description

The CLUSAMP-II computer program provides estimation of population means and estimates of variances for eight estimators. All estimators are used with single-stage cluster sampling. The program requires, as input, data for an entire population of clusters. It selects samples from the population using simple random sampling, PPS sampling, and PPES sampling. One hundred independent samples of each kind are drawn, and appropriate estimates are computed for each sample. Program output is in the form of punched cards and printout. Punched-card output permits analysis of the empirical distributions of estimates of a mean and estimates of estimator variances.

Estimates of a population mean and estimates of estimator variances are provided as output for the following sampling and estimation methods:

1. Simple random sampling with unbiased estimation
2. Simple random sampling with ratio estimation
3. Sampling with probabilities proportional to cluster sizes (PPS) and conventional PPS estimation
4. Sampling with probabilities proportional to cluster totals for an auxiliary variable (PPES) and conventional PPES estimation
5. Simple random sampling with jackknife-unbiased estimation
6. Simple random sampling with jackknife-ratio estimation
7. Sampling with probabilities proportional to cluster sizes (PPS) and jackknife-PPS estimation
8. Sampling with probabilities proportional to cluster totals for an auxiliary variable (PPES) and jackknife-PPES estimation

PROGRAM INPUT

The first data card required by CLUSAMP-II contains the number of clusters in the population. The input format is I4, a four-digit integer field. The input format number is 100 (card 4 in the program listing).

The remaining program input consists of one data card for each cluster in the population. Each card must contain the average within a cluster of the variable for which the population mean is to be estimated, $\bar{y}_i$, the number of elements in the cluster, M_i, and the average within the cluster of an auxiliary variable used in PPES sampling, z_i, in that order. The input format is F8.5,F2.0,F8.5—an eight-digit floating-point field, a two-digit floating-point field, and an eight-digit floating-point field. The input format number is 200 (card 6 in the program listing).

PROGRAM OUTPUT

CLUSAMP-II first prints, in columnar form, the input data for each cluster in the population. Three columns are printed under the heading $DATA. The first column contains the identifier YMEAN = and lists the average within a cluster of the variable for which estimation of the population mean is desired. The second column contains the identifier SIZE = and lists the number of elements in a cluster. The third column contains the identifier ZMEAN = and lists the average within a cluster of the auxiliary variable used in PPES sampling.

Directly below the input data, the program prints, in columnar form, totals of the variable for which estimation is desired and totals of the variable used in PPES sampling in each cluster. The identifier in the first column is YTOT = and in the second column is ZTOT =.

The main printed output begins on a new page. To the right of the heading SAMPLE SIZE =, the program prints the number of clusters sampled. As presently written, CLUSAMP-II computes estimates for four sample sizes. The sample sizes used are one-fourth, one-half, three-fourths, and the total number of clusters in the population.

The heading ESTIMATES OF MEANS AND VARIANCES ARE AS FOLLOWS is printed next. Following that, the column headings UNBIASED, RATIO, PPS, OPTIMAL-PPS, JACK-UNBIASED, JACK-RATIO, JACK-PPS, and JACK-PPES are printed. Estimates of the mean and estimates of estimator variances are printed in columns below these headings. Two lines are used to print estimates resulting from each of the 100 samples drawn from the population. The first line contains estimates of population means, printed with E formats. The second line contains corresponding estimates of estimator variances, printed with F formats. A line is skipped between estimates resulting from each sample. Estimates of means and estimator variances shown in the attached sample program output should be read as follows:

UNBIASED	RATIO	PPS	OPTIMAL-PPES
0.67144E 02	0.67310E 02	0.72445E 02	0.66905E 02
32.3456	4.0797	2.9071	0.3384
0.76217E 02	0.69068E 02	0.67776E 02	0.67927E 02
29.8400	3.2914	7.2077	1.1515

Numbers in the first two lines were computed from the first sample drawn. The number 0.67144E 02, printed under the heading UNBIASED, is the unbiased estimate of the sample mean resulting from the first sample. It is read "67.144." The number immediately below it, 32.3456, is the estimate of the variance of the unbiased estimator, based on data from the first sample. Under the heading RATIO, the number 0.67310E 02 is the ratio estimate of the population mean, computed from data from the first sample. It is read "67.310." The number 4.0797, printed immediately below the ratio estimate of the mean, is the corresponding estimate of the variance of the ratio estimator. Pairs of numbers printed below the headings PPS and OPTIMAL-PPES are estimates of the population mean and estimator variances for the PPS estimator and the PPES estimator, respectively.

The next two lines of data, beginning with the numbers 0.76217E 02 and 29.8400, are estimates of the population mean and estimates of estimator variances, computed from data from the second sample. Numbers in the first column result from unbiased estimation, those in the second column result from ratio estimation, etc.

After printing estimates computed from data from all 100 samples and a sample size equal to one-fourth the number of clusters in the population, the program prints estimates resulting from a sample size of one-half the clusters in the population. Whenever the sample size is changed, printout begins on a new page and the headings SAMPLE SIZE and ESTIMATES OF MEANS AND VARIANCES ARE AS FOLLOWS are repeated.

In addition to printed output, CLUSAMP-II provides punched-card output. Cards are punched in a standard format and may be used as input to programs such as the University of California BMD series. Output for each sample drawn from the population is punched on three successive cards. The first card contains estimates of the population mean for eight estimators in the order unbiased, ratio, PPS, PPES jackknife-unbiased, jackknife-ratio, jackknife-PPS, and jackknife-PPES. The format of the first card is 8F9.4. The second card contains jackknife estimates of the estimator variances in the order jackknife-unbiased, jackknife-ratio, jackknife-PPS, and jackknife-PPES. The format of the second card is 4F11.4. The third card contains conventional estimates of estimator variances in the order unbiased, ratio, PPS, and PPES. A set of three cards is punched for each of the 100 samples drawn and for each value of sample size in the order data are listed in printed output.

PROGRAM MODIFICATION

CLUSAMP-II can easily be modified to conform to the special needs of users.

If it is desired to read a value for sample size, rather than using sample sizes computed from the size of the population, the program should be modified as follows: Card number 3, which now reads "READ 100, N," should be changed to read "READ 100 N, NSAMP." Card number 4, which now reads "100 FORMAT (I4)," should be changed to read "100 FORMAT (2I4)." Cards numbered 60, 61, 62, 63, 64, 65, and 301 should be removed.

If it is desired to generate estimates for fewer than 100 samples or more than 100 samples, card number 72, which now reads "DO 1221 KOUNT = 1,100," should be changed to read "DO 1221 KOUNT = 1,*x*," where *x* denotes the number of samples for which estimates are desired.

SUBROUTINES INCORPORATED

The CLUSAMP-II program uses two subroutines. The first, named RANDU, is a pseudo-random number generator. Each time it is called, it provides a simulated random number between 0 and 1.0. RANDU uses the multiplicative congruential method of random number generation.

The second subroutine is named ZAKNIF. ZAKNIF computes jackknife estimates of population parameters and jackknife estimates of estimator variance. It requires as input the results of conventional estimation applied to an entire set of data and applied to *k* subsets of data. A jackknife estimate of the population parameter and a jackknife estimate of the variance of the jackknife estimator are subroutine outputs.

(Program listing and sample output follow.)

Program Listing

```
$WATFOR
C THIS PROGRAM COMPUTES ESTIMATES OF MEANS FORMED THROUGH SINGLE STAGE
C CLUSTER SAMPLING. JACKKNIFE ESTIMATES AND JACKKNIFE VARIANCE ESTIMATES
C ARE ALSO FORMED.
C FOUR TYPES OF ESTIMATES ARE FORMED:
C         1. UNBIASED
C         2. RATIO
C         3. PPS
C         4. OPTIMAL - PPES
C THE PROGRAM ACCEPTS THE ENTIRE POPULATION AS INPUT, AND COMPLETES
C NECESSARY SAMPLING. THE FIRST CARD INPUTS POP.SIZE,A CONTROL VARIABLE
C (J=1 IF PPES SAMPLING,2 IF NOT) AND SAMPLE SIZE. FOLLOWING CARDS
C INPUT CLUSTER AVERAGES,SIZES AND PPES VARIABLE AVERAGES.
      REAL Y(150),Z(150),YPPS(150),ZTOT(150),X(150),ZTOT1(150),RANDY(50)
      C,YPPES(150),XPPES(150),YSRS(150),YSUM(150),YTOT(150)
      INTEGER M(150),MM(150),LL(150),        MPPS(150),IZTOT(150),IZTOT1
      C(150),ZZ(150),WW(150),MSRS(150),MSUM(150),IJK(150)
C THIS BLOCK READS CLUSTER MEANS,SIZES, AND OPTIMAL PPES VARIABLES
      READ 100,N
  100 FORMAT(I4)
      IF(J2.EQ.2) GO TO 300
      READ 200,(Y(I),M(I),Z(I),I=1,N)
  200 FORMAT(F7.4,F2.0,F7.4)
      PRINT 201,(Y(I),M(I),Z(I),I=1,N)
  201 FORMAT(' ','YMEAN=',F7.4,'SIZE=',I3,'ZMEAN=',F7.4)
      DO 210 I=1,N
      ZTOT(I)=Z(I)*M(I)
  210 CONTINUE
  400 DO 410 I=1,N
      YTOT(I)=Y(I)*M(I)
  410 CONTINUE
      PRINT 411,(YTOT(I),ZTOT(I),I=1,N)
  411 FORMAT(' ','YTOT=',E14.5,'ZTOT=',E14.5)
C THIS BLOCK CREATES NEW ARRAYS FOR PPS SAMPLING
C M(I) IS SIZE OF THE I-TH CLUSTER,N IS NUMBER OF CLUSTERS,NSAMP IS
C DESIRED SAMPLE SIZE
      LL(1)=0
      MM(1)=M(1)
      LL(2)=M(1)+1
      NN=N-1
      DO 10 I=2,NN
      J=I-1
      J6=I+1
      MM(I)=MM(J)+M(I)
      LL(J6)=MM(I)+1
   10 CONTINUE
      MM(N)=MM(NN)+M(N)
C THIS BLOCK ELIMINATES STARTING TRANCIENCY IN RANDU
      IX=813261543
      DO 2222 I=1,50
      CALL RANDU(IX,IY,YFL)
      IX=IY
      RANDY(I)=YFL
 2222 CONTINUE
      SUMZ=0.0
```

```
      DO 4000 I=1,N
      SUMZ=SUMZ+ZTOT(I)
4000  CONTINUE
      DO 4100 I=1,N
      X(I)=ZTOT(I)/SUMZ
4100  CONTINUE
      DO 4300 I=1,N
      IZTOT(I)=INT(ZTOT(I))
      IZTOT1(I)=IZTOT(I)+1
      ZTOT1(I)=ZTOT(I)+0.5
      IF(ZTOT1(I).GE.IZTOT1(I)) GO TO 4200
      GO TO 4300
4200  IZTOT(I)=IZTOT1(I)
4300  CONTINUE
      ZZ(1)=0
      WW(1)=IZTOT(1)
      ZZ(2)=WW(1)+1
      NN=N-1
      DO 4310 I=2,NN
      J=I-1
      J6=I+1
      WW(I)=WW(J)+IZTOT(I)
      ZZ(J6)=WW(I)+1
4310  CONTINUE
      WW(N)=WW(NN)+IZTOT(N)
      IN(1)=N/4
      IN(2)=N/2
      IN(3)=3*N/4
      IN(4)=N
      DO 1331 IZ=1,4
      NSAMP=IN(IZ)
      PRINT 12,NSAMP
  12  FORMAT('1','SAMPLE SIZE=',I6)
      PRINT 580
 580  FORMAT('0','ESTIMATES OF MEANS AND VARIANCES ARE AS FOLLOWS')
      PRINT 590
 590  FORMAT('0','UNBIASED        RATIO         PPS           OPTIMAL-P
     CPES   JACK-UNBIASED  JACK-RATIO   JACK-PPS      JACK-PPES')
      DO 1221 KOUNT=1,100
C THIS BLOCK DRAWS PPS SAMPLES
      DO 3300 K=1,NSAMP
      CALL RANDU(IX,IY,YFL)
      IX=IY
      RANDO=YFL
      RAND=RANDO*MM(N)
      IRAND=INT(RAND)
      RAND1=RAND+0.5
      IRAND1=IRAND+1
      IF(RAND1.GE.IRAND1) GO TO 3000
      GO TO 3100
3000  IRAND=IRAND1
3100  DO 3200 KK=1,N
      IF(IRAND.GE.LL(KK).AND.IRAND.LE.MM(KK)) GO TO 3110
      GO TO 3200
3110  YPPS(K)=Y(KK)
```

```
        MPPS (K)=M(KK)
 3200 CONTINUE
 3300 CONTINUE
C THIS BLOCK DRAWS PPES SAMPLES
      DO 4430 K=1,NSAMP
      CALL RANDU(IX,IY,YFL)
      IX=IY
      RAND= YFL*ZZ(N)
      IRAND=INT(RAND)
      RAND1=RAND+0.5
      IRAND1=IRAND+1
      IF(RAND1.GE.IRAND1) GO TO 4410
      GO TO 4420
 4410 IRAND=IRAND1
 4420 DO 4425 KK=1,N
      IF(IRAND.GE.ZZ(KK).AND.IRAND.LE.WW(KK)) GO TO 4424
      GO TO 4425
 4424 YPPES(K)=YTOT(KK)
      XPPES(K)=X(KK)
 4425 CONTINUE
 4430 CONTINUE
C THIS BLOCK SELECTS CLUSTERS WITH EQUAL PROBABILITIES
 9999 DO 47 I=1,N
      IJK(I)=I
   47 CONTINUE
      J=1
      MX=N
   50 CALL RANDU(IX,IY,YFL)
      IX=IY
      XYZ=FLOAT(N)/FLOAT(2**J)
      IF(FLOAT(MX).LT.XYZ) GO TO 52
      GO TO 53
   52 YFL=YFL/FLOAT(2**J)
      J=J+1
   53 RAND=YFL*N
      IRAND=INT(RAND)
      RAND1=RAND+0.5
      IRAND1=IRAND+1
      IF(RAND1.GE.IRAND1) GO TO 54
      GO TO 55
   54 IRAND=IRAND1
   55 IF(IRAND.GT.MX) GO TO 50
      IF(IRAND.EQ.0) GO TO 50
      ITEMP=IJK(IRAND)
      IJK(IRAND)=IJK(MX)
      IJK(MX)=ITEMP
      MX=MX-1
      IF(MX.GT.1) GO TO 50
      DO 5030 K=1,NSAMP
      L=IJK(K)
      YSRS(K)=YTOT(L)
      MSRS(K)=M(L)
 5030 CONTINUE
      IUMM=0
      DO 6000 I=1,N
```

```
      IUMM=IUMM+M(I)
 6000 CONTINUE
C THIS BLOCK PREPARES UNBIASED ESTIMATE FOR JACKKNIFING AND JACKKNIFES
      K1=NSAMP/10
      IF(K1.EQ.0) K1=K1+1
      L5=(NSAMP-K1)/K1
      L1=K1+1
      L6=L5+1
      DO 99 I=1,L6
      YSUM(I)=0.0
   99 CONTINUE
      DO 1 I=L1,NSAMP
      YSUM(1)=YSUM(1)+YSRS(I)
    1 CONTINUE
      DO 4 J=2,L5
      L2=(J-1)*K1
      DO 2 I=1,L2
      YSUM(J)=YSUM(J)+YSRS(I)
    2 CONTINUE
      L3=L1+L2
      DO 3 I=L3,NSAMP
      YSUM(J)=YSUM(J)+YSRS(I)
    3 CONTINUE
    4 CONTINUE
      L4=NSAMP-K1
      DO 5 I=1,L4
      YSUM(L6)=YSUM(L6)+YSRS(I)
    5 CONTINUE
      DO 7 I=1,L6
     YSUM(I)=((YSUM(I))*(FLOAT(N)))/((FLOAT(L4))*(FLOAT(IUMM)))
    7 CONTINUE
      YUNB=0.0
      DO 8 I=1,NSAMP
      YUNB=YUNB+YSRS(I)
    8 CONTINUE
      YBAR=YUNB/FLOAT(NSAMP)
      ZSUM=0.0
      DO 666 I=1,NSAMP
  666 CONTINUE
      XXXX=FLOAT(N*(N-NSAMP))/FLOAT((NSAMP-1)*NSAMP*IUMM**2)
      VARUNB=ZSUM*XXXX
      YUNB=(YUNB*FLOAT(N))/((FLOAT(NSAMP))*(FLOAT(IUMM)))
      CALL ZAKNIF(YUNB,YSUM,L6,UNB,VUNB)
C THIS BLOCK PREPARES RATIO ESTIMATE FOR JACKKNIFING AND JACKKNIFES
      DO 101 I=1,L6
      YSUM(I)=0.0
      MSUM(I)=0
  101 CONTINUE
      DO 110 I=L1,NSAMP
      YSUM(1)=YSUM(1)+YSRS(I)
      MSUM(1)=MSUM(1)+MSRS(I)
  110 CONTINUE
      DO 140 J=2,L5
      L2=(J-1)*K1
      DO 120 I=1,L2
```

```
      YSUM(J)=YSUM(J)+YSRS(I)
      MSUM(J)=MSUM(J)+MSRS(I)
  120 CONTINUE
      L3=L1+L2
      DO 130 I=L3,NSAMP
      YSUM(J)=YSUM(J)+YSRS(I)
      MSUM(J)=MSUM(J)+MSRS(I)
  130 CONTINUE
  140 CONTINUE
      L4=NSAMP-K1
      DO 150 I=1,L4
      YSUM(L6)=YSUM(L6)+YSRS(I)
      MSUM(L6)=MSUM(L6)+MSRS(I)
  150 CONTINUE
      DO 160 I=1,L6
      YSUM(I)=YSUM(I)/FLOAT(MSUM(I))
  160 CONTINUE
      ESSRS=0.0
      MOSRS=0
      DO 170 I=1,NSAMP
      ESSRS=ESSRS+YSRS(I)
      MOSRS=MOSRS+MSRS(I)
  170 CONTINUE
      ESRAT=ESSRS/FLOAT(MOSRS)
      ZSUM=0.0
      DO 667 I=1,NSAMP
      ZSUM=ZSUM+(YSRS(I)-ESRAT*MSRS(I))**2
  667 CONTINUE
      CALL ZAKNIF(ESRAT,YSUM,L6,RAT,VRAT)
C THIS BLOCK PREPARES PPS ESTIMATE FOR JACKKNIFING AND JACKKNIFES
      DO 301 I=1,L6
      YSUM(I)=0.0
  301 CONTINUE
      DO 310 I=L1,NSAMP
      YSUM(1)=YSUM(1)+YPPS(I)
  310 CONTINUE
      DO 340 J=2,L5
      L2=(J-1)*K1
      DO 320 I=1,L2
      YSUM(J)=YSUM(J)+YPPS(I)
  320 CONTINUE
      L3=L1+L2
      DO 330 I=L3,NSAMP
      YSUM(J)=YSUM(J)+YPPS(I)
  330 CONTINUE
  340 CONTINUE
      L4=NSAMP-K1
      DO 350 I=1,L4
      YSUM(L6)=YSUM(L6)+YPPS(I)
  350 CONTINUE
      DO 360 I=1,L6
      YSUM(I)=YSUM(I)/(FLOAT(L4))
  360 CONTINUE
      ESPPS=0.0
      DO 370 I=1,NSAMP
```

```
      ESPPS=ESPPS+YPPS(I)
370   CONTINUE
      ESPPS=ESPPS/FLOAT(NSAMP)
      ZSUM=0.0
      DO 668 I=1,NSAMP
      ZSUM=ZSUM+(YPPS(I)-ESPPS)**2
668   CONTINUE
      XXXX=FLOAT(NSAMP*(NSAMP-1))
      VARPPS=ZSUM/XXXX
      CALL ZAKNIF(ESPPS,YSUM,L6,PPS,VPPS)
THIS BLOCK PREPARES PPES ESTIMATE FOR JACKKNIFING AND JACKKNIFES
      DO 501 I=1,L6
      YSUM(I)=0.0
501   CONTINUE
      DO 510 I=L1,NSAMP
      YSUM(1)=YSUM(1)+(YPPES(I)/XPPES(I))
510   CONTINUE
      DO 540 J=2,L5
      L2=(J-1)*K1
      DO 520 I=1,L2
      YSUM(J)=YSUM(J)+(YPPES(I)/XPPES(I))
520   CONTINUE
      L3=L1+L2
      DO 530 I=L3,NSAMP
      YSUM(J)=YSUM(J)+(YPPES(I)/XPPES(I))
530   CONTINUE
540   CONTINUE
      L4=NSAMP-K1
      DO 550 I=1,L4
      YSUM(L6)=YSUM(L6)+(YPPES(I)/XPPES(I))
550   CONTINUE
      ESPES=0.0
      DO 560 I=1,NSAMP
      ESPES=ESPES+(YPPES(I)/XPPES(I))
560   CONTINUE
      ESPES=ESPES/FLOAT(NSAMP*IUMM)
      ZSUM=0.0
      DO 669 I=1,NSAMP
      ZSUM=ZSUM+((YPPES(I)/XPPES(I))-IUMM*ESPES)**2
669   CONTINUE
      VARPES=ZSUM/(XXXX*FLOAT(IUMM**2))
      DO 570 I=1,L6
      YSUM(I)=YSUM(I)/FLOAT(L4*IUMM)
570   CONTINUE
      CALL ZAKNIF(ESPES,YSUM,L6,PES,VPES)
      PRINT 710,YUNB,ESRAT,ESPPS,ESPES,UNB,RAT,PPS,PES
710   FORMAT('0',E14.5,1X,E14.5,1X,E14.5,1X,E14.5,1X,E14.5,1X,E14.5,1X,E
     C14.5,1X,E14.5)
      PRINT 711,VARUNB,VARRAT,VARPPS,VARPES,VUNB,VRAT,VPPS,VPES
711   FORMAT(' ',8F15.4)
      PUNCH 380,YUNB,ESRAT,ESPPS,ESPES,UNB,RAT,PPS,PES,VUNB,VRAT,VPPS,VP
     CES,VARUNB,VARRAT,VARPPS,VARPES
380   FORMAT(8F9.4/4F11.4/4F11.4)
1221  CONTINUE
1331  CONTINUE
      RETURN
      END
```

```
      SUBROUTINE ZAKNIF(ESTAL,A,K,EST,VAR)
      DIMENSION A(30),B(30)
      DO 10 J=1,K
      B(J)=K*ESTAL-(K-1)*(A(J))
   10 CONTINUE
      TSUM=0.0
      TSUMSQ=0.0
      DO 20 J=1,K
      TSUM=TSUM+B(J)
      TSUMSQ=TSUMSQ+(B(J)**2)
   20 CONTINUE
      EST=(1/FLOAT(K))*TSUM
      VAR=(TSUMSQ-(1/FLOAT(K))*(TSUM**2))/(FLOAT(K-1))
      VAR=VAR/FLOAT(K)
      RETURN
      END
```

```
      SUBROUTINE RANDU(IX,IY,YFL)
      IY=IX*65539
      IF(IY) 5,6,6
5     IY=IY+2147483647+1
6     YFL=IY
      YFL=YFL*.4656613E-9
      RETURN
      END
```

Sample Output

```
                        $DATA
        YMEAN=71.8889 SIZE=27 ZMEAN=61.6296
        YMEAN=71.6875 SIZE=32 ZMEAN=59.5000
        YMEAN=69.5667 SIZE=30 ZMEAN=56.5333
        YMEAN=66.9333 SIZE=30 ZMEAN=56.3000
        YMEAN=78.5000 SIZE=12 ZMEAN=60.5833
        YMEAN=59.5000 SIZE=26 ZMEAN=51.1538
        YMEAN=70.2069 SIZE=29 ZMEAN=53.5517
        YMEAN=66.0938 SIZE=32 ZMEAN=55.2500
        YMEAN=68.8788 SIZE=33 ZMEAN=57.0303
        YMEAN=67.5454 SIZE=33 ZMEAN=53.7879
        YMEAN=69.2414 SIZE=29 ZMEAN=55.0000
        YMEAN=74.5000 SIZE=28 ZMEAN=59.9643
        YMEAN=63.5000 SIZE=18 ZMEAN=51.2222
        YMEAN=73.3030 SIZE=33 ZMEAN=57.8485
        YMEAN=77.7857 SIZE=28 ZMEAN=61.8929
        YMEAN=65.2414 SIZE=29 ZMEAN=53.8965
        YMEAN=62.1923 SIZE=26 ZMEAN=49.5385
        YMEAN=70.4828 SIZE=29 ZMEAN=56.8621
        YMEAN=53.3125 SIZE=32 ZMEAN=47.9375
        YMEAN=74.5172 SIZE=29 ZMEAN=59.9655
        YMEAN=54.3333 SIZE=15 ZMEAN=51.8000
        YMEAN=74.9355 SIZE=31 ZMEAN=62.5806
        YMEAN=62.1154 SIZE=26 ZMEAN=49.6923
        YMEAN=64.9688 SIZE=32 ZMEAN=54.2813
        YMEAN=61.2857 SIZE=28 ZMEAN=52.9286
        YMEAN=67.5454 SIZE=33 ZMEAN=57.4545
        YMEAN=64.6842 SIZE=19 ZMEAN=54.9474
        YMEAN=66.6923 SIZE=13 ZMEAN=56.6923
        YMEAN=67.6250 SIZE= 8 ZMEAN=56.0000
        YMEAN=82.2069 SIZE=29 ZMEAN=63.8276
        YMEAN=70.7200 SIZE=25 ZMEAN=57.4000
        YMEAN=69.6800 SIZE=25 ZMEAN=55.5600
        YMEAN=72.3333 SIZE=30 ZMEAN=60.5333
        YMEAN=67.7742 SIZE=31 ZMEAN=54.9032
        YMEAN=77.5555 SIZE=18 ZMEAN=64.4444
        YMEAN=66.5454 SIZE=11 ZMEAN=56.8182
        YMEAN=71.2200 SIZE=50 ZMEAN=58.3400
        YMEAN=74.6400 SIZE=25 ZMEAN=62.8000
        YMEAN=59.0000 SIZE=23 ZMEAN=53.3043
        YMEAN=49.7083 SIZE=24 ZMEAN=42.9167
        YMEAN=72.0000 SIZE=23 ZMEAN=58.4348
        YMEAN=62.6538 SIZE=26 ZMEAN=53.4615
        YMEAN=66.7273 SIZE=22 ZMEAN=56.9545
        YMEAN=53.9130 SIZE=23 ZMEAN=48.4783
        YMEAN=75.7273 SIZE=22 ZMEAN=62.6364
```

```
YTOT = 0.19410E  04ZTOT = 0.16640E 04
YTOT = 0.22940E  04ZTOT = 0.19040E 04
YTOT = 0.20870E  04ZTOT = 0.16960E 04
YTOT = 0.20080E  04ZTOT = 0.16890E 04
YTOT = 0.94200E  03ZTOT = 0.72700E 03
YTOT = 0.15470E  04ZTOT = 0.13300E 04
YTOT = 0.20360E  04ZTOT = 0.15530E 04
YTOT = 0.21150E  04ZTOT = 0.17680E 04
YTOT = 0.22730E  04ZTOT = 0.18820E 04
YTOT = 0.22290E  04ZTOT = 0.17750E 04
YTOT = 0.20080E  04ZTOT = 0.15950E 04
YTOT = 0.20860E  04ZTOT = 0.16790E 04
YTOT = 0.11430E  04ZTOT = 0.92200E 03
YTOT = 0.24190E  04ZTOT = 0.19090E 04
YTOT = 0.21780E  04ZTOT = 0.17330E 04
YTOT = 0.18920E  04ZTOT = 0.15630E 04
YTOT = 0.16170E  04ZTOT = 0.12880E 04
YTOT = 0.20440E  04ZTOT = 0.16490E 04
YTOT = 0.17060E  04ZTOT = 0.15340E 04
YTOT = 0.21610E  04ZTOT = 0.17390E 04
YTOT = 0.81500E  03ZTOT = 0.77700E 03
YTOT = 0.23230E  04ZTOT = 0.19400E 04
YTOT = 0.16150E  04ZTOT = 0.12920E 04
YTOT = 0.20790E  04ZTOT = 0.17370E 04
YTOT = 0.17160E  04ZTOT = 0.14820E 04
YTOT = 0.22290E  04ZTOT = 0.18960E 04
YTOT = 0.12290E  04ZTOT = 0.10440E 04
YTOT = 0.86700E  03ZTOT = 0.73700E 03
YTOT = 0.54100E  03ZTOT = 0.44800E 03
YTOT = 0.23840E  04ZTOT = 0.18510E 04
YTOT = 0.17680E  04ZTOT = 0.14350E 04
YTOT = 0.17420E  04ZTOT = 0.13890E 04
YTOT = 0.21700E  04ZTOT = 0.18160E 04
YTOT = 0.21010E  04ZTOT = 0.17020E 04
YTOT = 0.13960E  04ZTOT = 0.11600E 04
YTOT = 0.73200E  03ZTOT = 0.62500E 03
YTOT = 0.35610E  04ZTOT = 0.29170E 04
YTOT = 0.18660E  04ZTOT = 0.15700E 04
YTOT = 0.13570E  04ZTOT = 0.12260E 04
YTOT = 0.11930E  04ZTOT = 0.10300E 04
YTOT = 0.16560E  04ZTOT = 0.13440E 04
YTOT = 0.16290E  04ZTOT = 0.13900E 04
YTOT = 0.14680E  04ZTOT = 0.12530E 04
YTOT = 0.12400E  04ZTOT = 0.11150E 04
YTOT = 0.16660E  04ZTOT = 0.13780E 04
```

SAMPLE SIZE = 11

ESTIMATES OF MEANS AND VARIANCES ARE AS FOLLOWS

UNBIASED	RATIO	PPS	OPTIMAL-PPES
0.67144E 02	0.67310E 02	0.72445E 02	0.66905E 02
32.3456	4.0797	2.9071	0.3384
0.67217E 02	0.69068E 02	0.67776E 02	0.67927E 02
29.8400	3.2914	7.2077	1.1515
0.74453E 02	0.68657E 02	0.70029E 02	0.67805E 02
39.9810	3.8685	0.9410	0.8434
0.67839E 02	0.66842E 02	0.69451E 02	0.68164E 02
21.2357	4.9270	2.9163	0.4477
0.73824E 02	0.68516E 02	0.67755E 02	0.66680E 02
54.4541	4.2418	6.6443	1.4236
0.58924E 02	0.65204E 02	0.66041E 02	0.68677E 02
23.8152	1.9432	5.4332	0.4732
0.65965E 02	0.68025E 02	0.68214E 02	0.67576E 02
15.8313	4.1378	8.3101	0.2316
0.82579E 02	0.69067E 02	0.67025E 02	0.69620E 02
32.5481	4.3017	1.9314	0.5739
0.71586E 02	0.69347E 02	0.68713E 02	0.66821E 02
21.0028	4.9944	3.5828	0.4460
0.7337E 02	0.68954E 02	0.67168E 02	0.68127E 02
42.6059	3.2774	5.3537	0.3797
0.70706E 02	0.71630E 02	0.70252E 02	0.67389E 02
33.1540	1.7476	2.6112	1.6637
0.66518E 02	0.65541E 02	0.66765E 02	0.67005E 02
26.8071	2.8343	4.9357	0.3215
0.70932E 02	0.65832E 02	0.72559E 02	0.67230E 02
49.5080	2.9732	2.6406	0.9116
0.76949E 02	0.69184E 02	0.69208E 02	0.69015E 02
48.6779	3.1359	1.6944	0.4997
0.70863E 02	0.69347E 02	0.67142E 02	0.68651E 02
23.9794	3.8000	6.4809	0.4417
0.59949E 02	0.66853E 02	0.66457E 02	0.67096E 02
22.8371	5.5647	2.9117	1.0975
0.63591E 02	0.70099E 02	0.69276E 02	0.66064E 02
27.6848	2.2859	2.6414	0.6576
0.75569E 02	0.66898E 02	0.68578E 02	0.66968E 02
39.8470	4.4676	3.7380	1.4174

JACK-UNBIASED	JACK-RATIO	JACK-PPS	JACK-PPES
0.67144E 02	0.67301E 02	0.72445E 02	0.66905E 02
42.8103	5.5299	2.9070	0.3384
0.67217E 02	0.69132E 02	0.67776E 02	0.67927E 02
39.4942	4.6876	7.2077	1.1514
0.74453E 02	0.68696E 02	0.70029E 02	0.67805E 02
52.9158	4.3005	0.9410	0.8434
0.67839E 02	0.66884E 02	0.69451E 02	0.68164E 02
28.1060	6.4639	2.9162	0.4476
0.73824E 02	0.68598E 02	0.67756E 02	0.66680E 02
72.0709	4.7506	6.6443	1.4235
0.58924E 02	0.65206E 02	0.66041E 02	0.68677E 02
31.5201	3.1448	5.4332	0.4731
0.65965E 02	0.67973E 02	0.68214E 02	0.67576E 02
20.9532	6.0047	8.3100	0.2316
0.82579E 02	0.69071E 02	0.67025E 02	0.69620E 02
43.0778	3.9664	1.9315	0.5739
0.71586E 02	0.69295E 02	0.68713E 02	0.66821E 02
27.7978	6.3498	3.5830	0.4460
0.73337E 02	0.68910E 02	0.67168E 02	0.68127E 02
56.3902	3.9186	5.3537	0.3798
0.70706E 02	0.71633E 02	0.70252E 02	0.67389E 02
43.8803	2.4333	2.6113	1.6637
0.66518E 02	0.65554E 02	0.66765E 02	0.67005E 02
35.4798	3.7150	4.9358	0.3215
0.70932E 02	0.65893E 02	0.72559E 02	0.67230E 02
65.5255	3.5697	2.6406	0.9116
0.76949E 02	0.69233E 02	0.69208E 02	0.69015E 02
64.4262	3.3426	1.6944	0.4996
0.70863E 02	0.69394E 02	0.67142E 02	0.68651E 02
31.7375	4.7984	6.4810	0.4417
0.59949E 02	0.66811E 02	0.66457E 02	0.67096E 02
30.2254	9.5578	2.9117	1.0974
0.63591E 02	0.70143E 02	0.69276E 02	0.66064E 02
36.6417	3.6870	2.6413	0.6575
0.75569E 02	0.66982E 02	0.68578E 02	0.66968E 02
52.7381	4.5944	3.7381	1.4174

Answers to Exercises

Chapter 2

1. The most difficult problem in defining a target population of "health care facilities" would be in delineating the class of health care facilities. Public and private hospitals are obvious choices for inclusion. How about emergency medical facilities that are not associated with hospitals? Are the offices of individual physicians to be called health care facilities? Very often, facilities for the elderly such as rest homes and retirement homes have a registered nurse on duty who regularly dispenses medication from an office in the rest home. Are such offices to be counted as separate health care facilities? Mental health clinics and the offices of individual psychiatrists and, perhaps, clinical psychologists could be called health care facilities. Also, most public schools have a school nurse's office, and most colleges and universities have an infirmary. Are these to be counted?

One approach to delineating the target population would be to impose some constraints on facilities to be included. For example, you might require that the facilities be open to all residents of the city. This would eliminate many of the facilities suggested above. In any case, your operational definition should be consistent with the research questions that motivated your survey.

2. The most logical method of obtaining a sampling frame of parents of pupils enrolled in the public schools of a small school system would be to consult the administrative records of the school system. Virtually all school systems maintain cumulative record folders on all pupils, and the names and addresses of parents are almost always requested at the beginning of each school year. No other records are more likely to include the names and addresses of the vast majority of pupils' parents. However, the records available from the school

system are likely to contain both underregistration and overregistration and to be out of date to some degree. Records are likely to be retained for some pupils who are no longer enrolled in the school system. Other pupils will have transferred into the school system since the beginning of the current school year, and records might not be available for some of them. Other records will be incomplete, with parents' names and addresses missing. Still other records will be inaccurate due to deaths, divorces, remarriages, or changes of residence.

Another potential source of a sampling frame would be a city directory that listed all addresses in the city. This source would contain considerable overregistration, since many addresses would house individuals or families without children enrolled in the system's schools. If the sampling frame were assembled from a city directory, a screening question would have to be used to eliminate addresses that housed no enrolled pupils.

A third potential source of a sampling frame would be a local telephone directory. This source would facilitate a telephone interview survey but would contain both overregistration and underregistration. Many listed numbers would have no associated parents of enrolled pupils. Many parents of enrolled pupils would have no telephones or would have unlisted telephone numbers.

In an actual survey, it would probably be best to begin developing a sampling frame from data available in school records, and to supplement the school system's listing of parents' names and addresses with information from the telephone directory and a city directory. School system records are often incomplete or out of date, necessitating the use of supplementary sources of data.

3. Of the three examples given, only the first (a) is a probability sampling procedure. Although the use of a physical process for randomizing a list of names is not recommended, if the slips of paper were stirred thoroughly between each drawing and 20 percent of the slips were drawn without looking at them, each slip should have a nonzero probability of entering the sample on each draw. This physical process would approximate simple random sampling. Example b is not a probability sampling procedure because members whose names were not among the first 20 percent of the alphabetized list would have no chance of entering the sample. Example c is not a probability sampling procedure because the population is not well-defined, there is no sampling frame, and it is impossible to enumerate the potential samples that might be drawn.

4. If an estimator has a bias of -3.0 and a variance of 16.0, it will have a mean square error of $16.0 + (-3.0)^2 = 25.0$.

5. The mean square error (and variance) of estimator A is equal to $8.0 + 0.0^2 = 8.0$. The mean square error of estimator B is equal to $6.0 + 0.5^2 = 6.25$. Estimator B is therefore more efficient than estimator A.

Chapter 3

1. *Food stamp allotments*

a. Since the total dollar amount of food stamp allotments for the 100 sampled families equals $12,000, an estimate of the average monthly allotment per family is easily calculated by dividing the total allotment by the number of sampled families: $12,000/100 = $120 per family. Since the sample mean is an unbiased estimator of the population mean when based on data collected using simple random sampling without replacement, our estimate of $120 per month per family satisfies the requirements of the problem.

b. The standard error of the sample mean is equal to the square root of the estimated variance of the sample mean. It is given by

$$s_{\bar{y}} = \frac{s}{\sqrt{n}} \sqrt{1 - f}$$

$$= \frac{25}{\sqrt{100}} \sqrt{1 - \frac{100}{10,000}}$$

$$= 2.5(0.995) = 2.487$$

In this equation, the sample standard deviation, $s = \$25$, was calculated from the sum of squared deviations by using the formula

$$s = \sqrt{\frac{\sum_{i=1}^{n} (y_i - y)^2}{n - 1}}$$

$$= \sqrt{\frac{61,875}{99}} = \sqrt{625} = 25$$

c. A 90 percent confidence interval on the population mean can be calculated by using Equation (3.13) as follows:

$$\hat{Y}_U = 120 + \frac{1.64(25)}{\sqrt{100}} \sqrt{1 - \frac{100}{10,000}}$$

$$= 120 + 4.1(0.995) = \$124.08$$

$$\hat{Y}_L = 120 - \frac{1.64(25)}{\sqrt{100}} \sqrt{1 - \frac{100}{10,000}}$$

$$= 120 - 4.1(0.995) = \$115.92$$

d. If the population standard deviation of monthly food stamp allotments per family was equal to $25, the number of families that would have to be sampled in order to estimate the average monthly allotment to the nearest $1 with 90 percent confidence would be given by

$$n = \frac{[1.64(25)/1]^2}{1 + \frac{1}{10,000}[1.64(25)/1]^2}$$

$$= \frac{1681}{1.1681} = 1439.09$$

which rounds up to 1440 families out of the population of 10,000.

2. *Salivating white rats*

a. When simple random sampling without replacement is used, the sample proportion is an unbiased estimator of the corresponding population proportion. Therefore, an estimate of the population proportion of white rats that would salivate at the sight of pellet food is simply $\frac{28}{50} = 0.56$.

b. The most conservative possible value for the population proportion of white rats that would salivate at the sight of pellet food would be 0.50. Using this value, the number of rats that would have to be sampled to estimate the proportion that would salivate at the sight of pellet food within ± 0.10 with 95 percent confidence is given by Equation (3.25):

$$n = \frac{(1.96/0.10)^2(0.50)(0.50)}{1 + \frac{1}{1000}\left[\left(\frac{1.96}{0.10}\right)^2(0.50)(0.50) - 1\right]}$$

$$= \frac{96.04}{1.095} = 87.70$$

which rounds up to 88 white rats that must be sampled.

3. *Tuition income*

a. An estimate of the total income from students' tuition that could be expected during the coming semester is equal to $1,615,800. This figure was calculated as follows: First, the average number of credit hours all sampled students intend to take was calculated by taking a weighted average of the figures reported for students in each class:

$$\bar{y} = \frac{14.3(20) + 12.5(25) + 12.6(30) + 14.8(25)}{20 + 25 + 30 + 25}$$

$$= \frac{1346.5}{100} = 13.465 \text{ credit hours}$$

Second, this average was multiplied by the $100 cost per credit hour and the total size of the population to yield

$$y = 13.465(100)(1200) = \$1,615,800$$

b. A 90 percent confidence interval on the total income to be expected from students' tuitions is given by Equation (3.14):

$$\hat{Y}_U = 1,615,800 + \frac{1200(1.64)(250)}{\sqrt{100}}\sqrt{1 - \frac{100}{1200}}$$

$$= \$1,615,800 + \$47,105 = \$1,662,905$$

to the nearest whole dollar.

$$\hat{Y}_L = 1,615,800 - \frac{1200(1.64)(250)}{\sqrt{100}}\sqrt{1 - \frac{100}{1200}}$$

$$= \$1,615,800 - \$47,105 = \$1,568,695$$

to the nearest whole dollar.

c. Assuming $S = s$, the number of students who would have to be sampled in order to estimate the college's income from students' tuition payments during the coming semester within $25,000 with 95 percent confidence is given by Equation (3.17):

$$n = \frac{[1200(1.96)(250)/25,000]^2}{1 + \frac{1}{1200}[1200(1.96)(250)/25,000]^2}$$

$$= \frac{553.19}{1.4610} = 378.6$$

which rounds up to 379. Thus $\frac{379}{1200}(100)$, or 31.6, percent of the students would have to be sampled. Note that the $250 figure for s in Equation (3.17) was found by multiplying the standard deviation of the number of credit hours, 2.5, by the cost of each credit hour, $100.

Chapter 4

1. The Dalenius and Hodges method was to be used to find optimal stratum boundaries. Four strata were to be constructed, and 20 intervals were to be used. The highest score on the classification variable is 50, and the lowest is 11. The range of scores is therefore 39, and each of the 20 intervals would have a width of $\frac{39}{20} = 1.95$. Using the data given, the 20 intervals and their associated frequencies, square root frequencies, and cumulative square root frequencies would be as follows:

Interval	Frequency	Square Root	Cumulative Root
11.00–12.95	12	3.46	3.46
12.95–14.90	28	5.29	8.75
14.90–16.85	45	6.71	15.46
16.85–18.80	54	7.35	22.81
18.80–20.75	80	8.94	31.75
20.75–22.70	96	9.80	41.55
22.70–24.65	102	10.10	51.65
24.65–26.60	92	9.59	61.24
26.60–28.55	78	8.83	70.07
28.55–30.50	62	7.87	77.94
30.50–32.45	58	7.62	85.56
32.45–34.40	57	7.55	93.11
34.40–36.35	50	7.07	100.18
36.35–38.30	49	7.00	107.18
38.30–40.25	39	6.24	113.42
40.25–42.20	35	5.92	119.34
42.20–44.15	27	5.20	124.54
44.15–46.10	24	4.90	129.44
46.10–48.05	16	4.00	133.44
48.05–50.00	9	3.00	136.44

Since the largest cumulative square root frequency, Z_L, equals 136.44 and $K = 4$ strata are desired, $Z_L/K = \frac{136.44}{4} = 34.11$, $2Z_L/K = 68.22$, and $3Z_L/K = 102.33$. From the tabulated figures above, the boundaries of the optimal strata are 11.00 to 20.75, 20.76 to 28.55, 28.56 to 36.35, and 36.36 to 50.00.

2. *Proportion of adults who want to outlaw "adult" bookstores*

a. An estimate of the citywide proportion of adults who, if interviewed, would answer "Yes" is given by

$$
\begin{aligned}
p_{st} = \sum_{k=1}^{K} W_k p_k = \ & \frac{6000}{25,000}(0.60) \\
+ \ & \frac{10,000}{25,000}(0.50) \\
+ \ & \frac{5000}{25,000}(0.40) \\
+ \ & \frac{3000}{25,000}(0.35) \\
+ \ & \frac{1000}{25,000}(0.50) \\
= \ & 0.476
\end{aligned}
$$

b. The variance of the estimator is given by $v(p_{st}) = 0.000464$.

c. A 95 percent confidence interval on P is given by

$$
\hat{P}_{U_{st}} = p_{st} + t\sqrt{v(p_{st})} = 0.476 + 1.96(0.0215) = 0.518
$$

and

$$
\hat{P}_{L_{st}} = p_{st} - t\sqrt{v(p_{st})} = 0.476 - 1.96(0.0215) = 0.434
$$

d. With simple random sampling, an estimate of the proportion of adults who would answer "Yes" is given by

$$
p = \frac{100(0.6) + 200(0.5) + 100(0.4) + 100(0.35) + 100(0.25)}{100 + 200 + 100 + 100 + 100}
$$

$$
= \frac{260}{600} = 0.433
$$

An estimate of the variance of this estimator is given by

$$
v(p) = \frac{N - n}{(n - 1)N} p(1 - p)
$$

$$
= \frac{25,000 - 600}{599(25,000)}(0.433)(0.567) = 0.0004
$$

The efficiency of stratified sampling relative to simple random sampling is therefore $(0.000400/0.000464)(100) = 86.2$ percent. Stratified sampling is less efficient than simple random sampling in this application because of the way sample sizes have been allocated to strata. The largest sample size has been allocated to the stratum with the smallest estimator variance and not to the stratum with the largest estimator variance.

3. The sample size needed to estimate the mean religiousity of adults in the city within one point, with 95 percent confidence, is

$$n = \frac{14.4^2}{\left(\frac{1}{1.96}\right)^2 + \frac{1}{25,000}(214)} = \frac{207.36}{26.039}$$

$$= 771.23$$

which would be rounded up to 772.

4. *Problem involving a population total*

a. An estimate of the population total is given by

$$y_{st} = 1000(10) + 1000(20) + 1000(30) + 2000(40) + 2000(50) + 2000(60)$$
$$= 360,000$$

b. To compute a 95 percent confidence interval on the population total, the variance of the estimated total must be estimated. $v(Y_{st}) = 13,600,000$. The confidence interval is given by

$$\hat{Y}_{U_{st}} = 360,000 + 1.96(3687.82) = 367,228$$

and

$$\hat{Y}_{L_{st}} = 360,000 - 1.96(3687.82) = 352,772$$

Chapter 5

1. Since the size of the population is 100 and the sampling interval is 5, the sample size is $\frac{100}{5} = 20$.

2. The means of the five linear systematic samples of size 20 that can be drawn from the population are as follows. They are listed in order of the first element that appears in each sample:

Sample 1: $\bar{y} = 50.85$
Sample 2: $\bar{y} = 67.90$
Sample 3: $\bar{y} = 58.30$
Sample 4: $\bar{y} = 62.80$
Sample 5: $\bar{y} = 60.90$

3. The mean square error (MSE) of the means of linear systematic samples equals 31.512.

4. If you wanted to estimate the population mean time to solve the maze within ±2 points with 95 percent confidence, a sample size of 20 would not be sufficient. This can be seen by noting that three of the five samples (60 percent of them) have means that are more than 2 points from the population mean of

60.15. (Since $N = nk$, linear systematic sampling is unbiased, and the population mean can be found by averaging the five sample means.)

5. The variance of the individual times that the children took to solve the maze is equal to $S^2 = 460.434$. Using this value, the variance of the means of simple random samples of size 20 is equal to 18.417.

6. The efficiency of linear systematic sampling relative to simple random sampling is equal to the ratio of their mean square errors, multiplied by 100. For the data in this problem, linear systematic sampling is only 58.44 percent as efficient as simple random sampling.

7. Estimates of the variance of the means of linear systematic samples of size 20, based on data from the first linear systematic sample, are as follows:

    a. $v(\bar{y}_{sy1}) = 17.408$
    b. $v(\bar{y}_{sy3}) = 12.486$
    c. $v(\bar{y}_{sy4}) = 13.656$

All the estimation methods underestimated the actual variance of the means of linear systematic samples (31.512), but the first estimation method came a bit closer to the actual value than the other two methods. Since these results are based on data from only one sample, we cannot put much faith on our identification of the most accurate estimation method.

Chapter 6

1. The means of the five linear systematic samples of size 20 that can be drawn from the population of 100 children arranged in increasing order of their WISC performance scores are as follows. They are listed in order of the first element in the ordered sampling frame that appears in each sample:

Sample 1: $\bar{y} = 57.80$
Sample 2: $\bar{y} = 58.55$
Sample 3: $\bar{y} = 58.85$
Sample 4: $\bar{y} = 61.30$
Sample 5: $\bar{y} = 64.25$

Notice that, as expected, the means of samples containing elements listed later in the sampling frame are larger than the means of samples containing elements listed earlier in the sampling frame. That is, the sample mean increases as a function of the initial sample element.

2. The mean square error (MSE) of the means of linear systematic samples is equal to 5.581.

3. The variance of the means of simple random samples selected from this population is 18.417 (Exercise 5 of Chapter 5). Therefore the efficiency of linear systematic sampling relative to simple random sampling is 330 percent. Arranging the sampling frame in increasing order of children's WISC performance scores resulted in a dramatic increase in the efficiency of linear systematic sampling (compare these results to those you found in Exercise 6 of Chapter 5).

4. The mean of the centrally located sample is 58.85. It is 1.30 points lower than the actual population mean of 60.15. The estimate is thus fairly good.

5. The means of the five balanced systematic samples are as follows:

Sample 1: $\bar{y} = 61.55$
Sample 2: $\bar{y} = 59.15$
Sample 3: $\bar{y} = 58.85$
Sample 4: $\bar{y} = 60.70$
Sample 5: $\bar{y} = 60.50$

Do not be concerned if the answers you found were not in the same order as those listed above. Note that all of these sample means are very close to the actual population mean of 60.15.

6. The mean square error (MSE) of the means of balanced systematic samples is equal to 1.015. This small value is consistent with our direct observation that all sample means are close to the value of the actual population mean.

7. The efficiency of balanced systematic sampling, relative to simple random sampling, is 1814.5 percent. Thus balanced systematic sampling is very efficient in this application and, for a sample size of 20, does a far better job than simple random sampling.

Chapter 7

1. The variance of the estimated mean weekly income of female heads of households in the 20 housing projects is equal to 20.2863, for a simple random sample of 10 housing projects and unbiased estimation.

2. The variance of the estimated mean weekly income of female heads of households in the 20 housing projects is equal to 1.2196, for a simple random sample of 10 housing projects and ratio estimation. The bias of the estimator is equal to -0.00069, indicating that for these data, the ratio estimator is essentially unbiased. The mean square error (MSE) of the estimated weekly income is, for all practical purposes, equal to the variance, since the square of the bias equals 0.0000005.

3. The variance of the estimated mean weekly income of female heads of households in the 20 housing projects is equal to 2.5349, for a sample of 10 housing projects selected with probabilities proportional to their total numbers of female residents and conventional PPS estimation.

4. The variance of the estimated mean weekly income of female heads of households in the 20 housing projects is equal to 1.1906, for a sample of 10 housing projects selected with probabilities proportional to the totals of the ages of their female residents and conventional PPES estimation.

5. The most precise sampling and estimation method is the PPES procedure, with the totals of ages of female residents used as an auxiliary variable. For this method, in order to estimate the mean weekly income of female heads of households within $5 per week with 95 percent confidence, it would be necessary to sample only 1.8295 housing projects. Of course, in practice, we would round this number up to 2. Just for your interest, if you wanted to estimate the female residents' mean weekly income within $2.50 per week with 95 percent confidence, it would be necessary to sample 7.318 (which rounds up to 8) hous-

ing projects out of the 20. If you wanted to estimate the female residents' mean weekly income within $2.50 per week but demanded only 90 percent confidence (instead of 95 percent), it would be necessary to sample 5.124 (which rounds up to 6) housing projects out of the 20.

6. If you selected a simple random sample of housing projects that resulted in housing projects numbered 2, 4, 6, 8, 10, 12, 14, 16, 18, and 20 being sampled and used ratio estimation of the mean weekly income of female heads of households, your estimates would be as follows:

a. The mean weekly income of female heads of households would be estimated as $67.55. This is a little over $1 lower than the actual mean weekly income of $68.76.

b. The estimated variance of the estimator would equal 2.7672.

c. An upper 90 percent confidence limit on the mean weekly income of female heads of households would be $70.28, and a lower 90 percent confidence limit would be $64.82. Notice that in this case, the confidence interval includes the true value of the population parameter.

References

Angoff, W. "Scales, Norms, and Equivalent Scores," in R. L. Thorndike, ed., *Educational Measurement,* 2d ed. Washington, D.C.: American Council on Education, 1970, pp. 508–600.

Babbie, E. *Survey Research Methods.* Belmont, Calif: Wadsworth, 1973.

Babbie, E. *The Practice of Social Research.* Belmont, Calif: Wadsworth, 1975a.

Babbie, E. *Practicing Social Research.* Belmont, Calif: Wadsworth, 1975b.

Bayless, D. and J. N. K. Rao. "An Empirical Study of Stabilities of Estimators and Variance Estimators in Unequal Probability Sampling ($n = 3$ or 4)." *Journal of the American Statistical Association* 65 (1970), pp. 1645–1667.

Berdie, D. and J. Anderson. *Questionnaire Design and Use.* Metuchen, N.J.: Scarecrow Press, 1974.

Bloom, B. *Stability and Change in Human Characteristics.* New York: Wiley, 1964.

Burkhead, J. *Input and Output in Large City High Schools.* New York: Syracuse University Press, 1967.

Campbell, P. *Phase I Findings, Educational Quality Assessment.* Harrisburg: The Pennsylvania Department of Public Instruction, 1968.

Cochran, W. "Relative Accuracy of Systematic and Stratified Random Samples for a Certain Class of Populations." *Annals of Mathematical Statistics* 17 (1946), pp. 164–177.

Cochran, W. "Comparison of Methods for Determining Stratum Boundaries." *Bulletin of the International Statistical Institute* 38 (1961), pp. 345–358.

Cochran, W. *Sampling Techniques,* 2d ed. New York: Wiley, 1963.

Cochran, W. *Sampling Techniques,* 3d ed. New York: Wiley, 1977.

Coleman, J., E. Campbell, C. Hobson, J. McPartland, A. Mood, D. Weinfeld, and R. York. *Equality of Educational Opportunity.* Washington, D.C.: U.S. Government Printing Office, 1966.

Cronbach, L. "Test Validation," in R. L. Thorndike, ed., *Educational Measurement,* 2d ed. Washington, D.C.: American Council on Education, 1971, pp. 443–507.

Cronbach, L. and P. Meehl. "Construct Validity in Psychological Tests." *Psychological Bulletin* 52 (1955), pp. 281–302.

Cronbach, L. and P. Suppes, eds. *Research for Tomorrow's Schools: Disciplined Inquiry for Education.* New York: Macmillan, 1969.

Curtin, R., ed. *Surveys of Consumers: Contributions to Behavioral Economics.* Ann Arbor: University of Michigan, Institute for Social Research, 1979.

Dalenius, T. *Sampling in Sweden, Contributions to the Methods and Theories of Sample Survey Practice.* Stockholm: Almqvist and Wicksell, 1957.

Dalenius, T. and M. Gurney. "The Problem of Optimum Stratification." *II. Skand. Akt.* 34 (1951), pp. 133–148.

Dalenius, T. and J. Hodges. "Minimum Variance Stratification." *Journal of the American Statistical Association* 54 (1959), pp. 88–101.

Davis, J. *Elementary Survey Analysis.* Englewood Cliffs, N.J.: Prentice-Hall, 1971.

Fetters, W. *The National Longitudinal Study of the High School Class of 1972.* Washington, D.C.: National Center for Education Statistics, 1974.

Gallup Poll. *The Gallup Report.* Princeton, N.J.

Hansen, M., W. Hurwitz, and W. Madow. *Sample Survey Methods and Theory,* Vols. I and II. New York: Wiley, 1953.

Hess, I., V. Sethi, and T. Balakrishnan. "Stratification: A Practical Investigation." *Journal of the American Statistical Association* 61 (1966), pp. 74–90.

Jaeger, R. "Designing School Testing Programs for Institutional Appraisal." Ph.D. dissertation, Stanford, Calif.: Stanford University, 1970.

Jessen, R. "Statistical Investigation of a Sample Survey for Obtaining Farm Facts." *Iowa Agricultural Experimental Station Research Bulletin* 304 (1942).

Jessen, R. *Statistical Survey Techniques.* New York: Wiley, 1978.

Johnson, F. "A Statistical Study of Sampling Methods for Tree Nursery Inventories." *Journal of Forestry* 41 (1943), pp. 674–689.

Kahn, R. and C. Cannell. *The Dynamics of Interviewing.* New York: Wiley, 1957.

Kendall, M. and B. Smith. *Tracts for Computers XXIV, Tables of Random Sampling Numbers.* London: Royal Statistical Society, 1960.

Kish, L. *Survey Sampling.* New York: Wiley, 1965.

Lahiri, D. "On the Question of Bias of Systematic Sampling." *Proceedings of the World Population Conference* 6 (1954), pp. 349–362.

Lord, F. "Test Norms and Sampling Theory." *Journal of Experimental Education* 27 (1959), pp. 247–263.

Madow, W. "On the Theory of Systematic Sampling, III." *Annals of Mathematical Statistics* 24 (1953), pp. 101–106.

Madow, W. Private communication, April 1970.

Madow, W. and L. Madow. "On the Theory of Systematic Sampling." *Annals of Mathematical Statistics* 15 (1944), pp. 1–24.

Martin, E., D. McDuffee, and S. Pressler. *Sourcebook of Harris National Surveys.* Chapel Hill, N.C.: University of North Carolina, Institute for Research in Social Science, 1981.

Mickey, M. "Some Finite Population Unbiased Ratio and Regression Estimators." *Journal of the American Statistical Association* 54 (1959), pp. 594–612.

Miller, R., Jr. "A Trustworthy Jackknife." *Annals of Mathematical Statistics* 35 (1964), pp. 1594–1605.

Mollenkopf, W. "A Study of Secondary School Characteristics as Related to Test Scores," Educational Testing Service Research Bulletin 56–6, Princeton, N.J., 1956.

Moses, L. and R. Oakford. *Tables of Random Permutations.* Stanford, Calif.: Stanford University Press, 1963.

Moser, C. and G. Kalton. *Survey Methods in Social Investigation,* 2d ed. New York: Basic Books, 1972.

Mosteller, F. and J. Tukey. "Data Analysis, Including Statistics," in G. Lindzey and E. Aronson, eds., *The Handbook of Social Psychology*, Vol. II. Reading, Mass.: Addison-Wesley, 1968, pp. 80–203.

Murthy, M. *Sampling Theory and Methods*. Calcutta: Statistical Publishing Society, 1967.

Neyman, J. "On the Two Different Aspects of the Representative Method; the Method of Stratified Sampling and the Method of Purposive Selection." *Journal of the Royal Statistical Society* 97 (1934), pp. 558–606.

Parzen, E. *Modern Probability Theory and Its Applications*. New York: Wiley, 1960.

Peaker, G. "A Sampling Design Used by the Ministry of Education." *Journal of the Royal Statistical Society* 116 (1953), pp. 140–165.

Quenouille, M. "Notes on Bias in Estimation." *Biometrika* 43 (1956), pp. 353–360.

Raj, D. *Sampling Theory*. New York: McGraw-Hill, 1968.

Rand Corporation. *A Million Random Digits with 100,000 Normal Deviates*. Glencoe, Ill.: Free Press, 1969.

Rao, J. and D. Bayless. "An Empirical Study of the Stabilities of Estimators and Variance Estimators in Unequal Probability Sampling of Two Observations per Stratum." *Journal of the American Statistical Association* 64 (1969), pp. 540–559.

Satterthwaite, E. "An Approximate Distribution of Estimates of Variance Components." *Biometrics* 2 (1946), pp. 110–114.

Sethi, V. "On Optimum Pairing of Units." *Sankhya* 27(B) (1965), pp. 315–324.

Shoemaker, D. *Principles and Procedures of Multiple Matrix Sampling*. Cambridge, Mass.: Ballinger, 1973.

Som, R. *A Manual of Sampling Techniques*. London: Heinemann Educational Books, 1973.

Sondquist, J. and W. Dunkelberg. *Survey and Opinion Research: Procedures for Processing and Analysis*. Englewood Cliffs, N.J.: Prentice-Hall, 1977.

Sukhatme, P. *Sampling Theory of Surveys with Applications*. Ames: Iowa State College, 1964.

Tschuprow, A. "On the Mathematical Expectation of the Moments of Frequency Distributions in the Case of Correlated Observations." *Metron* 2 (1923), pp. 461–493, 646–683.

U.S. Office of Education. *National Assessment of Educational Progress. General Information Yearbook*, Rep. no. 03/04-GIY. Washington, D.C.: National Center for Education Statistics, 1974.

Warwick, D. and C. Lininger. *The Sample Survey: Theory and Practice*. New York: McGraw-Hill, 1975.

Weisberg, H. and B. Bowen, *An Introduction to Survey Research and Data Analysis*. San Francisco, Calif: W. H. Freeman, 1977.

Yates, F. "Systematic Sampling." *Philosophical Transactions of the Royal Society of London* A241 (1948), pp. 345–377.

Yates, F. *Sampling Methods for Censuses and Surveys*, 4th ed. New York: Macmillan, 1981.

Index